NATURAL HISTORY OF TENERIFE

Published by
Whittles Publishing Ltd.,
Dunbeath,
Caithness, KW6 6EG,
Scotland, UK
www.whittlespublishing.com

© 2016 Philip and Myrtle Ashmole

*All rights reserved.
No part of this publication may be reproduced,
stored in a retrieval system, or transmitted,
in any form or by any means, electronic,
mechanical, recording or otherwise
without prior permission of the publishers.*

ISBN 978-184995-225-5

Printed by Severn, Gloucester

NATURAL HISTORY OF TENERIFE

Philip & Myrtle Ashmole

Whittles Publishing

*We dedicate this book to all our friends on Tenerife,
who have made the writing of it possible.*

*Boxed contributions by
Antonio Machado, José María Fernández-Palacios,
Juan Carlos Rando, Cristóbal Rodríguez Piñero,
Aurelio Martín, Rubén Barone, Pedro Oromí,
Manuel Marrero & José L. Martín*

Photographs by Philip Ashmole except where acknowledged

Diagrams and maps by Myrtle Ashmole

CONTENTS

Acknowledgements	ix
Preface	xi
Note on names and terms	xiii
Box: Historical snapshot of biological investigation on Tenerife	xv
Map: *Tenerife Featured Sites*	xvi

1 The physical environment and arrival of life — 1

Geography	1
Climate	3
Surrounding sea	6
Colonisation	8
Getting established	10
Origin of new species	11
Box: Colonisation and speciation in Macaronesia	14
The pristine island and the impact of humans	17
Box: The Long-legged Bunting and the Slender-billed Greenfinch	20
The richness of life on Tenerife today	29

2 Ecology of Tenerife — 31

Coastal and lowland shrubland – 'matorral costero'	33
Dry woodland remnants – 'bosque termófilo'	34
Laurel forest – 'monteverde'	35
Pine forest – 'pinar'	35
High mountain shrubland – 'matorral de cumbre'	36
Lava flows, cinders and caves	37
Freshwater habitats	38

3 Coastal and lowland shrubland – 'matorral costero' — 39

The coastal and lowland shrubland ecosystem	39

	Box: Plants and animals coping with sea salt	44
	Sites: *Malpaís de Güímar; Tabaibal del Porís and Punta de Abona; Acantilado de La Hondura; La Caleta del Río; El Médano and Montaña Roja; Malpaís de la Rasca; Teno Bajo; Northern coastal sites; Southeastern desert slopes; Masca Bay; Offshore islets.*	48
4	**Dry woodland remnants – 'bosque termófilo'**	83
	The dry woodland ecosystem	83
	Typical dry woodland plants	86
	Transition zones	92
	The main types of dry woodland	94
	Box: Dry woodland restoration project in the Teno Rural Park	98
	Cliffs and ravines	100
	Animals of the dry woodland	105
	Box: The White-tailed Laurel Pigeon restoration project in Gran Canaria	107
	Sites: *Sabinar de Afur; Roque de los Pinos; Roque de las Ánimas; Anaga beyond Chamorga; Valle Brosque, Anaga; Ladera de Güímar; Siete Lomas and the barrancos of the Güímar valley; Barranco del Infierno; Degollado de Cherfe and the Masca road; Montaña de la Mulata; Barranco de Cuevas Negras; Interián Cliffs; Genovés and El Guincho.*	108
5	**Laurel forest – 'monteverde'**	157
	The laurel forest ecosystem	157
	Laurisilva	161
	Fayal-brezal	163
	Other plants of the laurel forest	165
	Box: Ferns of the laurel forests of Tenerife	180
	Animals of the laurel forest	183
	Sites: *Laurel forest of Anaga; Chanajiga; Barranco de Ruiz; Monte del Agua.*	187
6	**Pine forest – 'pinar'**	199
	The pine forest ecosystem	199
	Dry and moist pine forest	201
	Box: Reconstruction of the pine forests of Tenerife – the 'repoblación'	204
	Animals of the pine forest	209
	Box: Fire ecology on Tenerife	212
	Sites: *Aguamansa; El Lagar; Montaña de Joco; Boca del Valle; Volcán de Chinyero and nearby habitats; Pine forest near Vilaflor; Pine forest and mountains near Ifonche.*	214
7	**High mountain shrubland – 'matorral de cumbre'**	237
	The high mountain shrubland ecosystem	237
	Box: Facilities in Teide National Park	239

	Box: Human use of El Teide and Las Cañadas	242
	Box: Two plant rescue projects in the Parque Nacional del Teide	252
	Animals of the high mountain shrubland	261
	Sites: *La Fortaleza trail, Teide National Park; Las Narices del Teide and Chavao; Los Roques de García.*	266
8	**Lava flows, cinders and caves**	275
	Box: The mesocavernous shallow substratum (MSS)	280
9	**Freshwater habitats**	285
10	**Birds, mammals and other vertebrates**	289
	Birds	289
	Box: Endemicity of Tenerife breeding birds	290
	Mammals	311
	Reptiles	320
	Amphibians	323
	Freshwater fish	323
11	**Butterflies, dragonflies and other invertebrates**	325
	Key to the dragonflies of Tenerife	336
	Other invertebrates	338
12	**Geology**	347
	The Canary Islands and Macaronesia	347
	Island formation and plate tectonics	349
	Box: Volcanic materials	350
	The life cycle of oceanic islands	352
	Box: Sea level, sediments, seamounts and islands	355
	Early development of Tenerife	358
	Rift zones and giant landslides	364
	Las Cañadas Edifice and the origin of Las Cañadas caldera	366
	The Teide-Pico Viejo stratovolcano	369
	Historical eruptions on Tenerife	374
	Box: Raised beaches on Tenerife	377
13	**Directions to featured sites**	378
	Further reading	386
	Index	389

Acknowledgements

Although we have compiled this book mainly from personal experience and published sources, we have been greatly privileged – over more than 30 years – to receive generous help and encouragement from numerous people on Tenerife. These include members of staff and postgraduate students in the University of La Laguna with whom we worked in the 1980s, in volcanic caves and on lava flows in several parts of the world. Our friends have generously supplied written contributions, suggestions, information and photos, as well as accompanying us on numerous field trips to places on Tenerife that we would have been unlikely to discover for ourselves. We cannot mention everyone, and we wish to acknowledge at the start all those who have helped us but who are not named here: we warmly remember their input.

Our primary support has always come from Pedro Oromí, entomologist and eminent biospeleologist, with whom we have shared so many hard-working days on so many islands, both above and below ground, and who has smoothed our path in so many ways. Pedro's superb photographs of specialised cave invertebrates and other tiny animals are a unique feature of this book.

We are much indebted to Antonio Machado, who has stimulated our thinking since 1983 with his writings and in relaxed discussions on island biology, as well as organising the launch of our first book and advising on distribution of this, our fourth one. We salute and thank Keith Emmerson, long-term expatriate naturalist with extraordinary knowledge of plants, animals and obscure roads to exciting places, who commented on many early drafts and advised on treatment of the major habitats. We are also deeply indebted to Aurelio Martín, whose encyclopaedic knowledge of the birds of the Canary Islands we have shamelessly exploited, and whose knowledge of the local book trade has been so helpful; to Juan José Bacallado, to whom we owe so much, who has provided most of the information on the Lepidoptera, as well as many beautiful photos; to José L. Martín Esquivel, who long ago introduced us to the delights of exploring lava tubes and volcanic pits, and who recently took us to see rare plants in remote parts of Las Cañadas; and to Cristóbal Rodríguez, who so often found time in a hectic schedule to take us to special habitats in Anaga and Teno, and who identified many unfamiliar plants in photographs.

Newer friends who have supported our work on the book include Juan Carlos Rando, who has helped us to understand the chequered fortunes of the vertebrate animals of Tenerife.

Rubén Barone has generously shared his wide knowledge of the plants, animals and current research relating to Atlantic islands, guided us to special places and provided crucial help in relation to book distribution in the Canary Islands. José María Fernández-Palacios, leader of the team of researchers who have illuminated so many aspects of the ecology of the Canary Islands, has reviewed each of the main chapters in draft and explained complex matters in the field, in spite of many competing demands on his time.

For comments on the geology chapter we are deeply indebted to Sergio Socorro and Francisco José Pérez Torrado, whose comprehensive understanding of Teide volcano has helped us to make sense of the complex and rapidly developing story of one of the most fascinating and intensively studied volcanoes in the world.

In addition to all the help we received during development of the book, many of these old and new friends have reviewed and corrected drafts of sections of the text, resulting in many improvements; special thanks for help at this stage are due to Rubén Barone and Pedro Oromí, who read almost all of the text and checked the identity of plants and invertebrates in illustrations.

This book might never have been published without the support of Keith Whittles and his team at Whittles Publishing. We owe a special debt to Moira Hickey, who gave us the confidence to make the initial approach. We offer warm thanks to Kerrie Moncur, who patiently worked on modifications to our original design and endured our technical shortcomings, and to Sue Steven, who has worked so hard at establishing contacts with the Canary Islands book trade.

Finally, we would like to thank Sue Allen for suggesting that we might revise our 1989 book: she planted a seed that grew into five years of work on an entirely new book, and we hope that she will read it somewhere on one of her voyages to distant parts of the world.

If there are errors in the book, they are of course our responsibility, and we shall be pleased to have them pointed out to us.

Preface

Oceanic islands have played a part in our lives since 1957-59, when Philip studied seabird ecology and sub-fossil bird bones on Ascension Island and St Helena, and the early 1960s, when we worked together on seabirds of islands in several parts of the Pacific.

Our first visit to the Canaries was in 1983, with time on La Palma and Tenerife; in 1984 we were based in La Laguna University for six months and visited La Gomera, Lanzarote and Graciosa; in 1985 we worked again on Lanzarote and spent the summer learning more about Tenerife; in 1986 we visited Fuerteventura and Gran Canaria and worked on Tenerife and La Palma; in 1987 we worked on El Hierro and visited Tenerife and also Madeira. In 1987, 1989 and 1991 we joined colleagues in caving expeditions from Tenerife to the Azores and the Galápagos Islands.

In subsequent years we spent many months on St Helena, working on volcanic desert and cloud forest habitats, and greatly increasing our understanding of the evolutionary, ecological and geological processes that generate the diversity of life on oceanic islands. On Ascension Island we found various new species of invertebrates in caves and barren volcanic terrain, as well as investigating the feasibility of restoring seabird populations by eradicating feral cats (subsequently achieved) and rats (not yet attempted).

Meanwhile on the Canaries, collaborative international research involving DNA analysis has shed new light on the evolution of the flora and fauna; the biodiversity of the Canaries has been fully documented in an extraordinary initiative by local biologists; studies of fossil pollen have demonstrated important historical changes in the vegetation; and the geology of Tenerife and other volcanic islands of the northeast Atlantic has been subject to intense investigation by research groups from the Canaries and around the world.

In the 1980s, however, this was still to come, and there was hardly any easily accessible information on the natural history of the Canaries. The classic works by European travellers were hard to find and in various languages, and the only guide books were mainly concerned with walking and scenic views rather than the fauna and flora. This lack of information stimulated us to write the book that we needed ourselves. In 1989 we published *Natural History Excursions in Tenerife: a guide to the Countryside, Plants and Animals,* which is long out of print.

In the last few decades the number of people visiting the Canaries has greatly increased, and numerous walking guides and books on plants, birds and geology have been published in German and English. There are now also many excellent locally published books in Spanish, although most of these have limited distribution and can be very hard to find, even on Tenerife.

Of course it is now also possible to access masses of information online, but much of it is fragmented and can be confusing. In this book we have therefore tried to provide within a single cover a coherent account of the natural history and geology of Tenerife. It is aimed at all speakers of English, but is written primarily from the viewpoint of a northwest European for whom the plants and many of the animals will be unfamiliar; we hope it will also be of use to the people of Tenerife and the other Canary Islands.

For visitors and residents, the variety of landscapes offers endless opportunities for exploring, and in this book we have tried to make such explorations more rewarding. After an outline of the geography, climate, and biological and human history of Tenerife we focus on the different types of vegetation on this extraordinarily diverse island. Separate chapters cover the semi-desert areas around the coast, the surviving fragments of rich dry woodland, the luxuriant laurel forests of Anaga and elsewhere, the pine forest zones of the north and south, and the austere volcanic landscapes around El Teide.

The accounts of these and other less extensive habitats include thumbnail descriptions and photographs of most of the conspicuous trees, shrubs and smaller plants that characterise them, emphasizing those that evolved on the Canaries and do not grow naturally anywhere else. In each main chapter we also describe particular places that we know and love, and at the end of the book we provide brief directions on how to find them. Although we should emphasise that this is not a detailed walking guide, we hope that the information provided will lead naturalists to some beautiful and uncrowded places overlooked by most visitors to Tenerife and unknown to many residents.

The systematic sections near the end of the book cover the birds and other vertebrates, butterflies, dragonflies and some other groups of invertebrate animals. With regret, however, we have omitted a systematic treatment of the plants, which became too long for inclusion and will have to wait for another book.

The fascinating geological history of Tenerife is treated in a separate chapter near the end. Knowledge in this field is advancing at great speed, with new ideas still under scrutiny by geologists around the world.

We generally assume that a visitor to the island has access to car transport, although the buses on Tenerife are excellent and there is a comprehensive system of walking routes, well signed and with helpful information boards. Nonetheless, it is essential to have a good map of the island, or even two, since topographic names often differ between maps, details of small roads vary, and signposting in the countryside is limited. A good walking guide is useful and some of these include GPS information.

We shall be content if this book helps some other people to gain even a fraction of the pleasure that we have had from our long involvement with an island characterised by Charles Darwin as *"perhaps one of the most interesting places in the world."*

Philip and Myrtle Ashmole

Note on names and terms

Books on natural history need to refer to many plants and animals by name. But which name? We find it impossible to be consistent. As biologists, we tend to cling to the scientific (Latin) 'binomial' (two-word) names, printed in italics to help them stand out from the text, and with the first word capitalised, as in *Quercus robur* (English Oak). Scientific names have the enormous advantages – not always appreciated – that the first word indicates the group (genus) to which the organism belongs, and that in combination with the second (species) name it provides a unique label that is understandable around the world. Latin names sometimes change annoyingly when taxonomists revise the classification of the group, but the changes are necessary to reflect increased understanding of evolutionary relationships.

In general, therefore, we give priority to Latin names, especially with plants; gardeners are used to dealing with them because of their obvious usefulness in bringing together related plants. Family names have similar value at a higher level, and we often refer to a plant in this way, for instance as 'a member of the laurel family (Lauraceae)'. For a book in English, however, it would be perverse not to use English names when these are well established, as in the case of mobile animals such as birds and butterflies, most of which occur in northwest Europe as well as the Canaries. For species that are not well known outside the Canaries we deprecate the use of the variable English names created mainly for tourist literature. We attempt to learn some of the local Spanish names, and have included many of these, especially in captions to photographs, and in the systematic sections near the end.

We have tried hard to avoid technical terms, but some are fundamental and need defining. The word **species** is now generally used to refer to a set of closely related organisms following an independent evolutionary pathway, adapting to their environmental circumstances under the influence of natural selection, but also subject to the vagaries of chance. They diverge genetically from other such groups and do not normally interbreed with them, although hybridisation is fairly common in some kinds of organisms. A **subspecies** is a group within a species, in which the individuals are distinguishable from the rest of the species and usually occupy a particular area. **Macaronesia** is a collective term used by biogeographers for the five groups of islands in the northeast Atlantic Ocean: Azores, Madeira, Salvages, Canaries and

Cape Verdes. It is sometimes extended to include a small part of the coast of northwest Africa that tends to share species with the Canaries.

In discussing the biodiversity of Tenerife, an essential concept is that of **endemism**: the occurrence of a species only in a particular restricted area such as an island or archipelago. In this book we refer to a species as **endemic** if it occurs only in one or more of the Canary Islands, while the term **Tenerife endemic** refers to a species known only from that island. We occasionally refer to a species as a **Macaronesian endemic** if it occurs only in one or more of the archipelagos mentioned above as forming part of Macaronesia.

Historical snapshot of biological investigation on Tenerife

Tenerife has been the flagship of Canarian natural history studies since Alexander von Humboldt described his ascent to the peak of Teide and wrote about the landscape and fantastic flora of the island. Even Charles Darwin confessed to having read these compelling descriptions and was deeply disappointed when he was not allowed to land on Tenerife on his trip to South America. Perhaps the Canaries could have replaced the Galapagos in the history of evolution and science if on that 6th of January 1832 the Beagle had not been requested to stay under cholera quarantine for twelve days in front of the harbour of Santa Cruz. Captain Fitzroy did not wait.

Being a stepping stone on the sailing routes from Europe to Africa and the Americas, it is not surprising that many educated travelers and adventurous naturalists were captivated by the landscapes and fascinating plant world of the island of Tenerife, so different and so close to Europe. Some of them decided to stay and study its "natural productions", as did Sabin Berthelot, who later wrote with Philip Barker-Webb a first comprehensive Histoire Naturelle des îles Canaries (1835-1850). Indeed, the foundations of Canary Islands' natural history, starting with Tenerife, were settled by this flow of European enlightenment explorers, especially from countries that enrolled in the dispute for prestige in science and endeavour around the world in the late 19th Century. British, German and French names fill the scientific pantheon of these early contributors: Louis Feuillée (botany and astronomy), Leopold von Buch (geology), Carl Bolle (birds), Ernest Haeckel (marine animals), Oscar Simony (zoology), Auguste Broussonet (plants), Thomas V. Wollaston (beetles and land snails), etc.

In the early 20th Century, after the heyday of natural history in Victorian times, the torch was taken up by the European academic institutions, which organized scientific expeditions to the Canary Islands, and eventually to the other Macaronesian archipelagos. Many endemic species of plants and animals of Tenerife were discovered and described in this period, and much more: The German Expedition to Las Cañadas in 1910 (to study tuberculosis), the first (1931) and second (1947-1951) Finnish expeditions; the Contribution a l'étude du peuplement des îles Atlantides by the French Société de Biogeographie (1959), the copious results of the Internationaler Forschungsprojekt Makaronesischer Raum promoted by the Senckenberg Institut, and so many others. Individual researchers did also their job, like David Bannerman (birds), F. Børgesen (algae) or Eric Sventenius (plants), and among them, also a few local and isolated scientists made relevant contributions (eg Ramón Masferrer in flora, Elías Santos in Diptera).

It was only long after the Spanish Civil War that the Canarian research institutions joined the study of the island biota in a systematic way. The establishment in 1969 of Biology studies in the University of La Laguna was a seminal initiative and influenced the consolidation of other previous institutions, among them the Museo Insular de Ciencias Naturales in Santa Cruz and the Jardín de Aclimatación de La Orotava, in Puerto de la Cruz (founded in 1792!). Some local scientific journals, like Vieraea (1970), started in this period. The number of publications related to the Canarian biota increased exponentially, and in the decade 1990-2010, an average of a new species was described from the Canary Islands every 6 days: a record in European biodiversity. Other research institutions followed, such as the Instituto de Productos Naturales y Agrobiología (1990) and the newly created Instituto Universitario de Enfermedades Tropicales y Salud Pública de Canarias (2011), which expanded their activities to other research fields and territories (eg Central and South America, and Africa).

In summary, Biological investigation of Tenerife is definitively alive, but now takes place mainly in Tenerife, headed by Spanish institutions. This fact does not prevent foreign scientists from forming an interest in the island biota and collaborating with local researchers. Some of them, like the authors of this book, have a long lasting commitment and fruitful results.

Antonio Machado

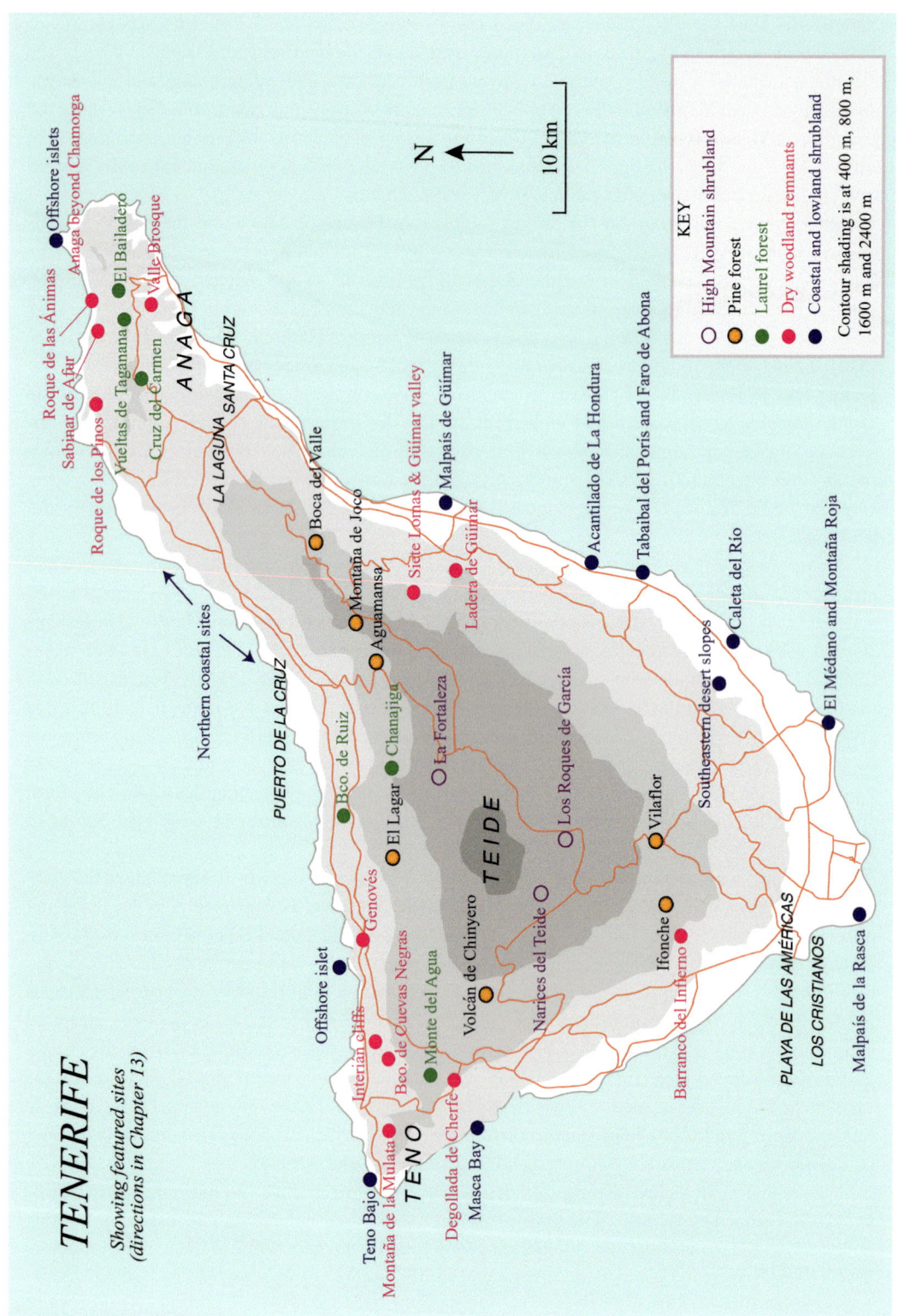

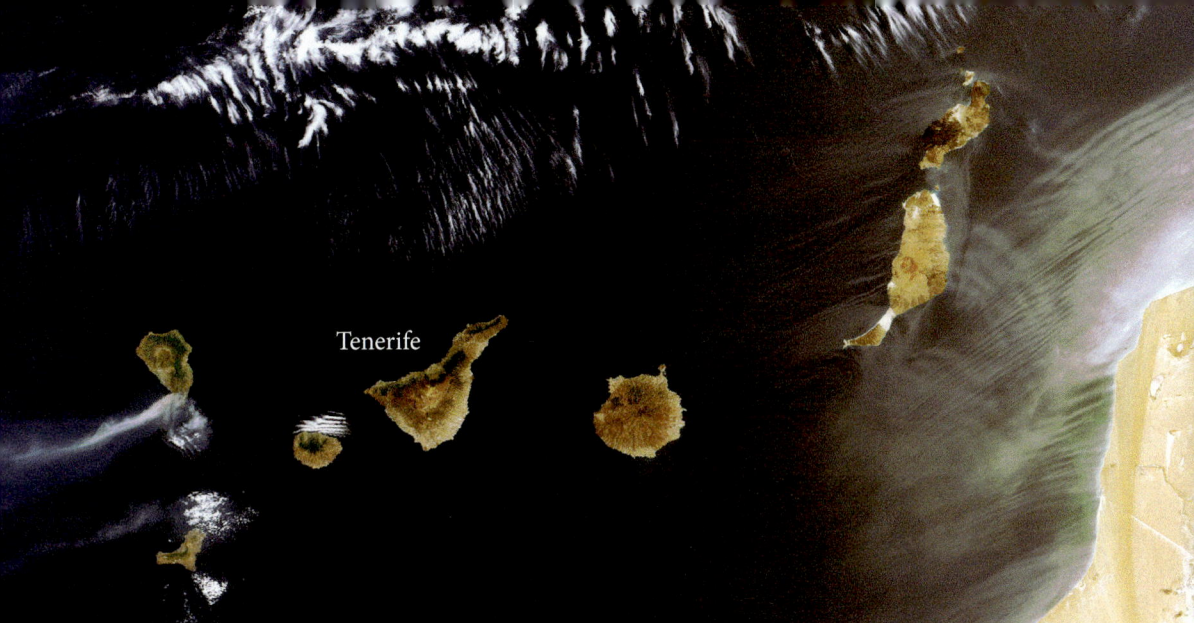

The Canary Islands Archipelago, 1st August 2009 (with fire on La Palma) ©ESA 2009

1

THE PHYSICAL ENVIRONMENT AND ARRIVAL OF LIFE

GEOGRAPHY

The Canary Islands Archipelago lies near the northwest coast of Africa, with the closest island – Fuerteventura – only 95 km offshore. This is a subtropical archipelago lying between 27.5–29.5°N and 13–18°W; it extends over 490 km from east to west, with a total area of 7,447 km². There are seven main islands, one small inhabited island and several islets lacking permanent human populations. Tenerife occupies a central position in the archipelago, with three islands (Gran Canaria, Fuerteventura and Lanzarote) to the east of it and three (La Gomera, El Hierro and La Palma) to the west. Tenerife is 27 km from La Gomera, 60 km from Gran Canaria and 284 km from the African coast.

Tenerife is the largest of the Canary Islands at a little more than 80 km long, with an area of about 2,034 km². The island is roughly triangular in shape and is dominated by El Teide in the centre, with a height of 3,718 m (over 12,000 ft), higher than any mountain in peninsular Spain. The peak rises from the northern edge of a vast 'caldera' – Las Cañadas – which is about 15 km across.

Tenerife tells a story of continual creation and destruction. As you journey round the island you can read parts of this story, but it is like a book with some of the text in a strange language, often overwritten or obscured and with pages torn out. The story starts around 12 million years ago, when the first of three separate shield volcanoes reached the surface of the ocean and became an island, now represented by the Conde massif near Adeje. The Teno and Anaga volcanoes followed, respectively about 6.1 and 4.9 million years ago, joining many older islands in the northeast Atlantic and followed by many younger ones. As the islands that would

Northeast ridge of Tenerife viewed from Teide, with the cloud sea enveloping the Orotava valley and the slopes beyond

form Tenerife increased in size with further eruptions, they were also being altered by the erosive action of rain, winds and waves, so that barrancos (ravines) were carved out, coastal cliffs were formed and debris spilled into the surrounding sea. In time, the ancestors of plants and animals found their way to these islands and many of them gradually evolved differences from their ancestors or split to form several groups of new species.

Then, starting about 3.5 million years ago, a new eruptive centre created a massive volcano between the three ancient islands, forming the foundation of the modern caldera. This, however, was not the volcano that now dominates the island, but an earlier version, largely destroyed by one of the catastrophes now known to be normal feature of the lives of oceanic islands. Like any tall pile of rubble, volcanic islands are unstable, and under the influence of a new eruption the flank of an island may collapse into the sea, sometimes giving rise to a tsunami.

It is now thought that around 200,000 years ago a collapse of this kind – somewhat similar to the Mount St Helens explosion of 1980 – removed about half of the Las Cañadas Volcano in a gigantic northward landslide. It was within the collapse scar that successive massive eruptions relatively quickly created the enormous piles of volcanic rock that we now know as the modern twin volcanoes of El Teide and Pico Viejo (the latter actually younger than Teide). Numerous smaller eruptions have continued sporadically to the present day, the latest eruption being little more than a century ago (see Geology chapter).

Extending northeast from Las Cañadas is a dorsal ridge of high mountains dissected by deep ravines leading towards the sea on either side. The geologically more ancient eastern and western peninsulas of Anaga and Teno are also mountainous; here the ridges and gorges are even more deeply eroded, and in some places there are sea cliffs over 500 m high. The southern tip of the triangular island is more complex; on the eastern side is a low semi-desert coastal plain stretching several kilometres inland towards the foothills of Las Cañadas and the dorsal ridge; to the west is the Conde massif, a complex landscape of steep mountains representing the third of the three ancient shield volcanoes.

Climate

The Canary Islands form a subtropical oceanic archipelago, only about 500 km north of the Tropic of Cancer. Oceanic islands generally have relatively mild winters and cool summers compared with inland areas at the same latitude, but in the case of the Canaries – and especially the eastern islands – the climate is also influenced by their closeness to the deserts of northwest Africa. In general, the climate is of 'Mediterranean' type, with hot, dry summers and fairly low rainfall spread over the rest of the year, and with the winters moderately warm and wet.

Teide and the north side of Tenerife, with the cloud sea persisting into late evening

Tenerife, at 28°N latitude, is within the influence of the northeast trade winds for most of the year. It lies near the northern edge of the Hadley Cell, an atmospheric circulation caused by the strong solar heating near the Equator. The heating produces rising air that flows polewards at high levels (10-15 km) but then sinks in the subtropics (in the northern hemisphere around 30°N) creating the subtropical high pressure belt. The air rising near the Equator produces a low pressure area, and air near the surface in the subtropics flows towards this, creating the Intertropical Convergence Zone (ITCZ). The converging surface winds

Cloud often streams over the summit ridge of Anaga and dissipates on the southern side

are turned westwards by the Coriolis effect resulting from the earth's rotation, and thus form the northeast trade winds. These low-level winds – 'los alisios' – acquire humidity as they blow over the cool waters of the North Atlantic, and then meet the obstruction of Teide and the rest of Tenerife, a mass of land large enough to create its own weather. The moist air rises as it strikes the warm lower part of the northern slopes of the island, and as it does so it expands and thus cools, leading to condensation and cloud formation, often followed by precipitation.

As the cool moist air rises, it meets a layer of warm dry air that comes with high-level northwesterly winds and is further heated by contact with the sun-warmed ground around Las Cañadas. The resulting temperature inversion, with warm air lying above a cooler

There is often cloud hanging over the higher parts of the Güímar valley, favouring woodland development

Las Cañadas, Teide and dorsal ridge from the air on a day of 'calima', with dust-laden warm air from the Sahara enveloping the island
Photo: Myrtle Ashmole

layer (the reverse of the normal situation) traps the clouds formed on the northern slopes against the mountainside, forming a dramatic 'mar de nubes' or 'cloud sea' along the north coast on most days in the year. The upper surface lies at a height of somewhere between 1700 m and 600 m (sometimes even lower) and is generally higher up in winter than in summer. It is around 500 m thick in winter and 300 m in summer, and tends to be thinnest in the morning and to thicken during the course of the day.

When the trade winds are blowing, therefore, the climate of the northern part of the island is quite different from that lying to the south of the main east-west ridge. The whole of the north coast may be under cloud for days on end while the south – and the high part of the Teide massif – basks in sunshine. If you are staying in Puerto de la Cruz, a journey up through the Orotava valley will almost always bring you into bright sunshine in Las Cañadas, and you will usually find a similarly striking transition as you go from the north to the south over the high pass near Erjos del Tanque, between Icod de los Vinos and Santiago del Teide.

The Anaga peninsula in the east of the island has its summit ridge at about 900 m, while the inversion layer that would trap the clouds is normally higher than this; there are thus often dense banks of cloud on the northern slopes of Anaga, but these continually flow over the ridge and then evaporate over the hot southern slopes. As a result, the Playa de Las Teresitas, northeast of Santa Cruz, is often just on the border of shade and sun.

More fragmentary cloud seas sometimes form over southwestern and southeastern parts of Tenerife, even though they are shielded from the trade winds by the Teide massif. These clouds seem usually to be caused by warming of the low southern slopes by the intense sun, causing the air to rise, expand and cool, forming cloud and generating an onshore breeze sucked in by the rising air. Cloud formed in this way can often be seen in the afternoon hanging against the steep southwestern slopes above Guía de Isora, while the somewhat lower land in the far south remains clear. The cloud sea in the north sometimes clears away for a while between November and January, especially when the trade winds give way to cyclonic north Atlantic weather – 'tiempo palmero' – with depressions approaching the islands from the northwest, bringing heavy rain, winds at high levels of up to 200 km/h, and snow on Teide. Occasionally a tropical wind from the southwest also brings heavy rain.

Brief respite from the cloud – offset by stifling heat – may also occur at any time of year when the trade winds are replaced by a gentler east or southeast wind – straight off the Sahara. When this 'tiempo sur' occurs in summer it may raise temperatures to 40°C. It brings with it a haze of dust – 'calima' – which may almost obscure the sun (see photos on pages 1 and 4) and leaves a fine layer of dust over cars and anything else left outside. The Saharan origin of the dust is sometimes emphasised by the simultaneous arrival of Desert Locusts *Schistocerca gregaria*, sometimes in swarms large enough to cause serious damage to crops. The hot air moves in at a fairly high level, leaving a layer below it which is kept cool by the sea; on some of these days Santa Cruz de Tenerife can be cooler than La Laguna, 500 m higher up. Calima was developing on the day we left Tenerife in June 2014, but as the photo shows, the cloud sea still persisted in the Orotava valley.

The weather in Tenerife, at 28°N, is strongly influenced by the seasonal (and also the less predictable) movements of the Azores (or Bermuda) High, a large and semi-permanent subtropical anticyclone lying further west in the North Atlantic and centred between 25°N and 35°N. It has tropical easterly winds on its equatorial side and westerlies on its poleward side. It moves polewards in summer, sometimes reaching 34°N to 38°N, but lies at lower latitudes in winter, when it is also less pronounced.

The relatively low latitude of Tenerife ensures that seasonal differences are less marked than further north in Europe, although there is a definite dry season and temperatures can be quite low in winter. August is the hottest month, with a monthly mean temperature of 25°C in Santa Cruz and 22°C in Puerto de la Cruz, both near sea level, but only 18°C at the mountain observatory of Izaña at 2367 m. The coldest month is January, with a daily mean temperature between 18°C and 4°C according to altitude. The difference between the highest and lowest temperatures in any one day is not more than 7°C in most places but often enormously more than this at high levels: in Las Cañadas there can be a difference exceeding 20°C.

Frosts and snowfalls may occur from December to March in the mountains, normally only above 1700 m, and small patches of snow can sometimes be found on El Teide up to the end of May. One consequence of the low latitude is likely to be noticed by visitors: in summer the days are shorter than further north in Europe (the longest day is 14 hours, as against 16½ in London) but in winter they are relatively long (the shortest day is 10¾ hours as against 7¼ hours in London). The average rainfall for a year varies greatly from place to place: it exceeds

800 mm in the highest parts of Anaga, is around 400 mm in Las Cañadas, and is below 100 mm in the extreme south of the island. Four-fifths of the rain falls between October and March, with virtually none between June and August.

The water resources of the island, however, are critically influenced by a phenomenon not taken into account by normal rainfall statistics: this is the capture of condensation by the foliage of trees, referred to locally as 'precipitación horizontal'. As the cloud drifts through the trees the minute water droplets in the air are caught on the cool, shiny surfaces of the leaves and drip off on to the ground below. The trees thus generate their own rainfall. In tests in the cloud zone at Aguamansa, at about 1000 m altitude, a rain gauge under trees recorded almost 20 times as much precipitation as a gauge a few metres away in the open. This effect was apparently well known by the indigenous inhabitants of the Canaries (the Guanches) and one of the famous stories of the Castilian conquest of the islands relates to the accidental betrayal by a local girl to her Castilian soldier lover, of the whereabouts of a providential tree, known as 'El Garoé' or 'árbol-fuente', which was thought to supply water for many of the inhabitants of the island of El Hierro, an island where surface water is especially scarce. More prosaically, one can often appreciate the effect when driving, from the wet patches on the dry road underneath trees at middle levels on the island, on days when cloud is down over the forest but it is not actually raining.

The Canary Islands are already being affected by changes in the global climate. Since the late 1970s trends have been towards some increase in temperature, a slight reduction in rainfall (with more dry years) and an increase in the number of significant weather events (but without clear increase in their severity). In the future, it is likely that the effects of climate change on oceanic islands – apart from low-lying ones threatened by rising sea level – will be less extreme than on continents. However, projections for the Canaries are for a roughly 1° degree rise in temperature by the end of the century, with perhaps a 10–15% reduction in rainfall. The most likely outcomes are an increase in forest fires, severe shortage of water, and heat stress leading to greater demand for air conditioning, with consequent strain on energy supplies.

Surrounding sea

The Canary archipelago lies immediately to the west of the path of the cool Canary Current flowing to the south-southwest at around 0.1–0.2 metres per second, roughly parallel to the coast of northwest Africa. Although the main part of the current passes well to the east of Tenerife, it brings cool water to the islands: in late summer the sea surface temperature reaches 23–25°C, but in late winter it drops to 16–18°C. As a result of these low temperatures the Canary Islands lack – in spite of their relatively southerly position – the extreme diversity of marine life often found in areas where sea temperatures remain in the twenties throughout the year. Coral reefs are not apparent here, although cold water corals are abundant offshore at depths of 200–1000 m.

Furthermore, the waters around Tenerife are poor in nutrients, so that populations of fish and other marine animals are not especially high. Some local enrichment results from turbulence caused by the flow of the current past the island and eddies formed on the downstream side. Calm and warm waters in the lee of the island probably have low productivity, but animals may be concentrated locally along boundaries between different water masses.

Another key factor relating to the seas near Tenerife is the lack of an extensive shallow submarine shelf around the coast, an environment that can be extremely productive of marine life. In this respect Tenerife is among the poorest of the Canary Islands, with an area of surrounding shelf (the zone less than 200 m deep) smaller than that of the island itself; only La Palma has a relatively smaller shelf. In contrast, Lanzarote (with its islets) has over 3 km² of shallow seabed for each one of land.

The intertidal zone is also very small, since the tidal range is only about one metre, though it may occasionally be as much as 2.5 m, especially in February and September. Waves generated by the trade winds have their main erosive effect on the northern and eastern coasts, but there are also periods with swell or 'mar de fondo' generated by distant storms, usually in the north of the Central Atlantic.

On a wider geographical scale the most significant phenomenon is the upwelling induced by the northeast trade winds along the coast of northwest Africa, which creates one of the major productive ecosystems of the world. It occupies an elongated zone over the continental shelf, and is most intense from July to September. The northern part of this rich zone, in the region of Cabo Juby, gives rise to large fish populations near the eastern Canary Islands of Alegranza, Lanzarote and Fuerteventura. These were exploited in the past by fishing vessels based in the eastern Canaries, but legal restrictions have led to a shift southwards.

The very high productivity along the African coast has little direct influence on the waters around the western Canaries, but it is now known that seabirds breeding on these islands are making use of it. In an elegant recent research project, Cory's Shearwaters *Calonectris diomedea* feeding chicks in burrows were fitted with satellite tracking devices. Birds from the western Canaries were shown to be collecting food for their young some 300-800 km to the south, in the largest area of high productivity. This extends from just south of Cabo Bojador to just south of Cabo Blanco (between 26° and 20° north latitude) and the birds are feeding mainly within 80 km of the African coast. This is precisely the area where the Canarian fishery for Pilchard *Sardinus pilchardus* is now concentrated.

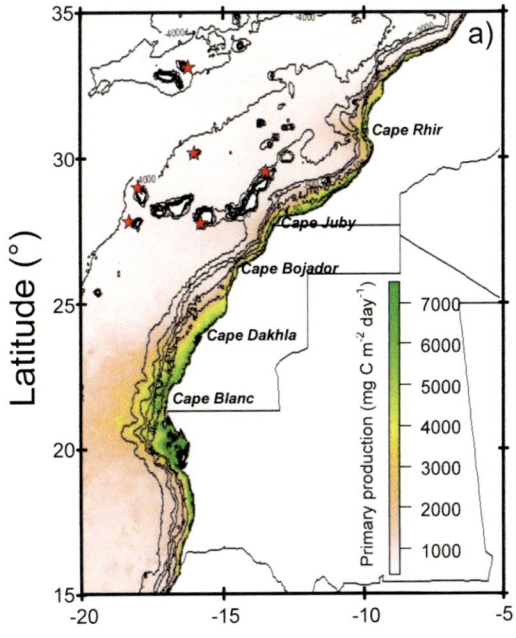

The NE trade winds along the coast of Africa cause upwelling of nutrient-rich waters and high productivity (scale shows annual mean production). Diagram from *Diversity & Distributions* (2013) 1-15, courtesy of Raúl Ramos

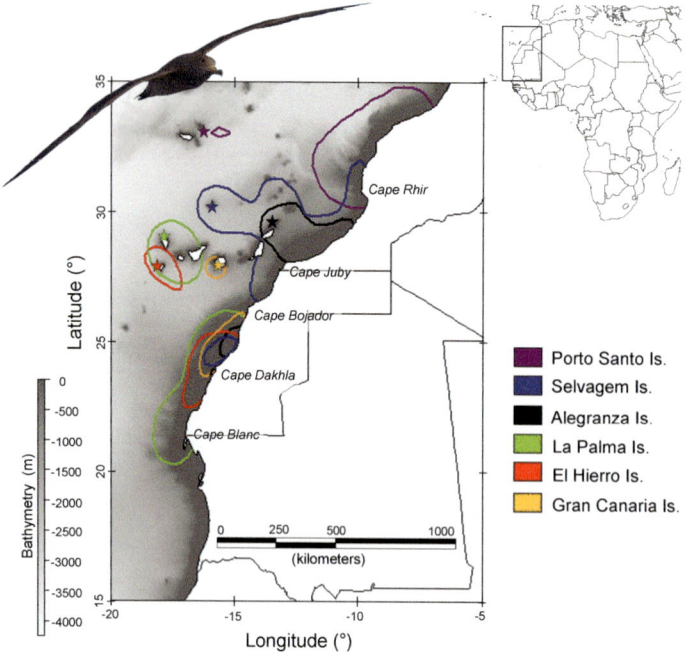

Cory's Shearwaters *Calonectris diomedea* breeding on the islands of the NE Atlantic collect food for their young off the African coast. Diagram from *Diversity & Distributions* (2013) 1-15, courtesy of Raúl Ramos

Global warming will undoubtedly lead to significant changes in the marine environment. A 20 cm rise in global sea level has been recorded since 1900, and a further increase of around 35 cm is predicted before the end of the present century. There is great uncertainty about these figures, but the effect on steep-sided volcanic islands like Tenerife will be slight. In contrast, predicted changes in sea temperature and acidity are known to have major effects on animal communities on the sea floor, and effects on fish and other marine animals will undoubtedly also be substantial.

COLONISATION

As soon as the volcanoes that became Tenerife emerged from the sea as chaotic and sterile piles of rock and ash, colonisation by microbes, plants and animals began. The first visitors were doubtless seabirds, but the first living organisms to establish permanent populations would have been bacteria capable of using hydrogen sulphide as a source of energy. They would have colonised the moist borders of the volcanic craters (as they did when Surtsey island broke the surface just off the coast of Iceland). Other pioneers would be free-living cyanobacteria (blue-green algae), primitive organisms probably arriving on the feet of seabirds. These can survive on volcanic deposits or lava flows even when no other living things are present, since they have the ability to use solar energy and nitrogen from the air; sometimes they form gelatinous masses in damp places, especially close to windward shores. Lichens, mosses, liverworts and ferns have reproductive stages that can be transported by air in great numbers over long distances, and it is likely that representatives of these groups soon became established in damp places among the volcanic rocks.

Higher plants adapted to life on beaches, with seeds that can survive in sea water, will also have begun to arrive at once. It is more difficult to visualise colonisation of new volcanic islands by plants of dry land and by animals other than birds or bats, although plants with small plumed seeds could also arrive by air. Many strong-flying insects such as dragonflies, butterflies and hoverflies undertake long distance migrations, and are inevitably liable to be blown off course. Animals that have evolved strategies for long distance dispersal would have occasionally reached the island. Some kinds of spiders go 'ballooning', supported by a long thread caught by rising air currents, while locusts and crickets form enormous swarms that also tend to drift downwind. This was brought home to us recently when we were in Anaga on a day of 'calima', with the air full of dust drifted from the Sahara by an easterly wind, when we came across a Desert Locust *Schistocerca gregaria* that had probably recently arrived, resting on a wire fence near Taborno.

Desert Locust *Schistocerca gregaria*, probably recently arrived from the Sahara

Tiny animals such as mites and pseudoscorpions often disperse by clinging to strong-flying animals. Accidental transport also occurs, perhaps mainly in the form of plant seeds or eggs of animals such as snails in mud on the feet of visiting shorebirds, or even as seeds of fruits in the guts of migrating birds from western Europe.

Crossing marine barriers by swimming or drifting in the sea is also a common means of dispersal. Many plants have seeds that can survive long immersion in seawater, and if they reach an island they can germinate close to the shore. Colonisation by most of the salt tolerant (halophytic) plants of Tenerife probably resulted from seeds drifting in the sea until they made a landfall on beaches of the island. Similarly, invertebrate animals such as isopod crustaceans, shoreflies, seashore crickets and springtails probably arrive mainly on driftwood, while some wood-boring insects may survive inside logs. The ease of inter-island travel for these plants and animals doubtless explains why most seashore species are widely distributed, also occurring on the nearby continents.

Rafts of tangled vegetation – mainly drifting out from the mouths of major rivers – are probably of key importance in the colonisation oceanic islands, especially for non-flying vertebrates, but rafts arrive so rarely that they are very difficult to study. In one case that we heard of, a biologist saw a natural raft with live lizards on it approaching the shore of an island, but it beached while he was fetching his camera and the chance to document the arrival was lost.

Some larger land animals are capable of dispersal to islands by swimming or drifting. The ancestors of the extinct Tenerife giant tortoise *Geochelone burchardi* probably arrived in this way, aided by the ability of members of this group to survive for long periods without fresh water. Although individuals would usually arrive singly and would rarely be carrying fertilised eggs, the long lifespan of tortoises would increase the chance of members of the other sex arriving in time for reproduction and the establishment of a new population.

Scenarios of this kind may seem implausible, but when millions of years are available, even extremely rare successful colonisation events can produce a diverse fauna and flora. For instance, a quick calculation suggests that the twelve million years or so since the first of the islands that formed Tenerife broke the surface would have been ample time for the modern flora to accumulate, even if only one new stock of plants became established every 10,000 years. In practice, the plants and animals with efficient means of dispersal probably arrived relatively soon, while more sedentary groups accumulated more slowly. At the extreme, however, there are plants with no mechanisms for long distance dispersal, and animals such as freshwater fish and amphibians, which cannot reach oceanic islands naturally.

Both the Northeast Trade Winds and the Canary Current now favour colonisation of the islands of the northeast Atlantic from the Iberian peninsula and northwest Africa, and it is therefore not surprising that most of the plants and animals (especially invertebrates) have their closest relatives in these areas. However, wind systems and ocean currents were probably very different at some times during the early history of the Canaries, and colonisation from the west may have been easier. Even today, meteorological disturbances such as tropical storms may occasionally bring species from unexpected directions.

Getting established

Arrival of animals, and of the spores and seeds of plants, begins as soon as land is available, and studies of Krakatoa and of Surtsey Island near Iceland have documented the accumulation of species on new islands. However, the first plant or animal species to arrive face extraordinarily austere conditions, and our experience on the Canaries, Ascension Island and elsewhere provides a fresh perspective on the process of establishment of invertebrate animals. We found that certain members of many animal groups (for instance pseudoscorpions, spiders, mites, woodlice, springtails, crickets, earwigs, psocids and even a few beetles and moths) can colonise recent barren lava habitats on islands before an extensive plant community has developed on them. These animals mostly shelter in cracks in the lava or among the cinders by day, emerging to forage at night. They are mainly scavengers, but there are also hunters such as spiders that prey on the scavengers.

On relatively old islands these lava communities depend on the input of food resources from elsewhere, such as aerially dispersing insects or plant fragments from productive vegetated areas in other parts of the island, but it is clear that these resources would not be available on a newly formed oceanic island. Intriguingly, however, when we sampled the barren 18th century lava on Lanzarote at different distances from the coast, we found that there were far more springtails and larger invertebrates just behind the beach and 200 m inland than there were 500 m or 4 or 6 km inland. This was in spite of the fact that the inland sites had a heavy growth of lichens, which were virtually absent near the shore.

Work on Ascension Island produced similar results, and it seemed clear that the pioneering animals near the shore lived by exploiting – directly or indirectly – material derived from the marine environment. Organic matter from sea foam, guano and scraps of food from seabirds and marine mammals, carrion, driftwood and the resources of the intertidal zone will all supplement the small primary production on land by colonies of cyanobacteria, mosses and scattered beach plants.

We suspect, therefore, that recently created volcanic islands may also tend to have a fringe of land close to the coast with a community of inconspicuous and mainly nocturnal invertebrate animals. As plants become established, however, their roots will speed up the weathering process, and dead leaves will promote the formation of soil, so that more familiar ecosystems can develop. Animals will then gradually spread inland and a greater range of species will be able to colonise the island.

Having initially established themselves on a volcanic island, populations of plants and animals are still faced with several threats to their long-term survival. The basic requirement is a place to live, and they will not survive long unless the island offers suitable habitat on a reasonable scale. There is also the obvious risk from volcanic activity, which is especially frequent during the early part of an island's life cycle (see Geology chapter). Whole populations are likely to be exposed to asphyxiating hot gases, molten lava or enveloping ash. Sometimes parts of an island will escape the worst effects, so that animals may recolonise the devastated area, but species restricted to one habitat must always be extremely vulnerable. In the case of the land tortoises mentioned above, which are known to have been present on Tenerife during the Pleistocene (within the last two and a half million years) volcanic activity may have been responsible for their apparent failure to survive to the time of arrival of humans.

Sampling invertebrates on Lanzarote, on barren lava close to the shore (above) and on lava covered with lichens 6 km inland (below)

The future of the population also depends on genetic issues. A very small colonising group will be subject to inbreeding (mating with relatives), which leads to genetic defects in the offspring and sometimes to extinction of the population. Even if it survives, the population may contain so little genetic variation that it will lack the capacity to evolve in response to future changes in its environment, or to survive threats posed by diseases to which it has no defence. However, if the colonising group has substantial genetic variation, its long-term future is good.

Origin of new species

Ever since the time of Darwin, biologists have viewed islands as places where natural experiments can help us to understand the origin of species. In any population, but especially if it is small, some genetic changes will occur as a result of chance; random mutation and the good or bad luck of groups of individuals will lead to unpredictable changes in the gene pool of the population. However, with the establishment of a colonising group on an island, natural selection will immediately swing into action, increasing the frequency of those genetic combinations best suited to the new environment, and starting the adaptive evolutionary process that will enable the species to flourish there over the long term.

Since conditions on the island inevitably differ from those where the colonising group originated, natural selection as well as chance effects will tend to lead to genetic divergence from the source population, and eventually to the evolution of a distinct new species. Such species are unique to the island where they evolved (at least initially), and are known as 'endemic' to the island. However, different kinds of organisms differ in their propensity to form endemic species.

Species that reach the island most quickly, such as seabirds and mosses and ferns with spores that disperse in the wind, are unlikely to diverge from the source population, since there will be frequent new arrivals from the mainland, diluting any genetic changes taking place in the new population and thus inhibiting divergence. This accounts for the presence in sandy coastal habitats on Tenerife of many plant species that also occur in northwest Africa and elsewhere, with adaptations for dispersal by sea. Species of this kind will also fail to diversify within islands, since there will be frequent movement of individuals (and thus genes) between local groups.

In contrast, natural arrival of reptiles and of most kinds of flowering plants from the mainland is rare, so that if a group does arrive and become established, the island population can evolve independently and is likely to become distinct from the mainland ancestors.

A key aspect of the biology of Tenerife is that it lies amongst an array of other islands. Recent research has shown that large, high islands – and many smaller ones – have been present in the northeast Atlantic continuously for up to 60 million years, a period much longer than the age of any of the modern islands. The archipelagos of the Azores, Madeira, Salvage Islands, Canaries and Cape Verdes as we see them today are simply the latest format of a highly dynamic pattern, but one in which changes occur on a timescale of tens of millions of years. Old islands shrink and disappear through erosion, while volcanoes build up from the sea bed thousands of metres below, eventually breaking the surface to create new islands.

This geographic dynamism has continually altered the nature of the ocean barriers that must be crossed by colonising plants and animals, so that during some periods crossings have been far easier than they are today, with now vanished 'stepping stones' available in some areas, which may have aided colonisation of islands now remote from other land.

The presence of many islands over long periods means that many continental stocks of animals and plants have been able to establish populations on a number of islands, each of which will follow more or less independent evolutionary pathways. This sets the scene not only for repeated arrivals of immigrants from the continents, but also for movement of plants or animals evolved on one island to other islands in the area, or even back to the continent. Furthermore, the flora and fauna of some modern islands may have been derived largely from far older islands that no longer exist. It is this continued process of island hopping that makes evolution on archipelagos so different from that on single remote islands.

1 The physical environment and arrival of life

When a colonising group from one island arrives on another and encounters a closely related population derived from the same continental stock, hybridisation and merger will often follow. However, if genetic divergence is sufficient to prevent interbreeding, the new arrivals may form a separate population, and the result is two closely related species coexisting on a single island, both of them endemic.

These fundamental processes of island biogeography have doubtless operated many times between the islands and the adjacent continents, and among the islands of the archipelago (see Box below). Furthermore, during the millions of years when Tenerife consisted of three islands, colonisation events must often have occurred between the Conde massif, the Teno massif and the Anaga massif (see Geology chapter).

The origin of new species (speciation) by isolation and evolutionary divergence is often most clearly seen in islands within an archipelago, but a similar process can occur within a single island (or larger land mass) if populations are sufficiently isolated. This situation may arise when animals or plants are separated by barriers such as mountains or hostile

Three of the more than 40 endemic species of the beetle genus *Laparocerus* on Tenerife: *L. inaequalis*, *L. crassus* (right) and *L. fernandezi*. Photos: Pedro Oromí

Colonisation and speciation in Macaronesia over its long and complex history

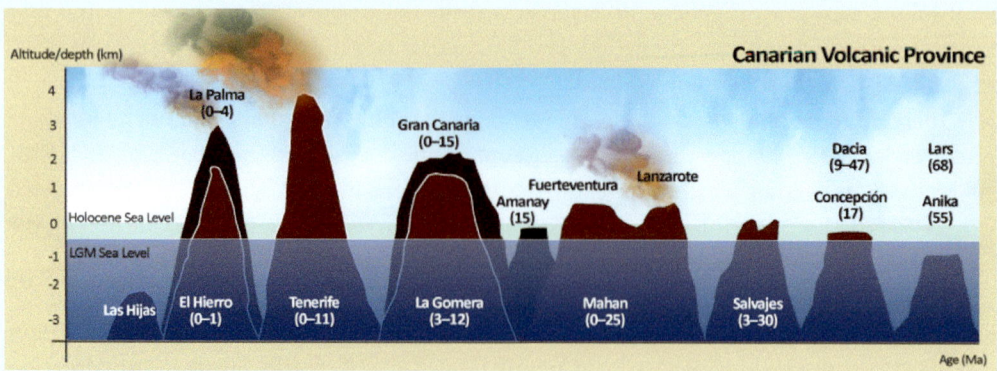

Macaronesia is a biogeographic region formed by five volcanic archipelagos (Azores, Madeira, Salvages, Canaries and Cape Verde) located in the Northeast Atlantic Ocean off Europe and North Africa. The region comprises about 40 islands larger than 1 km² distributed between 14.8°N (Cape Verde) and 39.7°N (Azores) and between 13.4°W (Canaries) and 30.9°W (Azores). The greatest distance between islands in Macaronesia is 2800 km, and Fuerteventura is only 96 km from the African coast. Although all the islands share a common volcanic origin, only three of the archipelagos have been active within the last few years (Canaries and Cape Verde) or decades (Azores).

Modern bathymetric maps of the world oceans have revealed the existence of enormous numbers of seamounts rising from the ocean floor. Many of them display a summit with a truncate cone shape (called 'guyots') and represent former islands that have been submerged below sea level due to erosion, subsidence or both. We know today that hot-spot volcanic islands (such as those in Macaronesia) go through a predictable pattern of growth and decline. They originate on the sea bottom due to the accumulation of magma, and if this is vigorous and prolonged they reach sea level and are born as islands. They continue growing to their maximum size, as is perhaps the case of La Palma and El Hierro, but if movement of the tectonic plate carries them away from the hot-spot they are no longer fed by magma. They then begin to subside and be destroyed by erosion, and in the process acquire maximum topographic complexity, as in La Gomera, Gran Canaria, Anaga and Teno.

Erosion continues until the highest parts of the islands are all removed, as in the case of Fuerteventura and Lanzarote, and eventually only the few last rocks remain above the surface, as in the Salvage Islands, or the island disappears entirely, as in the case of Amanay, a guyot off Fuerteventura. In the eastern Atlantic close to Africa, where the oceanic crust is old, cool and dense, so that subsidence is very slow, it is possible to have all the developmental stages within a single archipelago, as in the family portrait of the Canaries in the diagram.

The existence of so many seamounts in the expanse of ocean now lacking islands between Macaronesia and Iberia or North Africa, and between the archipelagos, has given rise to the idea that Macaronesia is actually much older than the oldest island still above sea level (Grand Salvage, 27 My), and that the isolation among the islands and the mainland (and among the archipelagos themselves) was much smaller in the past. It is now clear that islands existed in the Macaronesian region as early as 60 million years ago, and that for long periods there were islands much closer to Iberia than any are today (see maps). At that time the forests around the Tethys Sea (predecessor of the Mediterranean) were composed mainly of evergreen members of the laurel family and other trees with similar characteristics. The early islands of Macaronesia were therefore probably colonised, over a period of tens of millions of years, by an impoverished version (due to the water gap that made colonisation difficult) of this ancient 'laurisilva'.

Although many of the ancient islands are now lost forever, others survive as guyots with summits fairly close to the surface. In the course of the cycles of glacial and interglacial periods during the two and a half million years of the Pleistocene, sea level varied by as much as 130 metres. Some seamounts below the surface in the current Interglacial (the Holocene) but with summits at shallow depths, therefore existed as islands around the Last Glacial Maximum (LGM) 18-20,000 years ago, and also during earlier glacial periods. The emergence of these seamounts in low sea level periods enhances exponentially the connectivity of the Macaronesian islands, both among themselves and with the mainland, and the contrary will happen in high sea level periods.

The sea level cycles naturally caused important variation in the area of the islands, so that the Canaries had twice as much land exposed at the last glacial maximum as they do today. The cycles also led to merging and splitting of islands, as in the case of the eastern Canary Islands of Lanzarote, Fuerteventura and nearby islets (collectively called Mahan) which were joined and separated more than 20 times during the Pleistocene, alternately decreasing and increasing the isolation of the animal and plant populations.

The shifts in wind regimes and marine currents caused by the Glacial-Interglacial cycles affected the sources, routes and pulses of colonisation of the Macaronesian islands. There is even evidence for instances of back-colonisation of continents from islands: some of the lineages that originated on the mainland but had then colonised Macaronesia survived the glaciations only on the islands, where the ocean buffered the climatic changes and the height of the volcanoes permitted species to respond to change by altitudinal migration. This seems to have enabled a few lineages that became extinct on the mainland to recolonise it from the island refugia; possible examples include *Arbutus canariensis* and *Erica platycodon*.

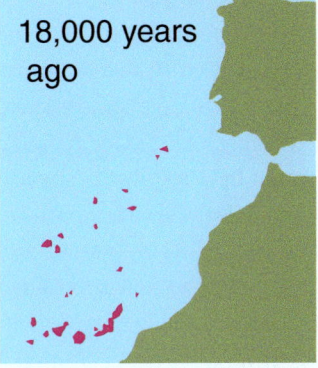
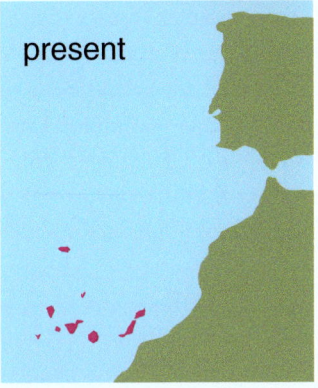

Islands of the Northeast Atlantic, past and presemt

Of course, all this outstanding dynamicity left an imprint in the genetic, ecological, and evolutionary configuration of the Macaronesian biota, and this region is today simultaneously a museum for Tertiary relicts (laurel forest palaeoendemisms), as well as an outstanding laboratory of evolution (Macaronesian neoendemisms).

José María Fernández-Palacios

expanses of fresh lava or cinders. It can also occur in subterranean habitats, as we discuss later, where isolation within distinct cave systems leads to the origin of a host of strange endemic invertebrates.

A key reason why single lineages of animals or plants sometimes diversify so quickly is that speciation is an exponential process; each time that one species splits into two, an additional species is added to the group, and every new species is itself capable of splitting. A classic example of the explosion of diversity sometimes generated in this way is that of the weevils (Curculionidae) of Madeira and other Macaronesian islands, studied by Thomas Vernon Wollaston, a coleopterist friend of Charles Darwin. In the Canaries the most notable weevil genus is *Laparocerus,* with over 40 described species on Tenerife alone (plus several subspecies probably on the way to becoming separate species) all of them endemic to the island or at least to the Canaries, and many of them confined to a particular part of Tenerife.

In plants a good example is the genus *Argyranthemum* (Asteraceae), large yellow-centred daisies with 10 species on Tenerife, as well as many subspecies; again all of these are endemic. The genus has representatives in almost all parts of the island, and it is fairly easy to see that the different species are adapted to the local conditions; for instance, those living close to the coast have swollen fleshy leaves, a common characteristic of salt-tolerant plants.

Fifty years ago evolutionary biologists and biogeographers could only infer these processes, but the techniques of molecular biology, and especially DNA analysis, have revolutionised the field, and now several new analyses appear each year. They can show how closely a species on one island is related to species on other islands or a continent, and indicate roughly when they last shared a common ancestor. This is enormously helpful in attempting to unravel the often complex history of colonisation events and subsequent adaptation and speciation.

One example is that of the skinks in the lizard genus *Chalcides* (family Scincidae) in which the molecular evidence suggests that there have been two colonisations of the Canaries from Africa. One was about 5 million years ago by a Moroccan stock that evolved into *Chalcides simonyi* on Lanzarote and Fuerteventura. The other, 5–7 million years ago, was when a different Moroccan group colonised the ancient western islands of Gran Canaria and Tenerife, producing *Chalcides sexlineatus*, endemic to Gran Canaria, and *Chalcides viridanus*, endemic to Tenerife. The skink on La Gomera is also now considered to be a distinct species, *Chalcides coeruleopunctatus*, whose ancestors reached that island long ago, possibly from Gran Canaria; it then colonised the much younger island of El Hierro in the last million years. The other young western island of La Palma seems to have only introduced skinks.

An endemic species is one that occurs only in a particular area such as an island, and endemic species of the kind discussed above originate when an island population evolves differences from the ancestral population on a

Tenerife Skink *Chalcides viridanus* (juv.)
Photo: Aurelio Martín

nearby continent or other island; it is then sometimes termed a 'neoendemic'. A different situation arises if a mainland population colonises an island (or island group) and survives there with little change, but the ancestral mainland populations fail to persist. Examples of this are well known among the plants of the Canary Islands, and include the laurel relative Viñátigo *Persea indica*, the holly Acebiño *Ilex canariensis*, and the olive relative Palo blanco *Picconia excelsa*. All were represented in deposits in southern France dating from about five million years ago, but are now found only in Macaronesia; they are thus endemic relicts of a species that once had wider distribution, and are sometimes referred to as 'palaeoendemics'.

THE PRISTINE ISLAND AND THE IMPACT OF HUMANS

During the last few million years the physical environment of the Canaries and the other archipelagos of Macaronesia was subject to massive climatic change. The start of the Pleistocene epoch 2.6 million years ago ushered in a series of glaciations (ice ages) in Europe, interspersed with warmer interglacials. Although the Canaries were too far south to be covered by ice, the climate changed and the level of the world's oceans alternately fell and rose, as ice ages were followed by interglacials. As a result, islands repeatedly expanded and diminished in size (with adjacent islands such as Lanzarote and Fuerteventura sometimes merging or separating) and shallow seamounts were alternately exposed and concealed.

Around 12,000 years ago the last ice age finished and the present interglacial – known as the Holocene – started. This meant that three thousand years ago, before humans arrived, the plants and animals of Tenerife had been evolving together in relatively stable conditions for many thousand years, apart from sporadic volcanic events. With the melting of the ice and warming of the oceans, global sea level had entirely recovered from its lowest stand, around 20,000 years ago, when it may have been 120 m below its level today. Furthermore, the climate of the Canary archipelago had remained relatively humid even while northwest Africa was becoming more arid. This stability, however, was the calm before the storm, since the last few thousand years have seen the ecology of the island altering in dramatic ways, mainly as a result of the arrival of people.

Surprisingly, there is still much doubt about the date of the original human colonisation of Tenerife. Recent archaeological studies elsewhere have shown that traditional radiocarbon dating methods are subject to large errors, with estimated dates often being much too old. For instance, dates of arrival of Polynesian people at some Pacific islands are now thought to be several hundred years later than previously estimated, suggesting that early attempts to date archaeological sites in the Canaries should be treated with great caution. However, it is likely that Berber people from Northwest Africa became established on the Canaries between 2000 and 2500 years ago. They developed large rural populations and highly organised society, but although they arrived by sea they apparently lacked any seafaring tradition, and contact between the people on the different islands seems to have been very slight.

The colonising people apparently brought with them domestic goats, sheep and pigs, doubtless along with various parasites and pathogens that may have affected native animals. Dogs may have arrived at the same time, or perhaps with seafarers somewhat earlier, but it is not certain that cats were present in aboriginal times; camels seem to have come with Moorish expeditions early in the 15th century. The house mouse *Mus musculus* probably arrived accidentally with the early colonists, reaching the eastern islands more than 1700 years ago and the western islands somewhat later. The Black Rat or Roof Rat *Rattus rattus* was introduced to Lanzarote more than 1000 years ago and reached the other islands after the Castilian conquest near the end of the 15th century. The Common Rat *Rattus norvegicus*, however, may not have reached Tenerife until the 18th century. The conquest brought many more changes to Tenerife, including the introduction of horses and cattle. Substantial human immigration from several Mediterranean countries followed quickly, and agriculture changed rapidly in some parts of the island, though Teno and other remote areas may have been little affected.

It is known from studies in many parts of the world that when humans and their fellow travellers (including disease organisms) first arrive on an island, massive ecological changes always follow quickly. The Canary Islands were no exception, and the nature of the changes is now becoming clearer. The initial impact of humans may have been greatest on the vertebrate populations, and some evidence is available from fossil bones. A Canarian biologist has written of the 'catástrofe silenciosa' that unfolded in the archipelago under the impact of humans. As with endemic species on other oceanic islands, those on the Canaries were rendered especially vulnerable by the absence of predatory animals during their evolutionary history. Although many of the animals concerned had their strongholds in the eastern islands, Tenerife also lost some species that had previously played a key role in the island's ecology.

When dogs, cats and people arrived, some naïve animals may have quickly succumbed, partly because they would have lacked fleeing responses to the approach of a land predator. One of the most vulnerable species was the Tenerife Giant Rat, *Canariomys bravoi*, with head and body length up to 30 cm and weighing up to one kilogram. It was primarily herbivorous, capable of climbing, and occurred in all regions of the island and from sea level up to Las Cañadas. It had evolved together with the local vegetation and had doubtless reached some sort of equilibrium. However, it was evidently unable to cope with the new threats, including predation by humans, who also hunted birds and the large lizards of the genus *Gallotia*, a group endemic to the Canary Islands. These lizards have been on the Canaries for some 20 million years and tended to evolve very large forms adapted to a mainly vegetarian diet. Their size would have made them attractive as prey for dogs and also for people, and when adult they would have been unable to escape into small crevices. The Tenerife Giant Lizard *Gallotia goliath*, up to

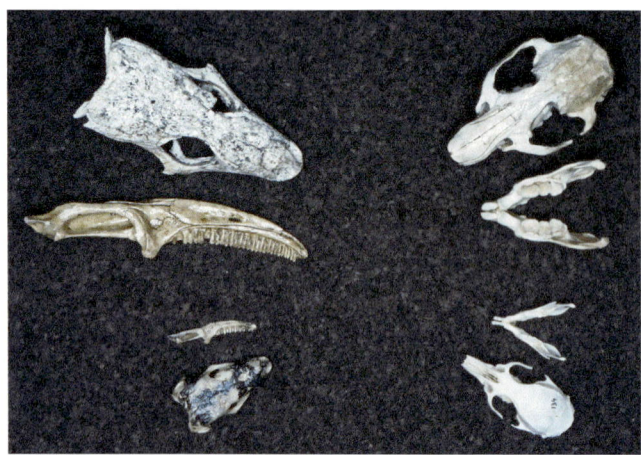

Left: Extinct Tenerife Giant Lizard *Gallotia goliath* above Tenerife Lizard *Gallotia galloti*. Right: Extinct Tenerife Giant Rat *Canariomys bravoi* above Black Rat *Rattus rattus*

1.5 m long, was probably once widespread in Anaga and the lower parts of the Teide massif. It seems to have survived until after the arrival of people, but is now apparently extinct, although small populations of large lizards have survived in precipitous places on other islands.

The abundant Tenerife Lizards *Gallotia galloti* can reach 40 cm and occur in almost all parts of the island. They are preyed on by the local Kestrel *Falco tinnunculus*, Buzzard *Buteo buteo* and Southern Grey Shrike *Lanius meridionalis koenigi,* but may not have been threatened by the arrival of people, since their small size makes them less attractive than the larger species and enables them to find refuges easily; they also reach sexual maturity in two years, while the giant lizards take several years more.

In 1996 there was an exciting discovery of small surviving populations of a third species of lizard, the Tenerife Speckled Lizard *Gallotia intermedia*, closely related to the extinct *Gallotia goliath* of Tenerife and the living giant lizards of the western Canaries; it is less closely related to *Gallotia galloti* of Tenerife. It is currently restricted to the extreme western and southern parts of the island. Tenerife is unique among oceanic islands in having had three species of primarily herbivorous lizards, one giant and two of medium size. It is possible, however, that the two medium species originally lived on separate parts of the island.

Skinks (*Chalcides*) and geckos (*Tarentola*) have been adapting to life on Tenerife for many millions of years. They may have been relatively little affected by human arrival, since they are small and secretive, and usually able to remain out of reach of predators under rocks and in narrow crevices.

An intriguing puzzle is presented by the extinct Tenerife Giant Tortoise *Geochelone burchardi*, known from fossils in Pleistocene ash-flow deposits around half a million years old, in the area around Adeje and apparently also near the Puertito de Güímar. Violent pyroclastic eruptions left ash-flow deposits over most of southern Tenerife in this period, and may well have killed all tortoises in that part of the island. However, if tortoises were present as far east as Güímar it is likely that they were also in Anaga, which has not experienced catastrophic volcanism in the last million years. There is currently no evidence that tortoises were still present on Tenerife when humans arrived, and it is certain that they did not frequent the high area around La Laguna. This is known because the lake sediments laid down around the time of human arrival do not contain spores of the 'coprophilous' fungi that thrive on tortoise dung in the Galapagos and elsewhere; on Tenerife these spores only appear in the sediments laid down after the introduction of cattle and horses by the Castilians. The puzzle about the role of *Geochelone* on Tenerife thus persists, since large tortoises are known to have

Gallotia intermedia, a recently discovered medium-sized lizard endemic to Tenerife
Photo: Aurelio Martín

The Long-legged Bunting and the Slender-billed Greenfinch: two extinct endemic forest birds from Tenerife

Reconstruction of Slender-billed Greenfinch. Art :A Bonner doi10.1371/journal.pone.0012956.g004

The only surviving endemic seed-eating passerine from Tenerife is the Blue Chaffinch *Fringilla teydea*. However, at least two extinct endemic species, the Long-legged Bunting *Emberiza alcoveri* and the Slender-billed Greenfinch *Carduelis aurelioi* lived alongside it, very probably until the arrival of humans. Bones of the Long-legged Bunting indicate that it was flightless and probably inhabited the dense underbrush of laurel forest, a habitat that offers protection from avian predators and also a great variety of seeds and insects. The Slender-billed Greenfinch was a weak flier with very curious anatomical traits. The weight was similar to that of the Greenfinch *Carduelis chloris* but it had smaller wings and longer legs. These data and the reduction of bones that support the most important muscles for flight strongly suggest that it was ground dwelling. Its more peculiar trait is the conical bill, which is similar to those of chaffinches (genus *Fringilla*) rather than the pyramidal shape in greenfinches (*Carduelis*). This bill suggests a more diverse diet with more invertebrates than in that of the Greenfinch; it is morphologically halfway between that of the Blue Chaffinch and the Common Chaffinch (*Fringilla coelebs*). Data from molecular divergences indicates that the former colonized Macaronesia 1.99 million years ago and that the second diverged on the Canary Islands 0.85 million years ago. These dates and the radiocarbon age of Slender-billed Greenfinch bones of more than 13,000 years indicate that these three species lived together in Tenerife at least from 0.85 million years ago until the extinction of the Slender-billed Greenfinch around 15,000 years ago. In this ancient situation three species with beaks of similar shape but of different sizes coexisted. Differences in size and shape of the bills of birds are directly related to the hardness of the seeds that they can eat, so the differences in size among the three finch species would allow them to specialize on different food sources (a case of ecological character displacement). The existence of morphological differences among extinct and extant finches in other Macaronesian islands suggests that these ancient interactions could have influenced patterns of divergence in beak size in other islands. In this situation, knowledge of the extinct finches is essential in order to understand the morphological differences in the species that have survived. The Long-legged Bunting and the Slender-billed Greenfinch, like many other endemic and naive insular species around the world, succumbed swiftly after human arrival and the introduction of alien species to the island.

Juan Carlos Rando

important effects on vegetation of other islands, but it is uncertain whether they were ever numerous enough on Tenerife to have played a major ecological role.

Rather little is known of the history of the avifauna of Tenerife. We have noted white deposits in several coastal areas on Tenerife that may be the remains of guano from large seabird colonies, but so far there is no confirmation of this. The Eastern Canary Islands evidently had large populations of seabirds, including at least two kinds of shearwater – *Puffinus olsoni* and *P. holeae* – that are now extinct. However, the finding on Tenerife of bones of the Houbara Bustard *Chlamydotis undulata* in the Cueva del Viento, near Icod, shows that a large terrestrial bird was present here in the past. The location suggests the possibility that these bustards once lived in open pine forests on Tenerife, rather than being restricted to arid plains at low altitude like those that they still inhabit in the eastern Canaries. An analogy is provided by the Stone Curlew *Burhinus oedicnemus*, which in El Hierro can be found in pine forest, though on Tenerife it currently occurs only in other types of vegetation at low levels. The Houbara Bustard feeds primarily on plant material, but also takes invertebrates, especially in the breeding season. The reason for its loss from Tenerife is not clear, but it is likely that predation by humans and dogs on eggs or adults was involved.

Another intriguing case is that of the Chough *Pyrrhocorax pyrrhocorax* and Alpine Chough *Pyrrhocorax graculus,* which currently coexist in Morocco. The former is a common bird on La Palma, but there are only rare sightings on Tenerife, although bones have been found in the Cueva del Viento, which is less than 170,000 years old, and also in a cave on La Gomera. The Alpine Chough does not now occur on any of the Canary Islands, but bones have been found in the Cueva de Cosme on Tenerife, and also in a cave on La Palma. Although both species normally nest in inaccessible rocky places, the absence of mammalian predators on Tenerife before humans arrived may have led to nesting on or near the ground, making them vulnerable to people and introduced predators such as dogs, cats and rats.

Songbirds on islands lacking predatory mammals tend to evolve reduced flying ability or to be completely flightless, thus saving the energy cost of flight muscles. It is now known that at least two such species were present on Tenerife in the past, but they probably did not survive long after the introduction of rats and cats (see Box). Since small birds are not often well preserved as fossils, it is very possible that other bird species with limited flying ability were also present but failed to leave a record. The most surprising apparent absence is of flightless rails (family Rallidae), since almost all oceanic islands had one or more species of rails before humans arrived, as we know from our experience on Ascension Island and St Helena.

Changes to vegetation are harder to pin down than the loss of animals that leave fossil bones, but it is now clear that the arrival of people had an equally profound impact on plants, eventually transforming several habitat types on the island. Burning of forests and felling of individual trees by people undoubtedly had significant effects, but the domestic herbivores that they brought with them were probably even more important. Although the Guanches seem to have cultivated barley and a few other crops in many scattered places, they were primarily pastoralists, herding goats and sheep. The goat is both a key resource for many people and a four-legged ecological disaster in much of the world, and it would be expected to have an especially heavy impact on vegetation that evolved in the absence of large grazing animals.

Direct evidence on the impact of human activities on vegetation can sometimes be obtained from plant pollen. In an exciting recent study, ecologists at the University of La

La Vega de La Laguna, once the site of a marshy lake, with the dorsal ridge of Tenerife running up towards Izaña, Las Cañadas and El Teide

Laguna analysed fossil pollen in a core two metres long through sediments from the now-dry lake that gave the city of La Laguna its name. The site is at an altitude of 560 m on the saddle between the north and south coasts of the island. To the east lie the surviving laurel forests of the Anaga peninsula, now above about 600 m, while somewhat further to the west at around 700 m are the pine forests near La Esperanza. This unique record is of immense value. It has yielded information on the dominant plants of the area around La Laguna over a period of several thousand years, massively increasing our understanding of the history of the vegetation of Tenerife and providing several significant surprises. One caution is necessary, however: we describe the results using the dates mentioned in the published account, but because of the previously mentioned problems with dating archaeological materials, these dates may be several hundred years too old.

In the oldest part of the sequence, some 4700–2900 years ago, the area around the lake was occupied by a mixed forest dominated by *Carpinus* (a species of hornbeam), *Quercus* (a species of oak), *Pinus* (presumably Canary Pine *Pinus canariensis*) and *Morella* (doubtless the modern Faya *Morella faya*). The presence of hornbeam and oak is especially intriguing, since neither had previously been considered native to the Canaries. Oak-hornbeam forest is a well-known forest type in southern Europe, but hornbeam survived the Pleistocene glaciations in only a few refuges, probably including the Pyrenees. It seems likely – but is not yet proved – that hornbeam and oak were already present in the Canaries during the Pleistocene and even earlier.

In the next part of the pollen sequence, between 2900 and about 2000 years ago, *Quercus* and especially *Carpinus* appear to have remained dominant, with *Morella* (Faya) somewhat reduced. There was an increase in trees typical of the laurel forest (species of Lauraceae along with *Picconia* and *Prunus*) while *Pinus* declined dramatically and *Ilex*, *Juniperus* and

Phoenix (Canary Palm) suffered some loss. The understorey tree *Viburnum rigidum* remained common, as it did to the end of the sequence. The change around 2900 years ago may have been due to a climatic shift towards more humid conditions.

A more drastic change about 2000 years ago undoubtedly reflects the activity of a growing human population. It was associated with a large increase in charcoal in the deposits, suggesting deliberate burning. It is likely that the Guanches used fire to clear forest for cultivation and to provide pastures, and both hornbeam and oak are valuable sources of timber for building as well as fuel; the foliage is used for fodder, and acorns are also eaten by pigs. The hornbeam and oak probably both lacked the ability to reproduce vegetatively, and so would have been especially vulnerable when animals were pastured in the forest, since palatable seedlings would be quickly eaten.

The pollen record shows that from about 2000 years ago up to the time of the Castilian conquest, *Carpinus* declined rapidly and *Quercus* and *Juniperus* more gradually. In contrast, *Morella* (Faya) showed a large increase, doubtless producing a precursor to the modern 'fayal-brezal' degraded woody heath now dominating large areas that once held laurel forest. It is curious, however, that pollen of Brezo *Erica arborea* (the other major component of fayal-brezal) increased hardly at all until the very end of the record 400 years ago, although grasses, composites (Asteraceae) and species of *Echium* were all more common during the last 2000 years. The dominance of *Morella* under human exploitation on Tenerife is no surprise given its ability to fix atmospheric nitrogen and its propensity to become invasive when planted in other parts of the world, especially where pristine forest on volcanic terrain is degraded by human activity. Intriguingly, the Canary Palm *Phoenix* showed a modest recovery in these Guanche times, possibly reflecting planting.

The Castilian conquerors, when they arrived in 1494, found that around the lake at La Laguna there was still dense forest, which seems to have had roughly the composition of the laurel forest a little further east in Anaga today, as described later. However, the forest close to La Laguna disappeared under the heavier exploitation that followed the conquest. Now, the vegetation is entirely secondary and the lake has disappeared.

The accelerating changes following the conquest affected most parts of Tenerife. The native pine and laurel forests, which once covered a large proportion of the island, were reduced to a fraction of their previous area. The most destructive initial exploitation was the establishment early in the 16th century of extensive sugar cane plantations. This crop needed not only land for growing the cane, but also huge amounts of timber for fuel used in the sugar extraction process, so it led to felling of extensive areas of pine forest. The industry declined late in the 16th century when large-scale sugar cane plantations were established in the New World. However, sugar cane was still grown on some estates in the mid 18th century, and hundreds of negro slaves were used on at least one of them. By this time in the north of the island the cultivation of maize had become common, plantations of Sweet Chestnut *Castanea sativa* were established, and there was even production of silk, with export of silk stockings to the Spanish colonies in the West Indies.

The pine forests that survived the early intensive exploitation were subject to continued harvesting in the centuries that followed, with the most accessible places suffering most; pines in the southwest were even exported to La Gomera for construction of buildings. However, extensive replanting, initially using alien species but more recently only Canary Pine *Pinus canariensis*, took place from the middle of the 20th century, as described later. The pine forests have also been used in a sustainable way by local people. Nowadays there are many recreation sites in forest zones, used mainly for barbecues at weekends, but a more ancient custom has been the collection of 'pinocha', the mass of fallen needles that accumulates in pinewoods unless there are frequent fires. In the past, the fallen needles were collected in most areas every few years, and more frequently along roadsides as a fire precaution; the practice continues, though less intensively than in the past. The needles are raked together and then taken down the mountains in trucks for cattle bedding and mulch. This removal of the fallen needles (and also the leaf litter in the laurel forest) destroys the habitat of the invertebrate animals that live on the forest floor, but does not endanger the forest as a whole.

Somewhat later in the 16th century there was extensive planting of grapevines *Vitis vinifera* along the north coast, primarily to produce the sweet white wines that were in demand in England and other European countries. This involved clearance of extensive areas, most of which would originally have been dry woodland and laurel forest. The industry was especially important in the 16th and 17th centuries and continued successfully into the 19th century, until fungal attack devastated the vineyards in the 1850s. Large areas were abandoned, but recently there has been a resurgence in the cultivation of vines, both on an industrial scale (sometimes on land previously used for other types of agriculture) and around individual houses. The latter is especially common in the remote fertile valleys of Anaga, where 'weekend' agriculture is still widely practised on a small scale: vines, patches of potatoes and other vegetables flourish here, as they probably have for hundreds of years, in close association with the laurel forest.

Taganana, centre of a major wine producing area for several centuries; old terraces are just visible on Roque de las Ánimas

Traditional wine press or 'lagar', repaired in 1965 when it was presumably still in use

The banana industry became important near the end of the 19th century, the plantations being at low altitudes. Establishment was partly in areas already ecologically degraded by other human activity, but bananas now cover over 4000 hectares (some 2% of the island's area) and must have led to substantial loss of natural habitat, mainly coastal and lowland scrub. In some places, the landscape changed rapidly. Whole hillsides were cut away and the soil and fine volcanic cinders from them were transported to the banana plantations, often created on raw lava fields where almost nothing grew before. Losses of this kind have continued recently with the establishment of plastic covered shelters for cultivation of tomatoes, cut flowers and other crops.

A more subtle influence on the native vegetation came with the introduction of alien plant species, of which the most conspicuous are prickly pear cacti in the genus *Opuntia* from the New World. Prickly pears were originally introduced for their fruit, which are still widely eaten on the islands, but in the 19th century they became the basis of a major industry, the production of carmine (cochineal) for use as a dye in the textile industry. The dye is extracted from

Separate ownership of different parts of a hill can lead to unsightly effects when cinders are excavated.

Introduced cactus *Opuntia maxima* suffering from late summer drought stress. Inset are Cochineal bugs *Dactylopius coccus* feeding on a healthier cactus pad.

females of the homopteran bug *Dactylopius coccus* (family Dactylopiidae) which feeds on the *Opuntia*. The industry in the Canaries grew from tiny beginnings around 1830 to a peak in about 1869, when – according to one source – six million pounds weight (over two and a half million kilos) of the bugs were exported to Britain, much of it to produce dye for the scarlet uniforms of the British army; around 70,000 bugs were needed for each pound weight. Production was greatest in the eastern Canary Islands, but was also significant on Tenerife, partly replacing wine production after the mid century fungal disease devastated the vines. Cochineal production declined rapidly late in the 19th century when aniline dyes were invented. Cochineal is still used as a food dye and in cosmetics, but is no longer produced on Tenerife.

The ecological significance of the cochineal industry results mainly from the fact that it involved planting of *Opuntia* in several parts of Tenerife. Eating of the fruit by birds (including Ravens *Corvus corax canariensis*) has aided long distance dispersal of seeds, while dispersal by lizards *Gallotia galloti* is effective over short distances. As a result, *Opuntia dillenii* (mainly near the coast) and *Opuntia maxima* (mainly inland) are now found in dry scrub habitats in most parts of the island, competing seriously with native plants.

Also conspicuous in these habitats is another invasive New World succulent, *Agave americana,* which was apparently introduced for fodder, while its relative *Furcraea foetida* is also present in some places. Introduced trees now growing wild in places include Sweet Chestnut *Castanea sativa* (in the north and southeast), two species of *Eucalyptus* (Myrtaceae), two species of *Casuarina*, the cypress *Cupressus macrocarpa*, three non-native pines (*Pinus*) and three non-native oaks (*Quercus*). Two alien and invasive shrubs, Castor Oil Plant *Ricinus communis* and Tree Tobacco *Nicotiana glauca*, are widespread in lower parts of the island, but mainly in waste ground and the beds of barrancos; several other species of *Nicotiana* are also present, but less prevalent. In damp barrancos several introduced plants are sometimes extremely abundant, the most widespread perhaps being Azucarera *Ageratina adenophora*, a white-flowered, purple-stemmed composite herbaceous perennial from Mexico, which can also often be seen along damp roadsides. In dryer places it is hard to avoid *Oxalis pes-caprae,* an invasive yellow-flowered relative of Wood-sorrel, with strange trifoliate leaves, each leaflet divided into two broad lobes by a fold in the midline. The numerous other alien plants are mainly concentrated around settlements and cultivated land, and most of them have had relatively little impact on natural habitats.

With the introduction of agricultural and ornamental plants, many associated insects and other animals arrived on the island. A large number of these are pest species, so insecticides are used for their control, with the inevitable detrimental affect on native insect species. The periodic influx of locusts is also combated by large scale spraying.

The vicinity of Santa Cruz de Tenerife is often cited as the place where a new introduction is first seen; this is the case for the scorpion *Centruroides gracilis* and the Turkish Gecko *Hemidactylus turcicus*. Other introductions by sea since the time of the Castilian conquest may have been here or at other anchorages or ports in the southwest or on the north coast.

An area of special interest in relation to plant introductions is Las Cañadas, with its unique community of endemic plant species. A recent study recorded about 30 species of flowering plants as successful invaders of Las Cañadas, half of them being grasses; another 50 or so alien species have been recorded there but are not established. Some of these invaders may have originally occurred naturally lower down on the island, while others have been introduced from other parts of the world. All of them, however, have probably been helped by humans in their invasion of Las Cañadas. At least half of the successful species are thought to have arrived as a result of goat herding, before it was stopped in the 1950s. A quarter of the invaders probably arrived as a result of tourism since that time, while the remainder may have come in either way. The species brought by goat herding probably mostly sprouted from seeds in the goat droppings or travelled on the feet or coats of the animals, but goatherds are also known to have spread seeds intentionally along the tracks to improve the grazing.

1 The physical environment and arrival of life

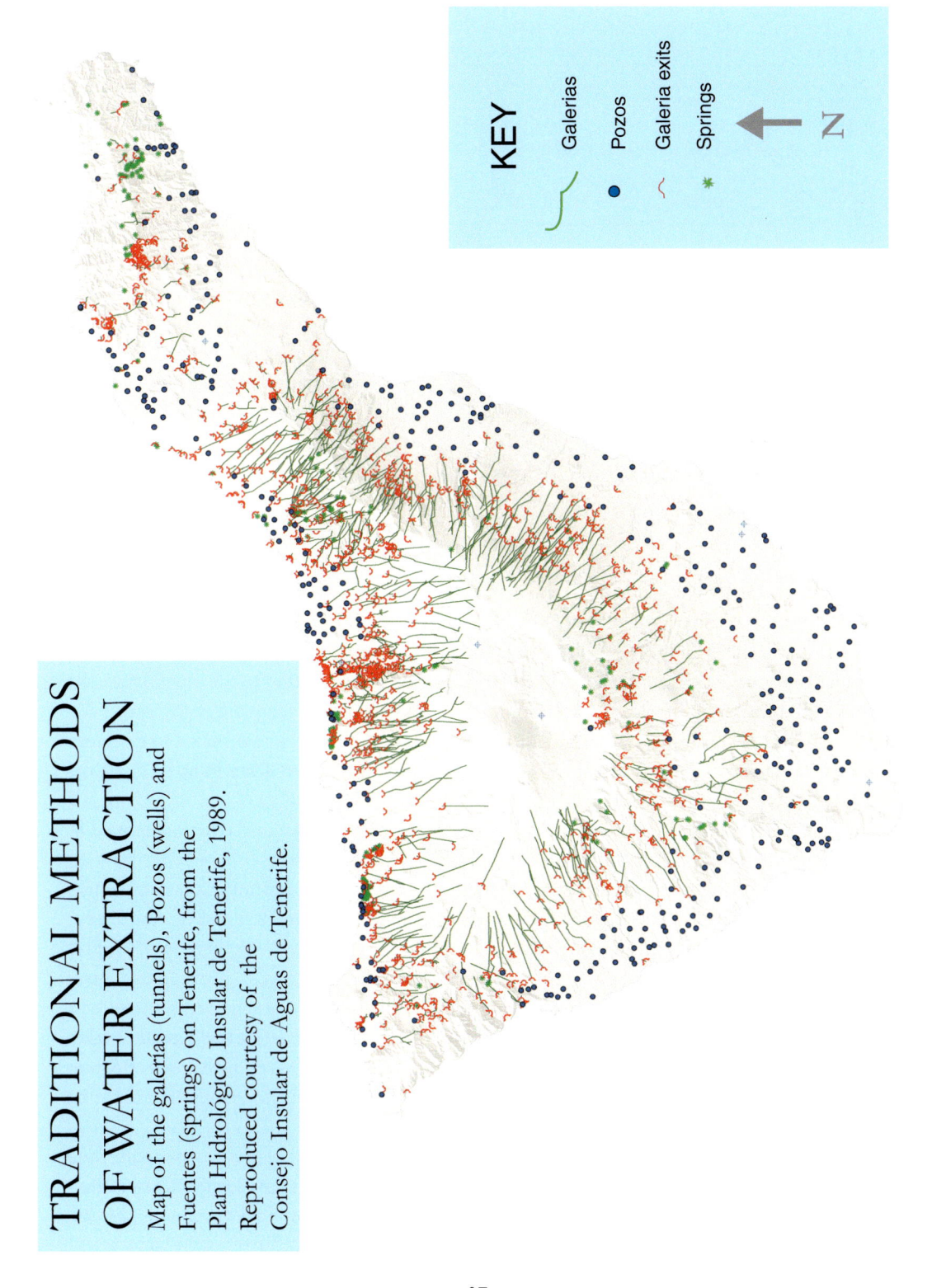

TRADITIONAL METHODS OF WATER EXTRACTION

Map of the galerías (tunnels), Pozos (wells) and Fuentes (springs) on Tenerife, from the Plan Hidrológico Insular de Tenerife, 1989. Reproduced courtesy of the Consejo Insular de Aguas de Tenerife.

KEY
— Galerias
• Pozos
⌒ Galeria exits
✳ Springs

← N

Disused water channel in Teno (1985)

The seasonal influx of herdsmen and their flocks to Las Cañadas took advantage of plant growth at high altitude in summer, when the whole of the south of the island was dry and unproductive; the custom evidently originated well back in Guanche times. In Teno in the far west and in the Anaga peninsula in the east seasonal movements were probably on a smaller scale, but there is little doubt that the herding of goats and sheep has been practised on a large scale for millennia, at least in the areas below the laurel and pine forests, with goat meat and cheese forming a key part of the aboriginal diet. The impact on the natural vegetation was profound.

Human activity has also altered the distribution of water on the islands. As early as the 18th century it is recorded that a system of wooden troughs was used to bring drinking water to the Puerto de la Orotava from a barranco a long way away, and a system of water channels (canales) became ever more complex in ensuing centuries. In addition to the pipes, channels and water tanks mentioned in the section on fresh water habitats, there are many very deep wells and over 1,000 almost horizontal tunnels (galerías) dug into the mountains to intercept underground water. The tunnels are 2 m x 2 m in section and up to 8 km long, with a total length of over 1,600 km (see map).

This proliferation of galerías took place in several stages. When construction started in 1850 they were placed near natural sources (of which there were about 135 on Tenerife) with the idea of capturing part but not all of the flow. After 1910 longer tunnels were excavated, starting near the springs but often penetrating more than 1000 m and attempting to increase the flow by intercepting the aquifer. Then, in the decades following 1945 there was a frenzy of activity, so that by 1973 nearly 1000 galerías and nearly 300 wells (pozos) had been constructed and the peak rate of extraction of subterranean water was achieved (200 million cubic metres per annum). However, by this time extraction from galerías had already begun to decline. For a while the deficit was made up by wells, but these were much less important on Tenerife than on Gran Canaria and Fuerteventura, producing only one eighth of the total volume of water extracted in 1973.

Later in the 1970s, with extraction declining, serious studies of water resources were carried out, but strategic planning developed only slowly. There was now serious competition among industries, especially banana growing and tourism, and the final decade of the 20th century saw the sale of water by landowners for use in the tourist industry, with consequent abandonment of cultivated land. By this time many galerías had been abandoned and permanent streams in barrancos had became intermittent, with serious effects on biodiversity in streamside habitats. At the present time, irrigated horticulture and tourist hotels continue to increase the demand for water, and bright green patches can be seen from the air in the semi-desert areas in the south, indicating the position of heavily watered golf courses.

The population of Tenerife now stands at over 800,000, with a continual flow of settlers and a high birth-rate leading to rapid growth. However, it is the relentless march of urbanization and development for tourists that has brought the greatest changes to Tenerife. Until the middle of the 20th century tourism was on a small scale and mainly in the north of the island, with relatively slight environmental impact. Since that time it has grown rapidly, and there are now around five million visitors per year, of which about 32% come from the United Kingdom, 23% from Spain and 11% from Germany. Tourist accommodation is mainly in the south and southwest, with apartment blocks spreading over large parts of the dry lowlands.

Tourist development inevitably makes great demands on the island's infrastructure, and there is no doubt that fragile ecosystems are sometimes threatened by it, especially in the region of Los Cristianos and El Médano. What to us may be an interesting and rare habitat worthy of preservation, to others is only an empty desert area useful for nothing but more facilities for tourists. In other parts of the island, however, the impact of tourists on natural ecosystems seems to be relatively small. Visitors to the Teide National Park are carefully controlled, and a fine system of footpaths ensures that most walks follow a limited number of routes.

The richness of life on Tenerife today

Charles Darwin arrived at Tenerife in HMS Beagle on January 6th, 1832, with eager anticipation, as he was well aware of the remarkable biodiversity on the island. When landing was prevented *"by fears of our bringing the cholera"* (there was an outbreak in England at the time) Darwin was devastated, lamenting *"Oh misery, misery ..... We have left perhaps one of the most interesting places in the world, just at the moment when we were near enough for every object to create, without satisfying, our utmost curiosity."*

Diversity of the fungi, plants and animals on Tenerife and the other Canary Islands is documented more thoroughly than in almost any other part of the world. The Canary Islands government, enlisting the help of biologists at the University of La Laguna and elsewhere, has published lists of all the species and subspecies known to be living wild on the island (marine species are dealt with separately). Data of this kind go out of date quickly, as naturalists discover new species and revise the classification, so we do not quote precise numbers, but the list provides an authoritative summary of the extraordinary richness of life on Tenerife.

Ignoring the microbes, over 9000 kinds of organisms live on Tenerife: more than 1100 fungi, over 900 lichens and lichen relatives, over 400 bryophytes (mosses and relatives), over 50 ferns and over 1400 flowering plants.

The island also has more than 5300 species of animals, of which 112 are vertebrates but the vast majority are invertebrates. More than 570 species of spiders and their relatives are present, as well as over 4100 insects and over 110 land molluscs such as snails and slugs.

The extraordinary biological richness of Tenerife is due to its great age, great area and great altitude

Humans are responsible for part of this diversity, since nearly 1200 of the species now living wild on Tenerife have been certainly or probably introduced by people – either intentionally or by accident. However, this leaves over 8100 different kinds of organisms thought to be native, having reached the island in the air or by sea, or having evolved from earlier colonists.

The high diversity has arisen partly because Tenerife is ancient (~12 million years) but partly because it has a large variety of habitats, resulting from its great area (2034 km^2) and its enormous altitudinal range (0–3718 m). The next richest island is La Palma, which is smaller and much younger than Tenerife but resembles it in being very high, so that condensation creates humid habitats with many lichens and mosses. Gran Canaria comes only third, in spite of size rivalling Tenerife and even greater age, probably because it is lower and less humid. The other islands have far fewer species. El Hierro and La Gomera are smaller and lower, and the former is also very young. The eastern islands of Lanzarote and Fuerteventura are very ancient, but are now relatively low and arid, lacking humid forested areas.

On islands and archipelagos it is particularly interesting to know how many of the species are endemic (occurring naturally nowhere else), since they reflect the course of evolution. Considering all non-microscopic terrestrial living things, 9% are endemic to Tenerife alone and nearly another 14% are endemic to Tenerife and one or more of the other Canary Islands.

Endemicity varies massively among taxonomic groups. Less than 2% of the species of mosses and ferns on Tenerife are endemic to the Canaries (although there are a good many endemic subspecies). In contrast, about 300 of the flowering plant species are endemic, which is over one fifth (21%) of the total.

Among the animals endemicity is even higher, with over 1700 species (32%) endemic to the Canaries. Here again, the variation among groups is great. The vertebrates have 112 species now living in the wild on Tenerife (a few more were present in the past). Only nine of these species are endemic, although several endemic subspecies have also been described (see Chapters 10-11). In contrast, over 1650 (33%) of the more than 5000 arthropods living wild on Tenerife occur naturally nowhere else in the world. Darwin had cause to be miserable!

DIAGRAMMATIC MAP OF PRESENT VEGETATION ZONES ON TENERIFE
Many remnants of natural habitats are too small to be shown

KEY
- Coastal and lowland shrubland
- Mainly urban and agricultural
- Dry woodland remnants
- Laurel forest
- Pine forest
- High mountain shrubland

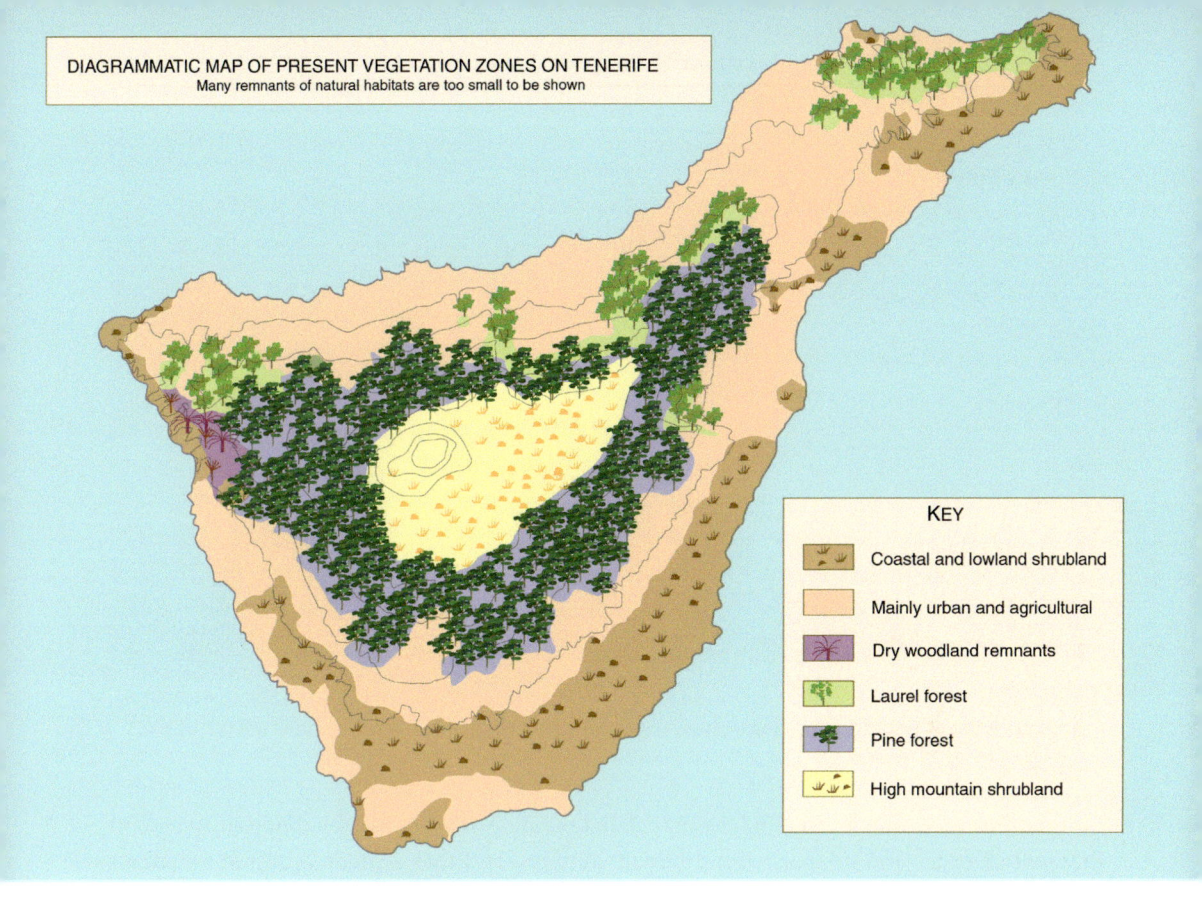

2

Ecology of Tenerife

The island of Tenerife can be viewed as a geological collage, created when the rise of a relatively young central volcano joined together three ancient volcanic massifs in the southwest, northwest and far east. Many millions of years were available for colonisation – of the separate components and then of the united island – by a huge variety of living organisms, originating both from the nearby African continent and from other islands in the archipelago. The plants and animals that established themselves continued to evolve, responding to pressures of natural selection generated by competition among individuals and with other species, fluctuations in climate and the impact of violent volcanic events.

Throughout this long evolutionary progression, members of each species become more elegantly adapted to the environment in which they live. Species in which individuals are mobile (either as adults or as seeds or larvae) are often generalists, and the populations tend to be genetically similar in different parts of the range. However, in sedentary species, such as snails and many plants, individuals mate within their local group and populations in different areas become genetically adapted to their local conditions, as well as changing by chance, and sometimes become distinct species. This process is rife on Tenerife, because the large size and extraordinary range in altitude provides a great range of different habitats. We therefore

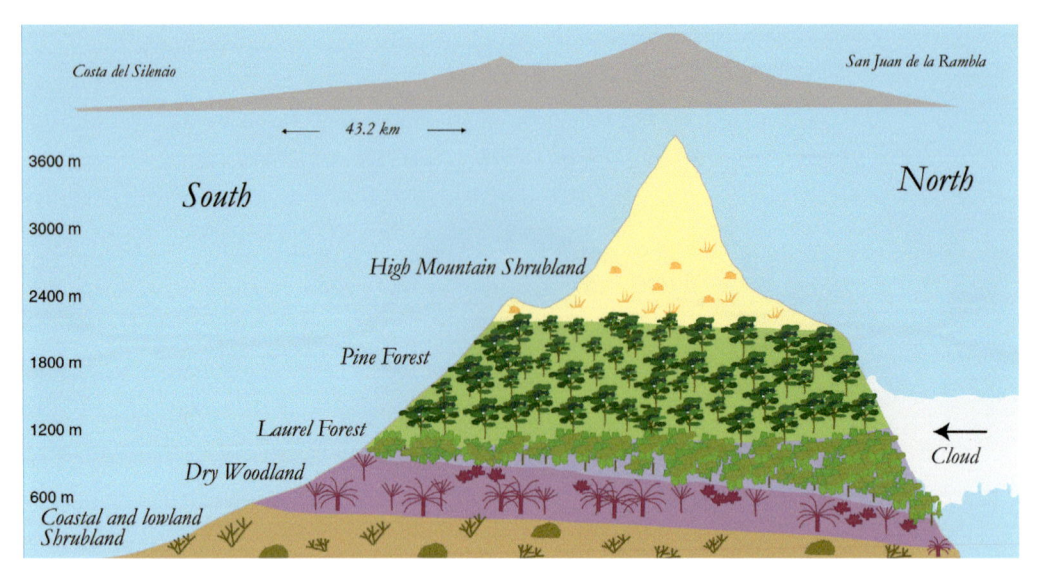

Diagram showing approximate vegetation zones on Tenerife in the absence of human disturbance. Vertical scale is amplified for clarity, but a real south-north profile of the island is shown above.

find, for instance, that a group of closely related plants is represented by different species at different altitudes and in areas with different climates, and that a single group of sedentary beetles has many different species living on different plants in different places.

Tenerife, as a result of its great topographic variety and great age, and its situation in an archipelago where additional species are always likely to arrive, is an island with a diversity of life that is extraordinary for a place outside the tropics. Some idea of this diversity is provided by the ecosystem map and the profile diagram above. The map is an approximate representation of the main vegetation types on Tenerife, as they are today. The natural vegetation has retreated under the successive onslaughts of subsistence agriculture, industrial (export-oriented) agriculture, and tourism, along with the introduction of many alien animals and plants, and massive increase in the human population. In this book, however, we focus mainly on those parts of the island that retain some of the natural vegetation.

A dominant factor in the ecology of Tenerife is the looming presence of El Teide and the Las Cañadas caldera from which it springs. The great height of the mountain and the adjacent volcanic massifs ensures that they largely control the weather conditions in many parts of the island and are responsible for the great variety of habitats. Teide is so high that the climate in Las Cañadas and on the peak is quite different from the climate nearer the coast, and the plant communities and the invertebrate animals that depend on them are also different. However, the mountains also create a north-south divide (see the section on Climate). The north is mainly green, the south dominated by greys and browns, with moist conditions in the north and arid conditions in the southeast and southwest. We therefore discuss the terrestrial habitats on Tenerife mainly in terms of altitudinal zonation on the windward (north) and leeward (southeast and southwest) sides of the island.

Local ecologists now normally describe the vegetation of Tenerife in terms of five main ecosystems, to which we have given English names: '**Coastal and lowland shrubland**', with succulent or hairy and periodically leafless shrubs exploiting every crack and pocket of soil among the salty coastal rocks and sand, as well as occupying the sun-baked lower slopes inland; '**Dry woodland remnants**', relics surviving in places inaccessible to goats, from a time when junipers, palms, dragon trees and other warmth-loving trees occupied their respective niches in the arid foothills and ravines; '**Laurel forest**', with about 20 kinds of tall, evergreen, broadleaf trees and giant heathers, moistened by the seas of cloud and competing to reach the light; '**Pine forest**', dominated by the majestic Canary Pine, damp and draped with lichens in much of the north, scarred by fire but enduring in the arid south; and '**High mountain shrubland**', with scattered brooms and other hardy shrubs living above the clouds, in Las Cañadas and in crevices on the raw rocky slopes of Teide's cone, where silvery hairs and narrow leaves aid survival in the baking days and freezing nights.

These five main ecosystems are briefly described below, with emphasis on their likely distribution on the island before humans arrived. In chapters 3-7 we give fuller accounts of each of them, accompanied by descriptions of some of the most interesting and accessible sites. Then follow accounts of two more specialized habitats that can occur at any level on the island: lava flows, cinders and caves, with their specialised fauna, and places with running or still freshwater, providing habitats for more species. This is a very simplified description of the island's ecology. Finer divisions are often necessary within the main ecosystems, and there are natural transitional zones that blur the distinctions between them. Also, several of the ecosystems have been substantially changed by many centuries of human activity.

Shrubland with dominant *Euphorbia* on the northeast coast

Coastal scrub on sand dunes and clinker on the south coast

Coastal and lowland shrubland – 'matorral costero'

This is a habitat formed by low-growing shrubs and herbs on dry sandy or rocky ground in areas with rainfall of less than 350 mm per year and mean annual temperature varying from 17–21°C according to the locality. The vegetation often has strong representation of shrubs in the genus *Euphorbia*. Close to the coast it is dominated by a suite of species tolerant of the influence of salt sea spray (halophytes) and in a few places there are sand dunes with plants more characteristic of the eastern Canaries and northwest Africa. Further inland the shrubland generally takes the form of patchy semi-desert vegetation, and near its upper limit there is a gradual increase in the presence of larger shrubs and trees. In the north of the island this limit is now generally around 300–400 metres above sea level, but in the past it may have formed only an irregular narrow band below 100–200 m, merging quickly into more wooded habitat. In the southeast and southwest a form of it can now be found as high as 800 m, but in the past its upper limit was probably around 400 m, with a very gradual transition to dry woodland.

Dry woodland remnants – 'bosque termófilo'

This habitat is often called thermophilous (ie warmth-loving) woodland by ecologists. It develops in places with annual rainfall around 250–600 mm per year, and occupies a zone with mean annual temperature of 14–19°C; it does not tolerate low temperatures. When fully developed it is an open woodland habitat with a rich understorey of shrubs and herbs. On Tenerife it once occupied the zone between the lowland scrub and the evergreen laurel or pine forests, but was largely destroyed during centuries of activity by humans and their accompanying animals.

In the north it may originally have extended down to sea level in some places, but only down to around 200 m in others, with a transition to coastal and lowland shrubland; its variable upper limit in the north was around 200–500 m, with a transition to laurel forest. In the south dry woodland probably once extended down to 300–500 m and up to around 700–900 m, with a transition to pine forest (or in a few places to a dry form of laurel forest).

After long human occupation only scattered remnants of dry woodland survive on cliffs and in a few remote areas, with degraded scrub taking over the rest. Originally, different tree species seem to have dominated the dry woodland in different parts of the island, and now that

The dry woodland at this site on the north coast is some of the richest surviving on Tenerife

goat grazing is reduced, some of them are spreading again in certain areas.

Laurel forest – 'monteverde'

This type of forest develops on Tenerife in areas with annual rainfall in the range of 500–900 mm, where misty conditions are frequent. It is made up of about twenty species of broad-leaved evergreen trees, several of them in the laurel family. They are capable of forming a high, dense canopy that largely inhibits development of a shrubby understorey but often supports an array of climbing plants ('lianas').

This habitat is now extremely fragmented, but on the northern slopes of the island laurel forest once occupied a zone between about 1200 m above sea level down to 400 m (even 200 m in a few places) where prevalence of the cloud sea ensured almost continuous high humidity and mild temperatures (mean annual temperature 12–16ºC). In the south of the island traces still remain of a relatively dry form of laurel forest that once occupied some barrancos and other frequently cloudy places on the lower fringes of the pine forest at around 800–1000 m.

Cloud is essential for full development of laurel forest

In favoured places a mature and diverse laurel forest or 'laurisilva' still survives in valleys and on steep slopes, with a variant on ridge tops where the giant heathers are more dominant. In many areas, however, agricultural activity has led to destruction of the natural forests or replacement by a more or less degraded form known as 'fayal-brezal', dominated by giant heathers but often including scattered broadleaf trees such as faya *Morella faya*, *Laurus novocanariensis* and *Ilex canariensis*.

Pine forest – 'pinar'

In its natural form the pine forest of Tenerife is composed almost entirely of *Pinus canariensis* (Canary Pine) with a sparse understorey. In the north of the island a moist form of pine forest occurs above the laurel forest, in a zone with annual rainfall of 500–800 mm and mean temperature 9–12ºC. A few small stands are as low as 450 m but the main extent is between

1200 m and 2000 m, with outposts in both north and south of Las Cañadas around 2200 m. In the south of the island a dry form of pine forest usually abuts directly on to the dry woodland at 700–900 m, but pines grow lower than this in a few places, and in others the transition involves some trees usually found in laurel forest.

Large areas of pine forest were felled in the two millennia following the arrival of humans. The 20th century saw massive replanting of pines, initially including alien species but later only with Canary Pine. This tree is not usually killed by forest fires, sprouting shoots from the burnt trunk, but fire is an increasing threat now that climatic extremes occur more frequently.

Moist pine forest with lichens and *Adenocarpus foliolosus* above the Orotava valley

Dry pine forest on recent volcanic terrain northwest of El Teide

High mountain shrubland – 'matorral de cumbre'

This vegetation is adapted to high altitudes, where the climatic conditions can vary from extreme heat to severe cold within 24 hours, and is dominated by a number of specialised shrubs belonging to only a few plant families. Annual rainfall is normally in the range 300–500 mm, but there are some drought years in which plant growth and flowering are largely inhibited; mean annual temperature is around 5–10°C. This shrubland is largely confined to the caldera of Las Cañadas, above 2000 m, although some of the characteristic species can be found at rather lower levels. The caldera offers a great variety of habitats, including dusty plains, cinders and raw lava, and precipitous rock walls. In the southwest Canary Pines have colonised the inner wall of the caldera, but have not spread on to the more level areas. Some of the shrubs manage to grow on the steep and exposed cone of El Teide, but there is also much exposed ground, and the slopes above about 3000 m are almost devoid of vegetation.

After the arrival of humans the caldera was used for seaonal grazing by the guanches, who took up herds of goats for the summer growing season. This intensive use inevitably had major effects on the vegetation. When the Teide National Park was established the seasonal grazing ceased, but the introduced rabbits and mouflon still exert much pressure on the plants.

Rich high mountain shrubland near the southern wall of Las Cañadas caldera

Lava flows, cinders and caves

Raw volcanic substrates on Tenerife give an overwhelming impression of barrenness. It is easy to see the gradual colonisation of these habitats by plants, but the animals living on lava, in cinder fields or in caves are harder to find, since they are mainly nocturnal or live in the dark. Investigations in the last 50 years have led to discovery of large numbers of insects and other invertebrates, most of them scavengers or carnivores, living on the lava or in underground lava tubes and pits. Many of the species are endemic, blind and with other adaptations to subterranean life.

Freshwater habitats

Wet places are rare on Tenerife, and many of them have dried up as the use of water by people has increased. However, there are still half a dozen permanent streams, and many other barrancos flow in times of heavy rainfall, leaving pools for much of the year. There are also many water tanks, often containing introduced fish, and a small number of more interesting ponds where waterbirds can be seen. In addition, there are some coastal lagoons, both freshwater and saline, used by migratory shorebirds.

Areas with pahoehoe (smooth) lava often also have lava tubes

Pools in the Barranco de Roque Bermejo

The Faro de Anaga, at the northeast point of Tenerife, with *Euphorbia* dominated scrub

3

Coastal and lowland shrubland – 'matorral costero'

The coastal and lowland shrubland ecosystem

The coastal and lowland shrubland consists of scrubby vegetation on various types of aerially deposited volcanic fragments (pyroclasts), cliffs, lava and coastal sand; it occurs mainly in arid areas exposed to the sun and often also to salt spray near the sea. It has been modified by centuries of heavy grazing by goats and other domestic animals, and by all the multifarious activities of humans. Species of *Euphorbia* are dominant in many places, but the least damaged areas also have a wide range of other plants, many of them found only on the Canary Islands, as well as diverse communities of invertebrates.

The shrubland is a semi-desert ecosystem, occupying areas where rainfall is normally less than 300 mm per year and mean annual temperature is around 18–21°C. Close to the coast the scrub is composed of a relatively small number of species tolerant of the influence of salt sea spray (halophytes) but further inland there is a greater diversity of low-growing shrubs and herbs, usually forming very patchy vegetation with much open space. Near the upper limit of the lowland shrubland, where humidity tends to be higher, there is a gradual increase in the presence of larger shrubs and trees, indicating the transition to one of the types of dry woodland.

The expanses of scrub close to the coast form one of the most characteristic types of vegetation on Tenerife, usually dominated by woody species of *Euphorbia* and giving rise to the local name 'tabaibal'.

Many other drought-adapted shrubs, bulbs and a few herbs also thrive on the arid slopes, dried by the sun and largely screened from the moistening effect of the trade winds by the bulk of the Teide massif. The shrubs typically grow to about 2 m in height, and in some of the most favourable places may create a closed canopy, while in more hostile sites the plants are small and sparse.

Unfortunately, several alien species have become established in the dry shrubland, competing with the native species. The most conspicuous are the prickly pears *Opuntia* spp., cacti imported from the New World, which are invasive in many of the unforested parts of the island. *Opuntia maxima* is the commonest prickly pear inland, but *Opuntia dillenii*, with vicious longer spines, is common in some coastal areas and needs to be given a wide berth.

Some coasts, like this at Masca Bay, are not easy places for plants

The current extent of the coastal and lowland shrubland on Tenerife is probably about half of the potential area, the rest having been lost to development of various kinds. Even where the shrubland survives, the natural ecology has often been more or less modified by modern development superimposed on the impact of earlier human activities, including centuries of grazing. Some places, however, show fairly natural conditions, and a few are described below.

The coastal and lowland shrubland can only persist in areas too salty, too exposed or too dry to permit development of woodland with trees and shrubs of greater stature. In the past the transition from scrub to dry woodland habitat doubtless occurred gradually and at different heights in the different parts of the island, since it was determined largely by the influence of the trade winds. Where the wind produced enough moisture for the growth of more substantial trees and shrubs, these excluded most of the species of the coastal and lowland shrubland.

Along much of the north coast of Tenerife the availability of moisture probably allowed woodland development as far down as the coastal cliffs, so the coastal and lowland shrubland may have formed only a very narrow and variable band above the shore. In a few places, however, there is low land near the coast that would have allowed scrub to persist in an extensive area below the woodland. This would have been mainly in the 'Isla Baja' (the low land around Punta de Teno and Buenavista), around Puerto de la Cruz, near Valle de Guerra and on Punta del Hidalgo. It is also likely that in those parts of the north of Anaga where the exposed coast is sloping and suffering rapid erosion, rather than having vertical cliffs, there was always a coastal strip dominated by salt-tolerant shrubs and herbs. In the current degraded situation the scrub tends to take over where there was once dry woodland, and it now reaches as high as 200 metres above sea level in some places.

In the drier southeast and southwest of Tenerife the natural upper limit of the scrub is probably around 400 m, with a gradual upward increase in the presence of larger shrubs and trees and originally a transition to one of the types of dry woodland that we describe later.

The most extreme form of the rocky coast habitat is found on the vertical faces of sea cliffs such as those near Los Gigantes, where the combined effects of hot sun, extreme aridity and salt spray prevent colonisation by all but a handful of plants that establish themselves in crevices. In less precipitous places, close to exposed coasts where there is often spray in the air, the plant community may be dominated by Tabaiba dulce *Euphorbia balsamifera*. This species is well adapted to coping with salty conditions and its low rounded form reduces wind stress in exposed situations; it is also remarkably successful at establishing itself in cracks in bare and exposed rock surfaces. This characteristic may explain the occasional occurrence of the species far inland on rocky slopes, for instance in Teno on the Montaña de la Mulata.

Coastal scrub dominated by Tabaiba dulce *Euphorbia balsamifera* is one of the most characteristic types of vegetation on Tenerife, here seen during a drought

Rapidly eroding unstable slopes exposed to sea spray are colonised by salt tolerant plants

Siempreviva de mar *Limonium pectinatum*

Leña buena *Neochamaelea pulverulenta*

Aulaga *Launaea arborescens*

Perejil de mar *Crithmum maritimum*

Magarza de costa *Argyranthemum frutescens*

Uva de mar *Tetraena fontanesii*

A few characteristic salt-tolerant plants accompany the *Euphorbia balsamifera* on rocky or gravelly slopes close to the sea. These include *Astydamia latifolia* with strikingly fleshy notched leaves and clusters of yellow-flowers in umbels, and the widespread composite *Argyranthemum frutescens* with its large 'daisy' flowers, *Mesembryanthemum crystallinum*, *M. nodiflorum* and *Launaea arborescens*. More typical of rocky places are *Crithmum maritimum* (familiar from northwest Europe) and the strange *Tetraena fontanesii*, with swollen yellow-green leaves looking like grapes. Other typical coastal plants include species of *Frankenia* and Siempreviva de mar *Limonium pectinatum*. Further from the coast Tabaiba dulce *Euphorbia balsamifera* is gradually replaced by its relative Tabaiba amarga *E. lamarckii*. There are many places where both species occur, but we have the impression that *E. lamarckii* has a competitive edge in relatively sheltered places where it can grow tall; its range extends far inland in some areas, especially those degraded by human activity. The third and most spectacular common *Euphorbia* is the tall, cactus-like Cardón *Euphorbia canariensis* with its clumps of succulent, candelabra-like stems, often forming impenetrable thickets. It colonises steep rocky slopes where there are few other large plants and is often the most conspicuous member of the plant community in such places, which are referred to as 'cardonales'.

Although species of *Euphorbia* dominate much of the coastal and lowland shrubland, the habitat also includes a variable number of taxonomically varied species adapted to the arid conditions. The *Euphorbia*-like composite Verode *Kleinia neriifolia* is widespread, as are the asclepiads Cardoncillo *Ceropegia fusca* and *C. dichotoma*, looking like sinister fat bits of jointed garden asparagus, though they occur in separate parts of the island and are numerous only in a few places. As well as these obviously succulent species, there is an array of shrubs with different adaptations for avoiding dessication. Some, such as Incienso *Artemisia* species and Salado blanco *Schizogyne sericea* (Asteraceae), Matorrisco *Lavandula* species (Lamiaceae) and Leña buena *Neochamaelea pulverulenta* (Cneoraceae) have leaves densely felted with silvery hairs, which insulate the plant and inhibit evaporation. A second group have small, cylindrical, succulent leaves: these include a number of Chenopodiaceae but also Romero marino *Campylanthus salsoloides* (Scrophulariaceae), Espino de mar *Lycium*

Tabaiba dulce
Euphorbia balsamifera

Tabaiba amarga
Euphorbia lamarckii

Cardón
Euphorbia canariensis

Plants and animals coping with sea salt

Land adjacent to the sea is influenced by sea salts arriving in the form of spray. In wet climates rain removes some of the salt, but on the coasts of Tenerife there is little rain to prevent its accumulation. As a result there is often a zone just above the high tide mark where high salt concentrations prevent the growth of most plants and only a few of the most salt-tolerant species (halophytes) are able to live. The width of this zone depends on the degree of exposure, with windward coasts receiving much more salt than those in sheltered parts.

Salty soil or sand presents a problem to plants because the process of osmosis tends to draw water out of anything in contact with it. Roots have semi-permeable tissues and those of ordinary plants are quickly sucked dry when exposed to salt. Halophytes can cope, however, since they have compounds in their tissues that give them a high osmotic pressure, enabling them to draw in pure water from moist salty sand or soil.

Halophytes are rarely found far inland, although it has been shown that they typically do not require salt, but simply tolerate it better than other species. Evidently where there is little or no salt, normal competition among the plants comes into play and halophytes generally lose out. One obvious exception on Tenerife is the attractive composite *Schizogyne sericea*, with clusters of brilliant yellow flowers and narrow, flattish silvery leaves. It flourishes in extremely salty areas but can also be found at an altitude of several hundred metres, for instance on the desert slopes below El Río.

Intriguingly, halophytes are sometimes found in severely exposed areas far from the coast, including mountaintops, suggesting that adaptation to salty conditions also fits them for life in situations where few kinds of plants can endure the dessicating winds.

Most halophytic species have wide geographic distribution, probably because their salt tolerance fits them for dispersal by sea, which leads to mixing of genes between the scattered populations and thus hinders genetic divergence and the formation of separate species.

---

intricatum (Solanaceae) and species of *Frankenia* (Frankeniaceae). A third and taxonomically diverse group have relatively long, thin, 'linear' leaves; these include species of *Asparagus* (Convallariaceae) and of *Atalanthus* (Asteraceae), *Reseda scoparia* (Resedaceae) and the small tree *Plocama pendula* (Rubiaceae). The last of these is especially interesting, since it is the sole member of its genus and occurs only in the Canary Islands, where it is capable of growing in some of the driest places. It is evergreen, with a strong and unpleasant smell, drooping branches, slender – almost threadlike – bright green leaves, and very tiny flowers. *Plocama* is a successful colonist, its seeds being distributed both by wind and by floodwater.

Related to *Plocama* and also capable of living in very dry places is the prickly, scrambling *Rubia fruticosa*, which is a Macaronesian endemic. Another scrambler is the asclepiad *Periploca laevigata*, a twining shrub that often envelopes other shrubs, especially Cardón, and has amazingly long, horn-like, twin seed pods that produce a mass of white fibres when ripe; its leaves are so variable that it is not always easy to recognise.

3 Coastal and lowland shrubland – 'matorral costero'

Salado *Schizogyne sericea*

Lechuga de mar *Astydamia latifolia*

Tasaigo *Rubia fruticosa*

Verode *Kleinia neriifolia*

Cornical *Periploca laevigata* with inset fruits

Vinagrera *Rumex lunaria*

Lone billy goat in Barranco de Roque Bermejo

Traditional sheep, still herded in parts of Anaga. Photo: B D Ashmole

The Bath White Pontia daplidice is largely green below

Several other kinds of plants sometimes found in the coastal and lowland shrubland seem to be more typical of the dry woodland above. A somewhat invasive species found mainly in disturbed areas is the endemic dock Vinagrera *Rumex lunaria*, a vigorous loose bush with very broad leaves. The genus *Echium,* so prominent in much of Tenerife, is not strongly represented in the lowland shrubland, but *E. aculeatum* is locally common low down in the Teno peninsula. Similarly, the diverse genera *Aeonium, Sideritis* and *Micromeria* occur mainly higher up, but include species also found in the lowland shrubland in some places. A few kinds of bulbs – *Scilla* spp., *Asphodelus* spp. and *Drimia maritima* (= *D. hesperia*) – are also present in the coastal and lowland shrubland, but are hard to identify when they are not in flower.

The desert plants of Tenerife - like those from the Mediterranean - are well adapted to aridity, but fewer of them are thorny, perhaps because there were few large herbivores on the pristine Canary Islands. Although large tortoises were present, their impact will have been much less than that of more agile goats brought by the Guanches, so the original dry land plants would have been under less selective pressure to evolve physical defences (the giant rat was probably mainly in moister habitats). One plant that does have vicious thorns is Espino de mar *Lycium intricatum*, related to the nightshades and more closely to the boxthorns, many of which also have toxic foliage. *L. intricatum* is not endemic and its protective adaptations were probably evolved in places where it was living with mammalian herbivores, but doubtless helped it to survive in Tenerife after the goats and sheep arrived.

Although most of the coastal zone in Tenerife is rocky, there are a few places in the extreme south, such as El Médano, where sand accumulates and may form dunes, though not on the scale of those on the eastern islands of Lanzarote and Fuerteventura, or even of Gran Canaria. On Tenerife the sandy habitats are small in extent, but they prove to contain interesting communities with plants not found elsewhere on the island, together with an array of beetles and other invertebrates associated with them. Sandy coastal habitats demand special adaptations in the plants and animals that live there, since they have to cope not only with the salty conditions, but also the loose sand where the surface is sporadically remodelled by the wind. Plants growing there need long roots, both to keep them in place and to reach the deeper layers of sand where there is moisture. Smothering by wind-blown sand is also a hazard, so the ability to grow upwards fast is also required. The species typical of sandy places are discussed further in relation to El Médano.

Invertebrates are hard to find in the lowlands in the heat of the day, but dragonflies and a variety of butterflies can be seen in suitable spots. At the eastern end of the Anaga peninsula, where there is rich scrub near the Faro de Anaga and along the Barranco de Roque Bermejo, one can find species such as the Small Copper *Lycaena phlaeas*, with brilliant orange, brown and red wings with dark brown blotches, and the Bath White *Pontia daplidice,* which is white above with black spots but has the underwing largely green. Plain Tiger butterflies *Danaus chrysippus* can also often be seen here; in this species both caterpillars and adults have the bright colours commonly found in poisonous insects, and in this case the poisons are derived from the food plants of the larvae, which are members of the family Asclepiadaceae, represented in this area by *Ceropegia dichotoma*. A similar case is the hawkmoth *Hyles tithymali*, whose caterpillars feed on the poisonous *Euphorbia lamarckii* bushes and signal their distastefulness by their gaudy colouration. Less conspicuous are the predatory mantids that lurk motionless on shrubs, in the hope that an unwary insect will settle nearby or walk within reach.

Caterpillar of *Hyles tithymali*, photo from eastern Canaries

Above and below: Caterpillar and pair of mating adults of the Plain Tiger *Danaus chrysippus*

Small Copper *Lycaena phlaeas* frequents open and sunny places

Malpaís with *Euphorbia balsamifera* and *Schizogyne sericea*

Malpaís de Güímar

The Malpaís de Güímar is a spectacular coastal lava field, in which the dominant plants are species of *Euphorbia* accompanied by some other salt-tolerant shrubs. Apart from these, the plant diversity is low, but there are some sandy patches with a few extra species. This site is a Special Natural Reserve and is one of only a few places at coastal level on Tenerife where raw and geologically recent lava has been protected, with its natural plant and animal communities almost intact.

We have always loved this site and it is evidently loved by local people. A notice that we found there in the 1980s read "Que la paz y el espíritu de este jardín vayan contigo" (May the peace and spirit of this garden go with you). The malpaís is a mass of jagged 'aa' lava between the sea and the base of Volcán de Güímar. Just to the north of the Puertito de Güímar a clear track leads past a number of semi-underground dwellings – which are still very much in use – and continues along the coast. It provides excellent views of this austere but fascinating area, with the jagged lava that covers most of the site and looks particularly attractive in low sunlight of early evening.

The plants living closest to the sea are mainly the silver-leaved and yellow-flowered composite shrub Salado *Schizogyne sericea*, but slightly further inland Tabaiba dulce *Euphorbia balsamifera* comes into its own, and the large dome-shaped clumps of this species, often with extensive patches of bare black rock in between them, define the character of the site. Near the path along the coast there are some sandy patches that provide habitat for three of the characteristic salt-tolerant plants, *Tetraena fontanesii*, *Astydamia latifolia* and *Frankenia* species. Another plant occurring very near the coast is *Forsskaolea angustifolia*, endemic to the Canaries.

Ratonera *Forsskaolea angustifolia* is at home on raw lava but also on waste ground

One is not allowed to go further inland without permission, but some other plants can be seen from the coastal path. Among the *Euphorbia balsamifera* one can find the strange Romero marino *Campylanthus salsoloides*, and also the evergreen Balo *Plocama pendula*. Near the entrance to the site there is a scattering of *Asparagus arborescens*, endemic to the Canaries and identifiable by its extremely long and threadlike leaf lobes. Also present in small numbers are *Allagopappus canariensis* (= *A. dichotomus*), *Ceropegia fusca*, *Lavandula canariensis* and the scrambler *Periploca laevigata*, as well as a species of *Frankenia*, while in shady places you may find the bulb *Scilla haemorrhoidalis*.

Asparagus arborescens is taller than any of the other shrubs on the malpaís

Further from the coast two other species of *Euphorbia* may be visible. The candelabra-like Cardón *Euphorbia canariensis* forms large clumps and Tabaiba amarga *Euphorbia lamarckii* grows as an erect shrub. All the *Euphorbia* species on the island have a milky white sap that drips out if you break them. It is very unpleasant and the sap from *E. lamarckii* is really dangerous if you get it in the eyes. Sap from this species and *E. canariensis* has been used as a narcotic for catching fish, and that from *E. balsamifera* to make a form of chewing gum; in the 1980s it was easy to see

Bushes of Romero marino *Campylanthus salsoloides* often look rather miserable, but have beautiful flowers

Long-horned beetle *Lepromoris gibba*. Photo: Pedro Oromí

Mantid *Pseudoyersinia subaptera*, with vestigial wings
Photo: Pedro Oromí

places where stems had been cut to collect sap. The alien prickly pear *Opuntia dillenii* is still rare in the core of the malpaís, in contrast to scrub areas in so many other parts of the island, and we hope that it can be kept at bay in the future.

Although the malpaís generally gives the impression of being raw lava with no soil, one sometimes finds small patches of fine, pale orange dust. This comes by air from the Sahara at times when strong winds set in from the east for a few days, producing a haze or 'calima'. The dust is not noticeable in places where there is already soil, but shows up on barren black lava – as well as on cars – and may accumulate in hollows, enabling small plants such as mesembryanthemums or bulbs to gain a foothold. Over the long term, this process is probably of great importance in the Canaries (and on many Mediterranean islands) in promoting soil formation in volcanic terrain.

Species of *Euphorbia* dominate this habitat, and in spite of their poisonous latex there are several endemic beetles and other invertebrates adapted to feeding on them. One such animal is the spurge hawkmoth *Hyles tithymali,* illustrated earlier, whose spectacular striped caterpillars can be seen on Tabaiba amarga *Euphorbia lamarckii*. Inside dead branches of this shrub one can sometimes find a particularly interesting insect, the endemic termite *Bifiditermes rogeriae*. It is a member of a genus that is otherwise restricted to the tropical zones of the Old World, and its ancestors presumably colonised the Canaries when the genus was also present in northwest Africa. In dead trunks of Cardón *Euphorbia canariensis* there are sometimes larvae belonging to the longhorn beetle *Lepromoris gibba*, the only member of a genus endemic to the Canaries, the adults of which are large and grey but not often seen. Another group of insects that are often difficult to see are the mantids (Order Mantodea) of which there are several species in many parts of the island, but mainly in open habitats. These slow moving insects lie in wait for their prey on shrubs and smaller plants; the mobility of their heads (unique in insects) and binocular vision enable them to watch prey on all sides, and the armed front legs can extend at high speed to make the capture. In contrast, a variety of spiders are very visible here, the most obvious being those of the communal-living orb-spider *Cyrtophora citricola*, whose webs festoon many bushes.

Judging from the number of droppings, rabbits also manage to live on the malpaís, and the Tenerife Lizard *Gallotia galloti* is common. Several species of birds also occur here, one of the more interesting being the Long-eared Owl, which has been recorded nesting on the ground inside large clumps of Cardón; you may also see Hoopoe and Barbary Partridge, as well as Spectacled Warbler. The Trumpeter Finch used to occur here but is now apparently lost from this site and also from the rest of the island.

Wind-pruned coastal scrub on an exposed coast at Porís

Tabaibal del Porís and Punta de Abona

The Tabaibal del Porís is a protected area with important coastal *Euphorbia* scrub. There are no information panels, but the paths are well-marked, apparently mainly used by fishermen, and the vegetation is fascinating in spite of its low diversity.

Some coasts are so exposed and salty that even Siempreviva *Limonium pectinatum* has trouble growing there

This site is an excellent example of coastal scrub fully exposed to onshore winds carrying salt spray. As a result, the rocks closest to the coast have deposits of crystalline salt and hardly any plants. Slightly further inland the rocky and relatively level ground has been extensively colonised by Tabaiba dulce *Euphorbia balsamifera*, but the degree of exposure is reflected in the way these normally dome-shaped shrubs have grown here as spreading mats. A few other salt-tolerant plants survive here, including *Limonium pectinatum*, *Schizogyne sericea*, *Lycium intricatum*, *Lotus sessilifolius*, *Frankenia capitata*, *Tetraena fontanesii* and *Aizoon canariense*.

Although the finest *Euphorbia* scrub is to the north of the town of Porís, the area immediately to the south is also well worth a visit. There is a large bay, and beyond this is the Punta de Abona with a Faro (lighthouse). Near to it the bay is backed by a steep sandy bank that provides a habitat for several shrubs with limited distribution. The widespread coastal umbellifer *Astydamia latifolia* occurs here, but there are also some clumps of *Crithmum maritimum*, a related halophytic (salt tolerant) shrub that occurs all along the north coast of the island but is very scarce in the south. Of equal interest is the presence of *Euphorbia paralias*, a species that occurs on many European coasts but in Tenerife is found only at Porís and a handful of other sandy sites near El Médano.

The hinterland of the bay is a dry plain with rocks scattered on coarse sand. It has been disturbed to some extend but has a diverse coastal shrub community of a type that is uncommon on Tenerife. Significant species here include the beautiful composite Piña de mar *Atractylis preauxiana*, a rare

Euphorbia paralias on the beach near the Faro at Porís

Chenoleoides tomentosa

Kickxia sagittata

and endangered shrub that forms compact clumps, with fleshy, pointed, silver-green leaves in rosettes. The pineapple-shaped buds appear in the centre of the rosettes and give rise to attractive white or mauve flowers and eventually to golden brown seedheads. Also to be found here are clumps of the yellow-flowered *Lotus sessilifolius* and the semi-prostrate succulent shrub *Chenoleoides tomentosa*, one of several members of the family Chenopodiaceae that occur on the beaches of Tenerife. A less common species is *Kickxia sagittata*, a scarce figwort-relative with yellow, spurred flowers, which grows here as a thin mat in open rocky places, but sometimes forms a dense bush.

Dry coastal scrub behind the beach near Punta de Abona

Piña de mar *Atractylis preauxiana*

Acantilado de La Hondura in a time of drought

ACANTILADO DE LA HONDURA

The protected site of Acantilado de La Hondura, just south of Fasnia, is interesting for both its geological features and its natural history. The name means 'cliff of the deep water' and the isobaths show that deep water comes unusually close to the coast here, a fact that has doubtless always been important to local fishermen.

Gymnocarpos decandrus

The Hondura sea cliffs are formed by a series of basaltic lava flows. However, interleaved with the lavas are some layers of white pyroclastic deposits, presumably the source of the pale grit among the black basaltic boulders, volcanic bombs and stones that cover the ground above the cliffs. This area is arid, exposed to the wind and much affected by salt spray, and is thus a challenging environment for plants and animals, including the non-native rabbit.

Seedlings of *Astydamia latifolia* emerging after the end of a drought

The dominant shrub at La Hondura is *Euphorbia balsamifera*, the member of the genus that is typical of salty coastal situations, but we were intrigued to come across a few specimens of *E. lamarckii*, with their green twigs, but growing in much the same form as *E. balsamifera* instead of their more normal upright stance. *Schizogyne sericea* is also common, along with *Astydamia latifolia, Limonium pectinatum*, some *Tetraena fontanesii* and at least a few *Salsola divaricata*. Also present here are *Lotus sessilifolius,* a striking blue-grey species with tiny succulent leaves and yellow pea-flowers, *Forsskaolea angustifolia, Micromeria hyssopifolia* and the mat-forming *Aizoon canariense*. We also found a few cushions of the succulent *Gymnocarpos decandrus,* in which the inconspicuous yellow flowers mature to a brilliant orange.

Varied salt-tolerant and drought-tolerant shrubs flourishing in a challenging environment

On our visit to the site on 2nd July 2012, at the height of the year-long Tenerife drought, all of the euphorbias had dropped their leaves and most of the other plants – for instance the *Astydamia latifolia* – were reduced to clumps of dry twigs. The only obvious exceptions were *Plocama pendula*, which seems capable of retaining its green needle-like leaves through severe droughts, and *Tetraena fontanesii*, in which some plants were apparently lifeless but others had remaining globular leaves, many of which were bright yellow or orange, probably as a result of stress. Two plants that still had a few flowers were a *Frankenia* species, with small pink flowers, and *Schizogyne sericea*, with its yellow groundsel-like flowers.

Corazoncillo *Lotus sessilifolius*

We went back to La Hondura in November, about four weeks after the first rains broke the drought, and found the area in process of transformation. The *Euphorbia balsamifera* were growing new leaves, the *Astydamia* plants had put up vigorous new growth and near some of them were masses of seedlings. Other species, however, had not yet recovered from drought, and we have not yet found the local population of the rare composite Piña de mar *Atractylis preauxiana,* discussed above in relation to Porís de Abona.

A bank with *Tamarix canariensis* separates the lagoon of La Caleta from the sea

La Caleta del Río

A small and unusual marshy site with a saline lagoon and silty sandbanks, offering space for opportunistic salt tolerant plants suited to sandy soils.

The Barranco del Río, a major ravine originating on the southern slopes of Guajara, the highest peak on the rim of Las Cañadas, reaches the sea about 10 km northeast of Montaña Roja, near the Urbanización Callao de El Río. Here there is an open area known as La Caleta del Río, with saline lagoons surrounded by damp banks of mud, sand and gravel that provide suitable habitat for a number of salt tolerant plants that also need moisture.

Some conspicuous plants here are invasive aliens such as Tartaguera *Ricinus communis* and Tabaco moro *Nicotiana glauca*, while others that may or may not be native include the crucifer *Hirschfeldia incana*, a thorn apple *Datura* sp. and a species of cudweed *Laphangium* sp. Native species include the succulent composite *Reichardia crystallina*, with blue-grey leaves covered with glandular papillae and so swollen that they almost appear to merge; its dandelion-type flowers are borne singly and seem disproportionately large. Several mat-forming members of the family Aizoaceae flourish here on the nutrient-rich silty flats left behind by floods. *Mesembryanthemum crystallinum* has highly succulent, pointed, dark green or partly reddish leaves and elegant white flowers reminiscent of sea anemones. In *Mesembryanthemum nodiflorum* the leaves are more cylindrical, with rounded ends, and are often red; in both species the leaves are covered with glandular papillae. *Aizoon canariense* has broadly oval, somewhat boat-shaped leaves, bright green with the edges tinged with red; its tiny yellow flowers give rise to striking pentagonal orange plate-like fruits that ripen to red. Also here are clumps of the pink-flowered annual sea heath *Frankenia pulverulenta*, rapidly spreading mats of the chenopod *Patellifolia patellaris* and taller clumps of *Polycarpon tetraphyllum*, a member of the Caryophyllaceae, and *Tetraena fontanesii* with its brightly coloured globular leaves.

3 Coastal and lowland shrubland – 'matorral costero'

Polycarpon tetraphyllum

Cerraja de mar *Reichardia crystallina*

Tebete *Patellifolia patellaris*

Patilla *Aizoon canariense*

Cosco *Mesembryanthemum nodiflorum*

Barrilla *Mesembryanthemum crystallinum*

El Médano and Montaña Roja

El Médano is the only substantial area on Tenerife with patches of wind-blown sand and dunes, with characteristic associated plants. The site is a Special Natural Reserve and the protected area includes Montaña Roja, a 171 m volcanic cone that dominates the coastline on the seaward side of the airport – Tenerife Sur.

El Médano (sand dune in Spanish) and Montaña Roja are heavily used by joggers and walkers, sometimes with dogs, and although the authorities have laid out a network of stone-bordered paths, the pressure of people results in much disturbance, especially to birds. Furthermore, the views from the two main car parks are largely of rather unprepossessing rough stony plains, so the site does not make a very favourable initial impression, especially in the heat of summer. However, this area is actually one of the most ecologically interesting in Tenerife, with great variety of habitats and an extraordinary richness of plant life. It is home to many species typical of the eastern islands of Lanzarote and Fuerteventura – and indeed of northwest Africa – which are tied to the sandy habitats found in only a few places on Tenerife.

3 Coastal and lowland shrubland – 'matorral costero'

The flat stony and sandy regions – now with much tourist development – along the coasts to the northeast of El Médano and to the west of Montaña Roja have a sparse – almost Saharan – vegetation that is at its best in winter and early spring, with many of the shrubs in flower.

One of the dominant plants here is the yellow-flowered composite shrub *Launaea arborescens,* which forms a dense grey-green spiny latticework up to 1 m high, decorated with yellow flowers with small centres and long, well-separated ray florets.

The sandy Playa del Médano itself, with a worldwide reputation for its strong winds, is now usually crowded with spectacular wind surfers and kite surfers. Behind the beach is a sandbar sheltering a saline lagoon.

The most extensive dune area is on the gradually steepening slope that runs up from the lagoon to Montaña Roja. On this slope a thin layer of windblown sand partially covers the rocks beneath and forms small dunes in places; further up – above a fairly definite horizontal line – the sand is more patchy and the lava and cinders become prominent. A patch of more substantial dunes is tucked in beside the coast road running south from the town of El Médano. This is partially sheltered from the winds off the sea that tend to erode the more exposed slopes. These dune areas are tiny, yet they have been colonised by a variety of dune plants and animals that are found hardly anywhere else in Tenerife.

A little further south along the road is the car park for the beach of La Tejita, the broadest natural sand beach on Tenerife. The scrubby area between the car park and this beach has some places with loose sand, with appropriate

Sheltered dunes near El Médano town, where Berthelot's Pipit can often be seen foraging

Playa de La Tejita from Montaña Roja, with straggly specimens of *Nicotiana glauca* at bottom left

Matabrusca negra *Salsola divaricata*

Balancón *Traganum moquinii*

Amuelle salado *Atriplex glauca*

plants, but there are not well developed dunes here. La Tejita beach has some tourist facilities, and the section closest to Montaña Roja is one of the main naturist beaches on the island.

At El Médano euphorbias do not form a major element in the scrub except in some rocky parts, and in contrast to the situation in so many parts of Tenerife, the plants here include many species that also occur on the nearby continent and some even on the coasts of western Europe. Although some endemics are present, widespread species are dominant. This is doubtless because the plants growing in the sandy coastal habitat have to be tolerant of salt, and many of them have reached the island by sea.

Several of the shrubs at El Médano are hard for amateurs to identify, mostly having simple succulent leaves and similar growth forms. Furthermore, the taxonomy of those in the Goosefoot family Chenopodiaceae is in a state of flux, so some scientific names have recently changed.

One of the commonest chenopods here is *Salsola divaricata*, which can form a large, loose bush; it has arching branches, sausage-shaped succulent leaves that often turn red, and striking papery flowers. Another chenopod is *Traganum moquinii*, which also occurs in North Africa and is almost entirely restricted to sand dunes, where the erect shoots enable the plant to stay partly in the light even in rapidly shifting sands; it has crowded, stubby, fleshy leaves and hairy flowers. Also in this family is *Atriplex glauca*, also present on continental coasts, with straight, prostrate silvery stems, fairly flat though succulent leaves and crowded cream-coloured flowers.

Two other noticeable succulent shrubs from the north African flora are the strange *Tetraena fontanesii*, which grows mainly in rocky places, with grapelike leaves that may be green, yellow or reddish, and the carnation relative *Polycarpaea nivea,* a shrublet that is common on the sand, forming prostrate clumps; it has silvery, swollen, racquet-shaped leaves that sometimes turn orange.

Several other salt tolerant plants here may be familiar to visitors from northern Europe. Two of these resemble *Traganum* in their upright growth form. The knotgrass *Polygonum balansae* forms clumps with rigid vertical stems and upward-pointing leaves that roll backwards at the edges so that they almost form a tube; the small whitish flowers are crowded along the stems among the leaves.

Sea Spurge *Euphorbia paralias* has more sinuous stems and almost scale-like fleshy leaves. In contrast the sea heath *Frankenia cf capitata* (in another group with many name changes) forms clumps; it has tiny fleshy leaves and white or pink flowers. The widespread crucifer *Cakile maritima* (European Sea Rocket) is common in some areas with loose sand; it has pale green, swollen, entire or irregularly toothed leaves, white four-petalled flowers and curious cone-nosed seed pods. On the fringes of the beach here and elsewhere in Tenerife are groves of *Tamarix canariensis,* a small tree with reddish brown bark, scale-like leaves and minute pink flowers.

On one visit to El Médano we came across a group of cylindrical steel mesh cages protecting tiny plants from the ubiquitous rabbits. It turned out that this was a conservation effort by the Cabildo (island government) and the plants were seedlings of the endangered composite *Atractylis preauxiana* (see Tabaibal del Porís).

On the southeast side of Montaña Roja is an intriguing area of windblown sand that has been consolidated by salt spray into fine-grain sandstone. It is a challenging substrate for plants to grow on, but a few species such as Tabaiba dulce *Euphorbia balsamifera* and the

Polygonum balansae

Saladillo blanco *Polycarpaea nivea*

Rábano marino *Cakile maritima*

yellow-flowered composite Salado *Schizogyne sericea* manage to gain a foothold.

Montaña Roja can be climbed from the dune area behind the lagoon by the track that starts low on the slopes and climbs gently up the ridge. From the summit it is clear that the mountain is actually the eastern part of a volcanic crater that has been eroded away from the west by the action of the waves. One can also see the similarity between the old accumulations of windblown sand on the southeast (now turned into soft rock) and the area with modern dunes behind the beach northeast of the mountain.

Looking down from Montaña Roja, the consolidated sand at centre right contrasts with the shifting sands at top left

A number of endemic or unusual shrubs grow on the slopes of Montaña Roja, although Tabaiba dulce *Euphorbia balsamifera* is dominant. *Artemisia reptans*, which on Tenerife occurs only in a few places on the south coast, presumably gets its Latin species name from the extending ground-hugging branches, reminiscent of octopus tentacles. Other species include the evergreen *Plocama pendula* with slender pendent branches, *Ceropegia fusca* with bare, cigar-shaped stems, and also the rare endemic *Kickxia sagittata*, a figwort relative with striking yellow spurred flowers, which forms compact clumps. In addition to these endemics there are unfortunately a few introduced species including the invasive prickly pear *Opuntia dillenii* and also a tobacco relative *Nicotiana glauca*, which is common on the steep western side of the mountain. Construction of a tourist resort and heavy use of the area by people and their dogs has reduced the in-

Ononis tournefortii

Volutaria canariensis

terest of El Médano for birdwatchers, but the area is still home to a tiny population of Kentish Plover, now on the edge of extinction, and an early morning visit in spring or autumn may be rewarded by views of migratory waders in the saline lagoon. The Little Ringed Plover breeds in some freshwater pools slightly inland.

The sandy habitat also has a special community of invertebrate animals. One member of this is a small pale yellow and black solitary wasp *Bembix flavescens* that lives in holes in the sand; it can sometimes be seen flying low over the sand and then landing and burrowing. You might also see on the ground the large (2 cm long) predatory bug *Reduvius personatus*, a Mediterranean species (which bites!). Slightly damp places here provide habitat for the mole cricket, *Gryllotalpa africana*, a fascinating insect that lives underground and has its front legs specially adapted for digging. Another insect living underground among plant roots is the large (>3 cm) and glossy black nocturnal carabid beetle *Scarites buparius*.

Artemisia reptans, Limonium pectinatum and *Schizogyne sericea* on the slopes of Montaña Roja

Some other notable plants (many other less obvious species have also been recorded here) *Allagopappus canariensis, Astydamia latifolia, Chenoleoides tomentosa, Fagonia cretica, Helianthemum canariense, Heliotropium ramosissimum, Herniaria canariensis, Lycium intricatum, Limonium pectinatum, Lotus sessilifolius, Ononis serrata, Portulaca* sp., *Reseda scoparia, Schizogyne sericea* and *Senecio glaucus*.

Espinosillo *Fagonia cretica*

Camellera *Heliotropium ramosissimum*

Malpaís de la Rasca

The Malpaís de la Rasca is a Special Natural Reserve comprising an extensive area of semi-desert that includes the southernmost point of Tenerife. The lower parts of the protected area consist of extremely rough, rocky ground from which several impressive cinder cones emerge. In the north there are some low-lying dusty and stony flats with vegetation reminiscent of El Médano.

This protected area has diverse scrub communities in which species of *Euphorbia* play a large role. Although the malpaís is extensive, walking is easy on well-marked paths. However, the southeast section is degraded, so we prefer the access from Palm-Mar in the north. From here one can quickly reach the unspoiled central area as well as the coast northwest of the Faro (lighthouse) de la Rasca.

The coast is clean black lava and is frequented by fishermen, but is still a good

Tomillo marino herbáceo
Frankenia pulverulenta

3 Coastal and lowland shrubland – 'matorral costero'

La Rasca has one of the largest areas of undeveloped, low-lying, arid land on Tenerife

place to see birds such as Whimbrel, Ringed Plover and Little Egret, especially in the early morning. Landbirds are generally scarce, apart from the ubiquitous Berthelot's Pipit, but there is a good chance of seeing or hearing birds such as Southern Grey Shrike, Hoopoe and Stone Curlew. Other species that used to occur at la Rasca include Trumpeter Finch, Lesser Short-toed Lark and Cream-coloured Courser.

Tenerife is not on major routes for bird migration in the way that some Mediterranean islands are, but many migrants do turn up in the appropriate seasons and la Rasca is one of the best places to look out for them. The Faro is also good for sea-watching, with a chance to see some of the locally breeding seabirds: Yellow-legged Herring Gull and Cory's Shearwater are the commonest, but Bulwer's Petrel, Little Shearwater, Manx Shearwater and Madeiran Storm-petrels and Storm Petrels can also sometimes be seen.

Migrant shorebirds like these Ringed Plovers share the coast with fishermen

Espino de mar *Lycium intricatum*

Gualdón canario *Reseda scoparia*

A feature of this area that intrigued us was the occurrence of white deposits that we suspected were phosphatic. These have not been analysed, but they might provide evidence for the past occurrence of substantial colonies of seabirds in the desert on the southern tip of the island, in contrast to the current situation where the presence of mammalian predators restricts the native seabirds to inaccessible cliffs and offshore rocks.

There are several notable archaeological sites at La Rasca, including accumulations of mollusc shells presumably harvested by the aboriginal inhabitants. Several rough stone shelters close to the shore just west of the Faro provide clearer evidence of past occupation. These have been the focus of close attention by archaeologists and are thought to have been constructed by the Guanche people in the period before the Castilian conquest of the island at the end of the 15th Century.

The plant communities at la Rasca are typical of scrub habitats in the arid south of Tenerife. The salt tolerant Tabaiba dulce *Euphorbia balsamifera* is dominant near the coast and extends well inland, but Tabaiba amarga *Euphorbia lamarckii* and Cardón *Euphorbia canariensis* are also present, as well as the superficially similar *Kleinia neriifolia*.

While the slopes of the cinder cones may appear from a distance to be clothed entirely by Tabaiba dulce, closer inspection of some of them shows a variety of interesting shrubs, including Romero marino *Campylanthus salsoloides* with its strangely beautiful flowers and the composite *Atalanthus* cf *capillaris* with stubby branches topped by heads of hair-like leaves and delicate, long-stalked, groundsel-like flowers. Also present at la Rasca is *Artemisia ramosa*, endemic to Tenerife and Gran Canaria, which occurs only at a few sites in the far south of the island. Smaller plants include at least one member of the Aizoaceae and the striking yellow composite *Asteriscus aquaticus*, with flowers flared at the rim like a pastry-baking dish.

Some animals can cope with the severe conditions on land in this arid region. One is the gecko *Tarentola delalandii*, which hides in relatively cool cavities under large rocks and emerges to hunt at night. Canary Lizards and Rabbits are also present. One noticeable insect is the Bath White butterfly *Pontia daplidice*, the larvae of which are probably feeding on the abundant *Reseda scoparia*, an upright broom-like bush with curiously shaped seed pods. A red darter dragonfly, *Sympetrum fonscolombii*, was also present during our last visit.

Some other notable plants
Argyranthemum frutescens, Ceropegia fusca, Frankenia laevis, Kickxia scoparia, Launaea arborescens, Limonium pectinatum, Lycium intricatum, Neochamaelea pulverulenta, Periploca laevigata, Schizogyne sericea.

Artemisia ramosa
Photo: Keith Emmerson

Tenerife Wall Gecko *Tarentola delalandii*

Dragonfly *Sympetrum fonscolombii*

Bath White *Pontia daplidice*

Teno Bajo

The gently sloping land at Teno Bajo (Lower Teno), which forms the extreme northwest point of Tenerife, has areas of *Euphorbia* dominated scrub in a relatively natural state. This protected area in Teno Natural Park is isolated from the rest of the island by the mountains of the Teno massif and the cliffs of Los Gigantes.

This area was spared from early tourist and agricultural development by its inaccessibility until 1965, when a tunnel was driven through the rock at Punta del Fraile, 3 km west of Buenavista. It was doubtless grazed for centuries by flocks of goats brought down from Teno Alto, the village high above it, but otherwise the impact of people was probably small.

When one emerges from the tunnel on the road to the Faro de Teno (Teno lighthouse) on the extreme western point of Tenerife, one is entering the stronghold of Tolda or Tabaiba parida *Euphorbia aphylla*, one of the strangest of the endemic plants of the Canaries. It is entirely leafless, as its Latin name implies, with much branched, fleshy, jointed, yellow-green stems, forming dense clumps 50–80 cm high. This *Euphorbia* is the dominant plant on a

3 Coastal and lowland shrubland – 'matorral costero'

couple of small rocky slopes immediately above the road on the south side, and also on the easternmost part of the relatively gentle seaward slope below the road. *Euphorbia aphylla* is typically associated here with the unrelated but similarly leafless *Ceropegia dichotoma,* which grows particularly well both here and also far to the east at the tip of the Anaga peninsula, forming grey-green clumps reminiscent of candelabras.

Below the road to the west are gentler slopes running down to a magnificent exposed coastline, where salt spray is often in the air. This explains the presence, up to almost 100 m above the sea, of several widespread salt-tolerant plants including the spiny composite *Launaea arborescens* and also the shore-line umbellifer *Crithmum maritimum,* whose light seeds are dispersed in the sea and can easily blow up slopes with onshore winds. Also common in this area is the salt-tolerant Tabaiba dulce *Euphorbia balsamifera,* though it is less dominant in Teno Bajo than in most places near the coast in Tenerife.

Continuing westwards, the cliffs and craggy gullies of the Teno massif to the south of the road have large clumps of the columnar Cardón *Euphorbia canariensis,* often accompanied by Tabaiba amarga *Euphorbia lamarckii* and also the much less common *Euphorbia atropurpurea* with noticeably broad leaves and striking red flowers, as well as the succulent composite *Kleinia neriifolia.*

Cardón *Euphorbia canariensis* dominates the rocky slopes above the road at Teno Bajo

Tolda *Euphorbia aphylla*

Further on there are some major agricultural structures, including two huge 'invernaderos' (plastic covered growing areas) and road construction has also damaged the vegetation. On the flatter coastal area towards the lighthouse are large expanses of disturbed land, as well as a house and the remains of buildings used for growing tomatoes in the past. However, this is still a very special space, well worth a leisurely visit.

Perhaps the best place to see most of the endemic plants characteristic of Teno Bajo is along track PR-TF51. This is a delightful easy path up the relatively undisturbed slope

Parolinia intermedia    *Cheirolophus burchardii*

opposite six wind turbines near the Luz Teno sheds. The variety of shrubs here can be somewhat bewildering, especially since many of them are endemic and thus unfamiliar.

One of the less common endemic species found along this track is *Ceballosia fruticosa*, a tall, loosely branching shrub in the family Boraginaceae (which also contains *Echium*) with long, curved, gutter-like leaves and small tubular white flowers. Another scarce species is *Justicia hyssopifolia*, an arching shrub with hooded white flowers. Even more restricted in its distribution is the pink-flowered crucifer *Parolinia intermedia*, endemic to the Teno massif and the southwest (and scarce even there); it is the only Tenerife member of a genus endemic to the Canaries, which has its closest relatives in East Africa.

Sadly, the prickly pear *Opuntia dillenii* is also present on the slopes of Teno Bajo and the steeper rocks further inland, competing with the special native plants. There is evidence of attempts to control it, in the form of piles of cut cactus pads in places, but we feel that the complete elimination of this invasive alien species from the unique plant communities of Teno Bajo should be one of the highest conservation priorities on Tenerife.

When going to Teno Bajo it is worth trying to pause near the entrance to the tunnel or the exit from it, where some unusual plants survive. On the approach from Buenavista, before reaching the tunnel, some unstable rubbly slopes provide conditions suitable for the endemic daffodil relative *Pancratium canariense*, which we have found in flower in November. In the same area one can find *Parolinia intermedia*, *Lavatera acerifolia*, *Ceropegia dichotoma*, *Euphorbia atropurpurea* and also *Sideritis cretica*, with silver-felted heart-shaped leaves.

Artificial embankments and accumulations of gravel by the road are favoured by the beautiful purple-flowered knapweed relative *Cheirolophus burchardii*, a species endemic to the northern slopes of the Teno massif. However, also present here is the dock *Rumex lunaria*, a rather invasive endemic species that colonises disturbed land, which may well have reached the area at the time of construction of the tunnel. From the road it is possible to see parts of the vertical cliffs of the Punta del Fraile, challenging habitats where there are some unusual species that can also be seen further inland in dry woodland sites.

One of these is *Vieria laevigata*, the sole member of its genus and entirely confined to the Teno massif. It gave its name (at a time when it was officially spelled differently!) to a local natural history magazine, 'Vieraea'. A rarer plant here is the tiny but beautiful composite

3 Coastal and lowland shrubland – 'matorral costero'

Hypochaeris oligocephala

Amargosa *Vieria laevigata*

Hypochaeris oligocephala, known only from this part of Tenerife. On damp cliff faces you may find *Aeonium tabulaeforme*, which forms a convex plate with leaves overlapping as tiles, while in crevices there are clumps of a succulent *Monanthes*, cliff-dwelling relatives of the aeoniums, with swollen, often ovoid leaves.

Some other notable plants
Aeonium canariense, Argyranthemum frutescens, Artemisia thuscula, Asparagus sp., *Astydamia latifolia, Atalanthus pinnatus, Campylanthus salsoloides, Chenoleoides tomentosa, Dicheranthus plocamoides, Echium aculeatum, Lavandula* sp., *Limonium pectinatum, Mesembryanthemum crystallinum, Micromeria varia, Monanthes pallens, Neochamaelea pulverulenta, Paronychia canariensis, Periploca laevigata, Plocama pendula, Polycarpaea carnosa, Rubia fruticosa, Schizogyne sericea, Scilla latifolia, Todaroa aurea, Withania aristata.*

Cardoncillo verde *Ceropegia dichotoma*

Lágrimas de Virgen *Pancratium canariense*

Northern coastal sites

A variety of beaches and coastal cliffs where access is possible, though sometimes complicated. The plants include a number of widespread species and also a few that have very restricted distribution.

Along most of the north coast of Tenerife to the east of the Teno massif, the dry woodland originally extended downwards almost to the shore, so there was never much scope for the coastal shrubland habitat. Furthermore, in some places where it may once have flourished, goats have grazed for centuries, while in other places banana plantations or urban developments occupy the land. The most extensive northern areas of surviving coastal shrubland are in Teno Bajo (described above) and on the northeast coast of Anaga. However, there are several headlands and small beaches scattered along the middle of the north coast and some of them have plants that are not present elsewhere on Tenerife. As some of these are sensitive sites, we describe them only in general terms.

In general, the coastal sites in Anaga – including some with wide sandy beaches – have rapidly eroding cliffs of friable rock or steep rubbly slopes, resulting in plant communities of fairly low diversity, with mainly widespread species that are both salt tolerant and able to cope with the shifting substrate; among the common plants are *Argyranthemum frutescens*,

3 Coastal and lowland shrubland – 'matorral costero'

Kunkeliella subsucculenta

Centaurium erythraea

Lotus tenellus

Frankenia capitata

Hyoscyamus albus

Limonium imbricatum

Frankenia ericifolia

Lantén coronado *Plantago coronopus*

Cerraja de costa *Reichardia ligulata*

Spergula fallax

Polycarpaea divaricata

Astydamia latifolia, *Crithmum maritimum*, *Schizogyne sericea* and *Lavandula canariensis*.

These species also occur on many small beaches, where they are often accompanied by the chenopod *Patellifolia patellaris*, *Frankenia ericifolia* and *Spergula fallax*. Sometimes also present is Henbane *Hyoscyamus albus*, which may be native; it is a hairy and toxic member of the nightshade family with cabbage-like leaves, which often produces long branches with twin rows of geometrically arranged flowers. Another distinctive plant is the composite *Reichardia ligulata*, with crinkled, fleshy, thorny leaves and all-yellow flowers. Somewhat further inland, where the slopes are slightly more stable, Tabaiba dulce *Euphorbia balsamifera* is often dominant. Firmer footing for plants is also available in a few places around the bases of intrusive phonolytic rocks; these enable rare endemic plants such as the scabious *Pterocephalus virens* and the white-flowered knapweed relative *Cheirolophus tagananensis* to survive on ledges out of reach of goats.

Coastal scrub on steep slopes above a sandy beach in Anaga, with *Pterocephalus virens* at front left and inset

Further to the west along the coast there are some cliffs of hard basaltic rock, which also provide refuges for unusual coastal species such as *Limonium imbricatum*, with impressive purple flowers and strongly winged stems; its better known relative *Limonium pectinatum* occurs at almost every coastal site. Where barrancos run down to the coast and cut down through the rock strata of the cliffs or rubble from ancient floods or landslides, they may open out onto small rocky beaches with coastal plant communities similar to those in Anaga, though sometimes also with shrubs often found in dry woodland, such as *Paronychia canariensis*, *Withania aristata* and *Aeonium arboreum*.

In a few places there are dry headlands on which more extensive coastal scrub can develop, with species such as *Plocama pendula*, *Frankenia capitata*, *Neochamelaea pulverulenta*, *Salsola divaricata*, *Polycarpaea divaricata* and *Micromeria varia*. In a few places there are also rarities such as the spectacular *Lotus maculatus*, with spiky gold and scarlet flowers, which is the subject of a restoration effort by the island government. Other beaches have its commoner relative *Lotus tenellus*, with entirely yellow flowers.

In one site there is also a small population of the strange *Kunkeliella subsucculenta*, in the hemiparasitic family Santalaceae, members of which carry out their own photosynthesis but get water and salts from other plants. The genus *Kunkeliella* is endemic to the Canaries and *K. subsucculenta* is one of three species endemic to Tenerife. It usually has fat, cylindrical, succulent stems and could be confused with *Euphorbia aphylla* of the Teno massif, but at some stages of growth it produces thinner, broom-like shoots.

On the edge of one cliff we found a stunted form of *Centaurium erythraea*, a pink gentian-relative that is thought to be introduced, although the remoteness of the site made us doubtful. Also present in a few sites is Estrellamar *Plantago coronopus*, a plantain that forms a striking rosette of radial prostrate leaves in a ring of tall 'rat-tail' inflorescences.

Balo *Plocama pendula* on a dry headland

Salt tolerant plants on the beach where Barranco de Ruiz reaches the sea after cutting a gorge through rubbly cliffs. Inset: *Atalanthus pinnatus*

An enclosure protects a planted group of the rare *Lotus maculatus*

Euphorbia dominated lowland scrubland near Chimiche

Southeastern desert slopes

Semi-desert scrubby vegetation patchily distributed on pale-coloured, porous, pyroclastic rocks that mantle the southeastern foothills of the Las Cañadas massif.

The old main road along the southeastern side of Tenerife runs close to the 400 m contour, high up above the coastal motorway. The lower part of the ravine Barranco del Río, which originates close to the summit of Montaña de Guajara on the rim of Las Cañadas, passes under the old road between the villages of El Río and Chimiche. As it makes its final descent to the sea across progressively less steep slopes, the ravine – which carries water intermittently – opens out and passes through a landscape of heavily eroded semi-desert slopes, mainly composed of porous consolidated pyroclastic rocks of the 'Bandas del Sur' (see Geology chapter) in which the scarce rainfall is quickly lost. Land of this kind extends for many miles both to the east and to the west, between the old road and the motorway, but is rarely explored by visitors.

Close to the coast Tabaiba dulce *Euphorbia balsamifera* is often dominant, though sharing the space (and water) with other especially salt-tolerant species such as *Schizogyne sericea*, *Launaea arborescens* and *Limonium pectinatum*. Further up the slopes there is greater variety of other shrubs including Tabaiba amarga *Euphorbia lamarckii, Plocama pendula, Kleinia neriifolia, Periploca laevigata, Artemisia thuscula, Atalanthus microcarpus* and *Argyranthemum frutescens.* There are also scattered clumps of small and apparently young Cardón *Euphorbia canariensis,* a species now more often seen on steeper ground, implying that this area of scrub, degraded by many centuries of grazing, may be recovering to some extent. However, goats do still graze there regularly, as well as the ubiquitous rabbits, and together they doubtless still have a large influence on the composition of the vegetation, which also includes the invasive prickly pear *Opuntia maxima, Nicotiana glauca* and a species of *Datura.*

The bluffs and relatively level areas along the ridges are mainly open stony ground where it is especially hard for plants to find enough moisture. *Ceropegia fusca* is noticeable here, along with *Rubia fruticosa* and *Reseda scoparia,* with some grasses and bulbs including *Asphodelus ramosus.* There are also smaller plants including *Lavandula canariensis, Micromeria hyssopifolia, Asteriscus aquaticus, Polycarpaea nivea, Helianthemum canariense* and *Heliotropium ramosissimum,* the latter capable of forming a substantial cushion. The large shrubs are concentrated in the hollows and other places where there is a little soil, and also along the bottoms and sides of the ravines, especially where water lingers.

Cardoncillo gris *Ceropegia fusca* with inset flowers

Heliotropium ramosissimum

Asteriscus aquaticus

Helianthemum canariense

Standing water is scarce on the southern slopes, but pools persist into summer in Barranco del Río

Vegetation soon disappears from goat enclosures

Up near the old main road we are close to the upper limit of the coastal and lowland shrub habitat; above this the plants typical of the areas close to the sea increasingly have to compete with those typical of the transition to the dry woodland habitat – now largely missing – that once occupied the foothills of the mountains above. One of these species, *Cistus monspeliensis*, seems especially prone to invade degraded areas.

Near the village of El Río the ravine itself is deep and precipitous, and the vegetation on parts of the cliffs less accessible to goats includes species such as *Aeonium urbicum* and *Carlina salicifolia*, and also dry woodland species such as *Convolvulus floridus* and the endemic olive *Olea cerasiformis*.

The birdlife of these stony desert slopes is varied, but birdwatching is always most productive near dawn and dusk rather than in the heat of the day. For us, the highlight of an early morning visit was a Stone Curlew, or Alcaraván, which has given its name to the nearby ridge 'Lomo de los Alcaravanes'. Other birds included a Barbary Partridge with a brood of very small but already flying young, Kestrel, Southern Grey Shrike on lookout from the bushes, Spanish Sparrows associated with a nearby goat farm, and the ubiquitous Berthelot's Pipit. The sheltered sides of the ravine, with more vegetation, have birds such as Tenerife Blue Tit, Spectacled Warbler and Canary Chiffchaff.

Barranco de Masca, lower section, with *Polycarpaea carnosa* (bottom) & *Vieria laevigata* (top)

Masca Bay

A visit to Masca Bay, on the steep southwest side of the island, provides a glimpse of one of the most challenging environments faced by plants on Tenerife. The Barranco de Masca is also a challenging objective for walkers starting from Masca village 700 m above, but the coastal section is easily accessed by boat from Los Gigantes. The whole area is protected, being part of Teno Rural Park.

The mouth of the Barranco de Masca has been much modified by human occupation over the years, with many introduced plants, but is still extremely interesting. The plants include many native species and some that are specialities of the Teno massif. One of these is the generically endemic composite *Vieria laevigata*, demonstrating its extraordinary ability to cling to barren rock walls, but almost equally impressive is *Polycarpaea carnosa*, which seems to favour rubbly cliffs – see inset photographs above. Water lingers here in some places, and we were intrigued to find a tree frog and water snails living in small pools.

Some other notable plants (additional cultivated plants are also present near the shore)
Adiantum capillus-veneris, Argyranthemum frutescens, Arundo donax, Ceballosia fruticosa, Datura sp., *Epilobium* sp., *Ephedra fragilis, Euphorbia atropurpurea, E. balsamifera, E. canariensis, E. lamarckii, Gnaphalium* sp., *Juncus* sp., *Kleinia neriifolia, Lavandula* sp., *Lavatera acerifolia, Periploca laevigata; Phoenix canariensis, Phyllis viscosa, Plocama pendula, Polygonum salicifolium, Rubia fruticosa, Rubus* sp., *Rumex lunaria, Schizogyne sericea, Sideritis brevicaulis.*

Offshore islets

These are small islands with important seabird colonies and some special animals and plants. They are protected sites with access rarely permitted.

The offshore rocks around the island of Tenerife, like those close to many other oceanic islands, offer sanctuary to some species – especially seabirds – that have fared badly on the main island since the arrival of humans and predatory mammals. The most significant islets are Roque de Fuera and Roque de Tierra, off the northeastern tip of the Anaga peninsula, and the Roque de Garachico towards the west of the north coast. The Roques de Anaga are a 'Reserva Natural Integral', with visits allowed only for research or conservation, while the Roque de Garachico is a 'Monumento Natural' where authorisation is needed for visits.

The Roques de Anaga are ancient intrusive plugs of phonolitic rock that once formed part of the main island. Roque de Tierra is only 200 m offshore and is almost connected by a rubbly ridge exposed at low tide; it has an area of 6.4 ha and an altitude of 178 m. The Roque de Fuera is much more isolated, being separated from the mainland by a channel 1450 m wide and reaching 20 m in depth. It would have been connected to the mainland during the last Ice Age, when global sea level was more than 100 m lower. Its area is 3.6 ha and it is 66 m high.

The Roques de Anaga have traditionally been used by people for fishing and shellfish collecting, and in the first half of the 20th century the Roque de Tierra was sometimes used for pasturing goats. Seabirds are the primary conservation interest, since the islands are the main breeding places on Tenerife for Bulwer's Petrel *Bulweria bulwerii*, Madeiran Storm-petrel *Oceanodroma castro*, Storm Petrel *Hydrobates pelagicus* and North Atlantic Little Shearwater *Puffinus baroli*. Cory's Shearwater *Calonectris diomedea* also has a substantial colony on Roque de Fuera, but its mainland population is much larger. The Yellow-legged Gull *Larus michahellis atlantis* also breeds in many places around the coast, but its largest Tenerife colony is on the Roques. The reptiles also hold special interest, since Roque de Fuera is home to an

Roque de Tierra (upper left) and Roque de Fuera, with *Gallotia galloti insulanagae*
Inset photo: Aurelio Martín

endemic subspecies of the Canary Lizard, *Gallotia galloti insulanagae*; this can reach a length of 40 cm and possibly much larger (a cave on the island is known as Cueva del Caimán!). It differs from the subspecies *G. g. eisentrauti* in the north of the main island mainly in its large size, a tendency towards melanism and the lack of the typical dorsal pattern. It has presumably diverged from the population of north Tenerife since the Roque de Fuera became separated, perhaps around 8000 years ago. The Roque de Tierra, however, has a population of the mainland subspecies. Both of the rocks also have populations of the West Canary Skink *Chalcides viridanus* and the Tenerife Wall Gecko *Tarentola delalandii*, but it is not known whether these differ from their main island counterparts.

Ringing a petrel on Roque de Fuera

The effect of isolation of the Roque de Fuera is also indicated among the invertebrates, by the presence of a distinct subspecies of land snail in the family Helicidae, *Hemicycla bidentalis inaccessibilis*. Both rocks have a variety of other invertebrates that also occur on the main island, including large populations of the tenebrionid beetle *Hegeter amaroides*, which apparently eats the excrement of both lizards and seabirds; but the beetle also forms part of the diet of the lizards.

The Roques de Anaga also have considerable botanical interest. The Roque de Tierra has many shrubs typical of the coastal zone, including several species of *Euphorbia*, but also some species that are more typical of dry woodland, including at least four old specimens of the Canary Olive *Olea cerasiformis*, and a hundred or so Dragos *Dracaena draco* near the summit on the north face of the island, where moist updrafts are doubtless most prevalent. Many young Sabinas *Juniperus turbinata* are present at middle levels, but there was evidently a well-developed sabinar (juniper woodland) in the past, before felling for charcoal-making in the 1930s. Five plants endemic to Tenerife - *Echium simplex, Lotus maculatus, Gonospermum revolutum, Lavandula buchii* and *Aeonium volkerii* - all have a foothold on Roque de Tierra, as well as *Asparagus arborescens* and the rare *Convolvulus* cf *fruticulosus*. The Roque de Fuera has apparently been subject to greater human influence and has greater accumulations of bird guano, as well as being more isolated. It has few plant species, but one Drago was present until the 1980s, and there are some *Euphorbia balsamifera* as well as other salt tolerant coastal plants, and weedy species including *Nicotiana glauca*, which was not found on Roque de Tierra.

The Roque de Garachico lies about 300 m offshore from the town of the same name, near the western end of the north coast of Tenerife. The islet is separated from the mainland by a channel with a depth of up to 12 m; it has an area of about five hectares and height of 77 m. It consists of a series of basaltic lava flows that also formed the adjacent part of the mainland. Marine erosion has both separated it and formed the cliffs and steep rubbly slopes that render access to the rock impractical except in calm weather. From a distance, there is little sign of human activity apart from the cross on the summit, but the rock has been used by local people for thousands of years, for fishing and collection of shellfish; goats have sometimes

Roque de Garachico, with view across to the town. Photo: Rubén Barone

been pastured there and rabbits were once introduced, although they soon all died.

The breeding birds of the islet include a colony of *Bulweria bulweri* second in size only to that in the Roques de Anaga. A small number of pairs of *Calonectris diomedea* breed on the rock, and it is highly probable that there are small colonies of *Oceanodroma castro* and *Puffinus baroli*. In addition there are about 100 pairs of Yellow-legged Gulls *Larus michahellis atlantis*, which are probably significant predators on the small petrels. The only land birds that breed regularly are the Rock Dove *Columba livia* and Plain Swift *Apus unicolor*, although the Kestrel *Falco tinnunculus* has bred occasionally; other species of raptors make occasional visits. The islet also provides secure roost sites for other species, especially Grey Heron *Ardea cinerea* and Little Egret *Egretta garzetta*.

Canary Lizards *Gallotia galloti*, geckos *Tarentola delalandii* and skinks *Chalcides viridanus* are all present on the Roque de Garachico, but in contrast to the situation in Anaga, the lizards are similar to those on the adjacent mainland. This lack of divergence is echoed in the landsnails: *Hemicycla bidentalis*, which has a distinct subspecies on the Roque de Fuera, is represented on the Roque de Garachico by specimens similar to those on the coast nearby.

The plants on the Roque de Garachico are much less diverse than on the Roque de Tierra in Anaga, which is similar in size, but they include the Canary endemics *Limonium imbricatum* and *Chenopodium coronopus*, both of which occur only at widely scattered sites around the coast of the main island. Several other plants typical of the lower zone on Tenerife are present, including species of *Euphorbia*; there are only a few clumps of Cardón *Euphorbia canariensis* but a large area is covered by dome-shaped bushes of Tabaiba dulce *Euphorbia balsamifera*, some of them partly covered by the sprawling *Asparagus umbellatus*. Unfortunately there are also a good many clumps of the invasive cactus *Opuntia dillenii*, the seeds of which are dispersed by lizards, whereas on the Roques de Anaga this species was very scarce when surveyed in the early 1990s. A conservation initiative to eliminate the cactus from the Roque de Garachico has been suggested, although the steepness of the island might make it impracticable.

Invertebrates on the Roque de Garachico include two species of ants, one of them the Canary endemic *Camponotus hesperius*. As on the Roques de Anaga the most abundant beetle is the tenebrionid *Hegeter amaroides*, which scavenges around seabird colonies. A notable lepidopteran is the Canary endemic moth *Amicta cabrerai* (Psychidae), whose larvae construct caddis-like cases that are conspicuous on the rocks. The females are wingless and wormlike, remaining in the cases for their entire lives, but the males are normal flying moths.

Dry woodland at Genovés, with *Echium giganteum* and *Osyris lanceolata* (bottom centre and right)

4

Dry woodland remnants – 'bosque termófilo'

The dry woodland ecosystem

Modern investigations suggest that before the arrival of the first human colonists between two and three thousand years ago, there was an almost continuous band of dry woodland around Tenerife. Along the north coast it occupied the space between small fragments of salt-tolerant *Euphorbia* dominated scrub near the coast and the moist laurel forest only a little higher up. In most of the southeast and southwest, however, it lay above a broad band of coastal and lowland shrubland, and extended upwards to make a transition to pine forest at a much higher level.

Whereas the laurel and pine forests are easily recognised ecosystems, the dry woodland occurred in various forms, dominated by different tree species. We use the term dry woodland for this set of habitats, but plant ecologists often refer to it as thermophilous (warmth-loving) woodland or 'bosque termófilo'. It can develop in places with annual rainfall as low as 100-400 mm, but it does not tolerate low temperatures. In its various forms, the dry woodland once covered more than 15% of the total area of Tenerife.

The modern situation is very different, since the dry woodland of Tenerife has disappeared more completely than any other natural ecosystem, because the zone which it once occupied is that in which much human agricultural activity – especially the pasturing of goats and sheep – has been concentrated. By the end of the 20th century only scattered remnants of

Sabinar – dry woodland dominated by Juniper – near Afur in Anaga

relatively intact dry woodland survived on Tenerife, covering less than 1% of the island, and almost all of the trees that once dominated it had become scarce.

In most places the undisturbed dry woodland would have been open enough to allow development of rich and varied communities of shrubs and herbs, in contrast to the closed-canopy laurel forest and pine forest at higher levels, which have little undergrowth. Partly because of this, the dry woodland ecosystem as a whole was the most diverse of all those in the Canaries, with large numbers of endemic plants and animals. However, in view of its generally degraded state, it is unsurprising that it contains the largest number of endangered species.

Until recently the various natural types of dry woodland in the Canaries received little attention from ecologists and conservationists, but the situation has now changed, since in spite of their currently lamentable state, the European Union has given special status to 'sabinares', 'palmerales' and 'acebuchales', respectively dominated by the Canary Juniper *Juniperus turbinata* subspecies *canariensis*, the Canary Palm *Phoenix canariensis* and the Canary Olive *Olea cerasiformis*. Furthermore, the EU recently funded a LIFE project on the Bosques termófilos of the Canaries. In connection with this, the Cabildo Insular de Tenerife purchased a 'finca' (estate) in Teno in order to undertake a major project in ecological restoration, to bring back part of the dry woodland that has so nearly vanished.

The dry woodland ecosystem of the Mediterranean and Macaronesia is thought to be relatively young in evolutionary terms, having developed mainly since the origin of 'Mediterranean' climates near the end of the Pliocene epoch, some three million years ago. Areas with this type of climate have dry, hot summers and cold but often wet winters, and plants living in them must withstand the stresses of both these extremes. Because of the strongly seasonal environment, the flowering and fruiting of most of the plants is restricted to certain times of year.

Many of the species typical of the dry woodland in the Canaries have affinities with the flora of the Mediterranean and North Africa. Ancestors of some of these trees and shrubs are thought to have lived in dry tropical areas close to the ancient Tethys Ocean, with adaptations such as thick, tough and usually small 'sclerophyllous' leaves that live for at least a year, resist water loss and are unattractive to herbivores. Examples are members of the genera *Asparagus*, *Jasminum* and *Olea*, but also the Canary Palm *Phoenix canariensis*. In contrast, some relevant plant groups such as *Pistacia*, *Ephedra* and *Artemisia* hail from the semiarid steppes of Central Asia, while others such as *Juniperus*, *Arbutus*, *Lavatera* and *Navaea* probably evolved closer to what is now the Mediterranean region. Two more species were recently added to this group, when researchers found fossil evidence of the presence of a hornbeam *Carpinus* sp. and an unidentified oak *Quercus* sp., neither of which was previously considered native to Tenerife. It is not yet clear, however, whether these two species played a part in the dry woodland.

We treat the remnant dry woodland habitats in some detail here, since although they are small in area, they include many interesting and little known species, and are important in terms of diversity and evolutionary history. In the Canaries archipelago as a whole one can still see good examples of several variants of the dry woodland, with different tree species dominant. We have gleaned information from some of the other islands, but have also tried to visit the sites on Tenerife where the best surviving remnants are to be found.

Dry woodland on coastal cliffs near Buenavista with *Artemisia*, *Marcetella* and *Pistacia* in foreground

Drago Dracaena draco

Acebuche Olea cerasiformis

Saquitero Heberdenia excelsa

Sabina Juniperus turbinata

Almácigo Pistacia atlantica

Typical dry woodland plants

The main characters in this chapter are the small number of trees and large shrubs that are illustrated and described here, which probably once dominated rather distinct types of dry woodland in different places. They were supported by a dazzling cast of smaller or less dominant trees and shrubs of all shapes and sizes, and together they formed a continuous band around the island, between the lowland scrub and the evergreen forests above.

Perhaps the most prevalent of the dominant trees was 'Sabina' *Juniperus turbinata*, a bushy conifer growing to 8 m, with much-branched scaly twigs and hard round fruits, initially green but ripening to bluish and then reddish black. The Canary Olive *Olea cerasiformis* occupied some of the driest areas and has survived mainly on cliffs; on open ground it forms large and almost spherical bushes with thin branches; its leaves arise in opposite pairs and are long, narrow, leathery and dark green. The Mt. Atlas Mastic Tree or 'Almácigo' *Pistacia atlantica*, which gave its name to a village in Anaga, also has thin branches but has pinnate leaves; it spreads out more than the olive, eventually forming a hemispherical deciduous tree, often noticeable from a distance when the leaves have fallen. 'Retama blanca' *Retama rhodorhizoides* is really a large broom; its twiggy branches are dull green for most of the year but erupt into a burst of dense white pea-flowers in early summer. The Canary Palm *Phoenix canariensis*, now reasserting itself in the wild as well as where it is planted, is an unmistakeable sturdy, thick-trunked variant of date-palm stock. Equally impossible to miss is the iconic tree of the Canary Islands, the Dragon Tree or 'Drago' *Dracaena draco*, with a straight trunk supporting a dense topknot of straplike

4 Dry woodland remnants — 'bosque termófilo'

Mocán Visnea mocanera

Palmera canaria Phoenix canariensis

Peralillo Maytenus canariensis

Retama blanca Retama rhodorhizoides

Marmulán Sideroxylon canariense

leaves; it is much planted in towns but also gives a special character to many remote areas, where scrawny specimens can be seen clinging to distant cliffs.

Other trees typical of the dry woodland but not normally dominant include two more that are now found especially on cliffs: the elusive Marmulan *Sideroxylon canariense* has long, boat-shaped leaves, while the rather spindly *Maytenus canariensis* has notched, broad leaves reminiscent of tennis rackets. Two additional trees occur in dry woodland but also in drier parts of the the laurel forest above: *Visnea mocanera* is a small tree or bush with rather flat, shiny and lance-shaped leaves, while *Heberdenia excelsa* in its dry-country form, often referred to as 'Saquitero', forms tall asymmetric trees that jut proudly out from largely deforested hillsides in a few places. Finally there is the small tree 'Orobal' *Withania aristata*, related to the nightshades and tomatoes, with striking hooded flowers and fruits, which is now often found around villages.

As well as the major trees, the dry woodland included a large suite of shrubs, including about 20 that are generally accepted as having been characteristic of this ecosystem on Tenerife; these are illustrated on the next three pages. One shrub that can almost always be found in the woodland remnants is *Hypericum canariense,* a large bush with dense clusters of yellow flowers often standing up above the highest leaves; they are followed by dark

Orobal *Withania aristata*, often seen around villages

Jócama Teucrium heterophyllum

Tepopote frágil Ephedra fragilis

Espinero negro Rhamnus crenulata

Lengua de pájaro Globularia salicina

Moralito Rhamnus integrifolia

brown fruit capsules, while the leaves tend to go yellow, making the bushes visible from far away. Another constant species is *Jasminum odoratissimum,* with dark green pointed leaves, tubular yellow flowers and fruits reminiscent of coffee beans. A third species almost always present is *Rhamnus crenulata,* with straight, stiff, thorny stems, small white flowers and pea-sized fruit that ripen from green to red to black. A rather inconspicuous but prevalent shrub in dry areas is *Globularia salicina,* low-growing and with crowded lance-shaped leaves and whitish flowers in the axils near the tips of the branches. Another widely distributed but less abundant shrub is *Echium strictum,* the only local member of the genus with broad, soft leaves, which can also be found in the laurel forest.

Marcetella moquiniana, known as 'Palo de sangre' for its often crimson vertical stems, forms an open bush with coarsely notched pinnate leaves; it turns up in many dry woodland remnants and occasionally forms almost pure stands. *Convolvulus floridus,* a large bush with long leaves similar to the olive, has unmistakeable clusters of large white flowers. A large shrub unfamiliar to most visitors to the Canaries is *Bosea yervamora*, a Canary endemic member of the Amaranthaceae, a family comprising mainly much smaller plants; it has greenish, arching stems, sprays of white flowers and red fruit, and sometimes forms curtains down damp cliffs in ravines. Another little known shrub is *Ruta pinnata,* a member of the *Citrus* family Rutaceae, which occurs at scattered sites around the island; it has yellow flowers and unusual pinnate leaves with 3-4 pairs of widely spaced leaflets and one at the tip.

Some of the dry woodland shrubs have more restricted distribution in Tenerife. The partially parasitic *Osyris lanceolata* occurs at only a few sites in the north, but is numerous in

Bayón *Osyris lanceolata*

Retama fina *Spartocytisus filipes*

Jasmín *Jasminum odoratissimum*

Oro de Risco *Anagyris latifolia*

Guaidil *Convolvulus floridus*

some of these and easily recognizable by its yellow green appearance and small, thick, lance-shaped leaves. *Anagyris latifolia,* named admiringly 'Oro de risco', is a leguminous shrub with broad leaves and large golden yellow flowers; it is now also restricted to a few scattered sites but might well flourish in many others if protected from goats. The broom *Spartocytisus filipes* is found at only a few sites towards the west of the north coast and near Masca; it is the more delicate dry woodland equivalent of Retama del Teide, the principal white-flowered bush of Las Cañadas. *Ephedra fragilis,* an unusual relative of the conifers, is another shrub with thin, straight stems and has similarly restricted distribution on the island. The same applies to *Rhamnus integrifolia,* recognizable by its long-stalked, stiff and dagger-like leaves, which now has a relict distribution including cliffs in Las Cañadas and deep barrancos in the southwest.

Bupleurum salicifolium is recognizable by its spreading growth and the narrow, pointed leaves that give rise to its specific name; it occurs mainly in dry places in the south at up to 1000 m, but also at a few moister sites in Anaga, where the leaves are noticeably broader. *Carlina salicifolia,* a large thistle, also has somewhat willow-like narrow leaves, but with prickles along the edges; it has yellow flowers but is more often seen with conspicuous dead golden seed heads. *Asparagus scoparius* has tall, upright, unarmed stems and very short needle-like cladodes; it is the member of its genus most clearly associated with dry woodland, but is now frequent only in the north; other species of the genus also sometimes occur in dry woodland.

Three of the most attractive species are left till last. The labiate *Teucrium heterophyllum,* with crimson flowers and silvery oval leaves that are sometime notched, is conspicuous and

- Norsa *Tamus edulis*
- Anis de risco *Bupleurum salicifolium*
- Yerbamora *Bosea yervamora*
- Granadillo *Hypericum canariense*
- Esparragón raboburro *Asparagus scoparius*
- Maloica *Carlina salicifolia*
- Navaea phoenicea
- Taginaste chico *Echium strictum*
- Palo de sangre *Marcetella moquiniana*
- Malva de risco *Lavatera acerifolia*
- Ruda canaria *Ruta pinnata*

beautiful when blooming, but scarce and easily overlooked at other seasons. The two native shrubs in the family Malvaceae, *Lavatera acerifolia* (endemic to the Canaries) and *Navaea phoenicea* (endemic to Tenerife), are dry woodland species, but both are hard to find. The former, which has pale mauve petals with darker bases, occurs in scattered sites around the island; the latter, with beautiful peach coloured flowers, occurs only in the Teno and Anaga massifs and is now very rare, and is still under pressure from goats at one of its important sites.

Goats still pose a threat to many rare plants on Tenerife

Apart from the typical shrubs mentioned above, several other species have an altitudinal range suggesting that they may once have been a normal part of the dry woodland communities, although some may never have been widespread. Among the strongest candidates are *Aeonium arboreum*, a spreading bush with striking yellow flowering heads, and *Echium giganteum*, with similar shape but white flowers. Other candidates are *Asparagus umbellatus*, *Carduus clavulatus*, *Descurainia millefolia*, *Echium simplex*, *Euphorbia atropurpurea* (typically occurring with *Retama rhodorhizoides* in the west), the rare *Limonium arborescens* and *L. macrophyllum*, *Pancratium canariense*, *Plantago arborescens*, *Salvia canariensis*, *Solanum vespertilio* and the scramblers *Smilax aspera* and *Tamus edulis*.

Other kinds of plants often occur in dry woodland sites but are either extremely widespread or are more typical of other ecosystems; some of these are mentioned below in the discussion of transition zones. Dry woodland sites also often have scattered alien shrubs *Opuntia maxima* and *Agave americana*, while roadsides and other disturbed areas are invaded by the notorious yellow-flowered weed *Oxalis pes-caprae*.

Bejeque arbóreo *Aeonium arboreum*, a spectacular shrub of the dry woodland

Tomillo salvaje *Micromeria varia*, an inconspicuous plant common in the north

Transition zones

Maps of the potential dry woodland of Tenerife may give the impression that it once formed an easily recognizable band around the island, but in reality the various kinds of dry woodland merged gradually – in broad transition zones – into lowland and coastal shrubland below and laurel or pine forest above. The lower transition zone is especially difficult to interpret, because of the overwhelming impact of humans and their animals. The transition originally occurred around the level – differing in the north and south of the island – where conditions become moist enough for growth of large bushes and small trees. In the north this transition occurs low down – below 100 m in some areas – so that in a few places the woodland once extended down to the top of the sea-cliffs, with only a narrow band of *Euphorbia* dominated scrub along the coast. Around the south of the island, however, the transition was usually at 300-500 m.

In both north and south the lower transition zone includes some shrubs more typical of the lowland shrubland. Two species of *Euphorbia* are especially relevant: Tabaiba amarga *Euphorbia lamarckii* now occurs at all levels within the potential dry woodland zone; it has been recorded as high as 1500 m and can be found among juniper and at the edge of regenerating pine forest, so there is a case for considering it as a typical species of dry woodland as well as of lowland shrubland. Cardón *Euphorbia canariensis* also extends far inland, often dominating the vegetation of dry cliffs but also extending on to more gentle slopes. Both these species may have been less common inland when the dry woodland was fully developed; they are moderately resistant to grazing by goats, and seem to be quick to recolonise when grazing pressure is reduced. This may also be true of some other shrubs that now occur in dry woodland remnants, such as *Artemisia thuscula* and the succulent *Kleinia neriifolia*. Other shrubs occurring in areas transitional between lowland shrubland and dry woodland remnants include *Allagopappus canariensis, Campylanthus salsoloides, Ceballosia fruticosa, Ceropegia dichotoma, Gonospermum fruticosum, Justicia hyssopifolia, Paronychia canariensis, Periploca laevigata, Rubia fruticosa* and *Teucrium heterophyllum*.

The upper limit of the dry woodland originally involved other transitions, again differing in the north and south. Along the north coast the change was related to the increased humidity in the zone just below the base of the prevalent cloud sea, at about 200-400 m; above this level the moisture promotes growth of a different vegetation type, the closed canopy

Canary Pine and Sabina *Juniperus turbinata* above El Río in a recovering transition zone

laurel forest composed of tall evergreen trees. The change is gradual and the transition zone is enriched in places by some trees more often found in the laurel forest but tolerant of relatively dry conditions, such as *Morella faya, Erica arborea, Arbutus canariensis, Apollonias barbujana, Heberdenia excelsa, Pleiomeris canariensis* and *Visnea mocanera*; smaller species found in these areas include the scrambler *Canarina canariensis* and *Dracunculus canariensis*.

On the southeast and southwest coasts the reduced influence of the trade winds and consequent greater aridity inhibits development of laurel forest; this allowed the dry woodland to extend further up the slopes, with a gradual transition to pine forest at around 700-900 m, where there may be communities with both pine and juniper in varying proportions. High up in the Güímar valley, where cloud is prevalent, the transition zone was more like a dry form of laurel forest: in some barrancos running up into the pine forest there are still substantial stands of *Arbutus canariensis* and *Bencomia caudata*, along with *Salvia canariensis* and other dry woodland shrubs. In Teno there is no transition to pine forest and degraded dry woodland can be found up to the highest levels, around 1100 m; here Retama blanca *Retama rhodorhizoides* is often dominant.

Grazing pressure has recently declined in many parts of Tenerife, so although much land has been developed for various purposes, some poor land that once held dry woodland has been abandoned by people and is being re-colonised by shrub communities. These degraded habitats often still lack the trees that would once have been dominant

Tacorontilla *Dracunculus canariensis*

Mocán *Visnea mocanera*, a tree typical of transition from laurel forest to dry woodland, with a heavy crop of fruit in an open site

in the various forms of dry woodland, and tend to contain large numbers of *Hypericum canariense, Cistus monspeliensis* and *Asphodelus ramosus*, with some *Euphorbia lamarckii*. Invasion by alien plants is also common, one of the most conspicuous of these being the succulent *Agave americana*, with enormous sharp-pointed, spear-shaped, silvery green leaves and tall, branched inflorescences. Equally invasive are several species of Prickly Pear or 'Tunera' in the genus *Opuntia*, which are cacti originating in the Americas; their seeds are spread both by humans and by various birds, including the Raven *Corvus corax*, and they are now found 'wild' in many parts of the dry lower zone, flourishing in disturbed ground and also invading wilder areas where they compete for water and space with native plants. However, the once-dominant native trees are also making a comeback in many places. We have found steep slopes with vigorous young junipers, barrancos dotted with palms only a few feet high, and dragon trees spreading on slopes that had once been devastated by goats. At one degraded site above the village of El Río in southeast Tenerife, now dominated by *Cistus monspeliensis* and *C. symphytifolius*, juniper *Juniperus turbinata* is apparently re-establishing itself around 650 m and pines are colonising the slopes immediately above.

The main types of dry woodland

It has been suggested that there were once about five more or less distinct types of dry woodland on Tenerife, dominated respectively by Sabina *Juniperus turbinata*, Canary Palms *Phoenix canariensis*, Retama blanca *Retama rhodorhizoides*, Canary Olive *Olea cerasiformis* and Drago *Dracaena draco*. Doubtless there were some areas where these species dominated, but we suspect that intimate mixtures of several of them were also widespread.

On Tenerife there are still a few substantial areas (around 440 ha in total) of 'sabinar', where the *Juniperus turbinata* ssp. *canariensis* is dominant. In the past juniper woodland was evidently much more widespread, occupying some of the highest dry areas in the south of the island between 300-500 and 700-900 m, immediately below areas of pine forest. It was also present in some parts of the north, from close to sea level up to several hundred metres, where it usually graded into laurel forest.

The Sabinar de Afur in the north of Anaga is perhaps the finest surviving example of juniper scrub on Tenerife, and is described later. Also in the north are good juniper sites near Punta de Anaga, on steep hilltops in the south of Anaga, near Genovés, and on an almost inaccessible steep slope between the village of Tigaiga near Los Realejos and Icod el Alto. In the west there are some extensive patches around 800 m near Guía de Isora, Chío and Tamaimo. In the south of the island there are few substantial surviving areas dominated by juniper, but there are also patches near Arico and several place names and scattered bushes testifying to the previous presence of sabinares, for instance near Arona, Fasnia and Güímar. We have also noticed many isolated juniper bushes in widely scattered precipitous parts of the island, and active regeneration seems to be in progress in many of these.

Reconstruction of the original distribution of Canary Palms *Phoenix canariensis* is particularly difficult because the areas where it once grew are almost all developed, and the species is often planted. Palms can survive on fairly dry slopes and there are significant groups in many parts of the island, so it is

Sabina *Juniperus turbinata* on intrusive rock in southern Anaga

easy to imagine palms dominating large areas in the ravines of the coastal fringe along the north coast of Tenerife and in Teno, and on both north and south coasts of the Anaga peninsula. On the other Canary Islands palms sometimes form pure stands, known as 'palmerales', but they may also grow with other trees typical of the dry woodland, while Canary Willows *Salix canariensis* often accompany them in wetter sites.

On Tenerife the place that perhaps gives the best impression of a palmeral is around Viña Vieja at 200 m in the Barranco del Cercado de Andrés. Here, in the heart of the Anaga peninsula and 3 km from the sea, there has been extensive recent palm regeneration from a tiny surviving group with possibly wild ancestry. Near Taganana in the north the process of regeneration can be seen at an earlier stage, with numerous healthy young palms in some barrancos. In the Teno massif regeneration is occurring in much drier and higher situations, with apparently wild groups of palms present in several places, for instance in a ravine at about 700 m above Los Carrizales near Masca. One of the most interesting stands that we have seen is on the cliffs at Interián, near Garachico, where palms growing in a ravine - with *Marcetella moquiniana* on the drier slopes nearby - provide a clue as to the vegetation that would have been natural on this site. Another intriguing modern situation is the presence of a number of scattered palms relatively high up in the sabinar at Afur in Anaga.

The broom Retama blanca *Retama rhodorhizoides* is capable of growing in extremely arid situations and still dominates large areas in parts of the Teno massif, up to the highest levels around 1100 m; this community is termed 'retamar blanco'. In the highest areas, and especially in the region around Masca, it is often accompanied by *Euphorbia atropurpurea*, a species which also occurs in other

Canary Palms and Palo de sangre *Marcetella moquiniana* at Interián

Retamar blanco below Los Carrizales with *Retama rhodorhizoides*, and *Salix canariensis* in the barranco

Woodland dominated by Acebuche *Olea cerasiformis* with a single Sabina *Juniperus turbinata* (top edge)

Perhaps the only wild individuals of Lentisco *Pistacia lentiscus* (dark green, centre) surviving on Tenerife

places where dry woodland once flourished, for instance near Buenavista in the north of Teno, high up above Güímar, and in the Conde massif near Adeje and Ifonche. In Teno Retama blanca often occurs in a degraded but perhaps recovering habitat that also includes species such as *Atalanthus capillaris, Bupleurum salicifolium, Carlina salicifolia, Sideritis brevicaulis, Echium virescens,* and *Globularia salicina*. Retama blanca has also colonised abandoned agricultural areas at lower elevations, which may once have been clothed with junipers or olives.

The 'Acebuche canario' or Canary Olive *Olea cerasiformis* is now very scarce on Tenerife, but the species may once have been important in some of the warmest and driest parts of the island, as it still is on Gran Canaria, where some 'acebuchales' persist. (European Olive, now treated as a separate species *Olea europaea*, has been planted in some places.) On Tenerife the nearest we have seen to woodland dominated by Canary Olive is at just over 500 m on the steep and rocky slopes of Valle Brosque in the south of the Anaga peninsula (see site account below). Scattered wild olives can be found in other dry rocky places and cliffs on Tenerife, mainly but not exclusively in the south. One especially good site for olive and other dry woodland species is on the cliffs of the Barranco del Río, above and below the village of El Río (see Cliffs and ravines).

The European Union defined its conservation habitat 'Acebuchales' in a broad sense to include not only areas with dominant *Olea cerasiformis* but also those with species of *Pistacia*. The Almácigo (Mt. Atlas Mastic Tree) *Pistacia atlantica* was widespread and probably common in the dry woodland of Tenerife in the past; in Gran Canaria it still forms almost pure stands of well-spaced, low-spreading trees in a few places. The village named Almáciga near the north coast of Anaga implies that the species was once common there, and it features in the coats of arms of Guía de Isora and Arona in the southwest. Scattered individuals can still be found in many remote places, including the steep slopes above Buenavista, Los Silos and Garachico, near Bajamar and a few sites further east in Anaga, and also in some barrancos in the southeast. This deciduous tree is very tolerant of drought and may have originally formed extensive stands in arid places below and among the olives. It has been suggested that the valuable wood of the Almácigo and the medicinal properties of its resin were the main causes of its almost total disappearance from the island. A close relative of the Almácigo is the Mastic Tree or Lentisco *Pistacia lentiscus*, which is evergreen. It now grows only in a semi-urban situation in Barranco Hondo near Santa Úrsula on the north coast, where there are a few individuals in a fragment of dry woodland on the vertical side of a ravine, but this tree may possibly have played a significant role on the island in the past.

Almácigo *Pistacia atlantica* near Buenavista in winter

Woodlands dominated by Canary Dragon Trees *Dracaena draco* 'dragonales' do not now exist in Tenerife, and it is possible that they never did. Planted Dragos flourish almost anywhere on the island up to middle altitudes, and they may have grown extensively – along with palms – on slopes at fairly low levels, as they do now below Chamorga in Anaga and in other places where grazing pressure has declined. However, the most striking quality of Dragos is their ability to gain a foothold in steep places where there is hardly any

Dragos on previously denuded hillside near Chamorga

Barranco Taburco de Adentro, site of the restoration project, with scattered Brezo *Erica arborea*. Photo: Cristóbal Rodríguez Piñero

Dry woodland restoration project in the Teno Rural Park

This ambitious project in the Parque Rural de Teno aims to recreate 'bosque termófilo' or dry woodland in a zone degraded by past human activity, by establishing conditions under which natural ecological processes can gradually restore the rich biodiversity that existed here in the past. The dry woodland is one of the Canaries ecosystems that has been most severely reduced, so it is inevitable that the restoration will be slow and complex. No large and diverse dry woodlands are available to act as seed sources for natural regeneration on deforested areas, so it is necessary to initiate the process by planting.

The project is an initiative of the Cabildo de Tenerife (the island government), which was concerned at the severe reduction in this kind of woodland and the poor understanding of its dynamics. The project has received financial support from the European Union by means of a LIFE project, and has also benefited from collaboration

Echium aculeatum, one of the surviving shrubs on the site

with the University of La Laguna, who undertook a study not only of the restoration site but also of the remnants of this type of woodland that still survive in Tenerife and some of the other islands.

The site chosen for this restoration project is a farm owned by the Cabildo, extending to 53.5 hectares (~130 acres) located in the Barranco de Taburco de Adentro. It has an altitudinal range of 500 to 900 metres, faces southeast, and was used for growing cereal crops in the past and more recently as pasture. The climatic conditions of the zone are appropriate for dry juniper woodland (sabinar) in the middle and lower parts of the barranco and for humid juniper woodland at the top. At the start of the project the vegetation of the restoration site consisted of degraded scrub dominated by shrubs such as *Euphorbia atropurpurea* and *E. lamarckii*, *Echium aculeatum*, *Cistus monspeliensis* and some alien species such as *Opuntia maxima* and *Agave americana*. However, in neighbouring areas there are fragments of dry woodland made up of species such as Sabina *Juniperus turbinata* ssp *canariensis* and Acebuche *Olea cerasiformis*, together with a number of trees and shrubs appropriate to this type of woodland, notably *Pistacia atlantica*, *Hypericum canariensis*, *Jasminum odoratissimum* and *Globularia salicina*.

Restoration work began in 2006 and 2007, with the collection of seeds of various species from nearby populations to ensure genetic integrity, which were propagated in one of the Cabildo's nurseries. The seedlings were kept in a greenhouse until they reached a height of 10-15 cm, after which they were hardened off before being planted out.

The planting density was 700 plants per hectare, implying a need for about 20,000 saplings; before starting the planting the exotic species present in the degraded scrub were removed. Shelters were installed to protect the plants from herbivores, especially to prevent damage by the herds of goats present in the area, as well as the rabbits. Subsequent monitoring has shown a mean survival rate of about 18%, with much variation among the species. Nonetheless, given that the plants which have survived these first years have a high chance of establishment, this can be considered as the start of a process of recovery of this type of woodland.

Cristóbal Rodríguez Piñero

Three of the species planted for the project: left to right,
Visnea mocanera, *Juniperus turbinata* and *Hypericum canariense*

Barranco del Río, with many kinds of dry woodland plants

soil. This doubtless always enabled them to maintain populations on cliffs and ridges where they could capture moisture from rising air currents, as they still do, for instance, on Roque de Tierra off the coast of Anaga and on seaward faces of the coastal cliffs. It also accounts for their frequent appearance on the skyline in some of the most precipitous inland parts of the island, such as Bco. de Masca in the Teno massif and Bco. del Infierno near Adeje.

Cliffs and ravines

Cliffs, ravines and steep rocky banks are considered here as part of the potential dry woodland habitat, although they are also mentioned in relation to high mountain, forest and coastal situations that support rather different plant communities. The massive inland cliffs that dominate the north side of Tenerife between Puerto de la Cruz and Garachico are particularly rich, as are the walls of the gigantic 'collapse embayment' of the Güímar valley (see Geology chapter). Cliff and rocky slope habitats are also provided by the deep erosion ravines or 'barrancos' that are prominent topographical features of the ancient massifs of Anaga, Teno and El Conde, and also by those that cut through the inland cliffs of the north and furrow the steep southeastern and southwestern slopes of Teide; all of these provide ecological links between the high ground around Las Cañadas and the dry coastal regions.

4 Dry woodland remnants — 'bosque termófilo'

Barranco de Añavingo, with Madroñero *Arbutus canariensis* and Canary Pines

Under the hot conditions in the south some very dry and vertical cliffs remain almost free of plants, as in parts of the Barranco de Masca and along the sea cliffs of Los Gigantes. In general, however, the sides of barrancos offer shadier and moister conditions than the surrounding open ground, and this can lead to development of rich communities of tall shrubs, while wet and deeply shaded cliffs, like the walls of the higher part of the Orotava valley, have vegetation more akin to laurel forest. In the humid north the upper parts of many barrancos are impenetrable, choked with lush vegetation, while lower parts are so prone to erosion that the plant communities are in a constant 'state of flux'.

Cliffs or steep slopes on islands in dry climates often play a key ecological role, especially if they are exposed to moist winds from the sea, since air rising against them produces condensation that can be captured by specialised cliff-dwelling plants that establish themselves on the rock faces. This is clearly the situation along the north side of Anaga, though at relatively low altitude, where north-facing sea cliffs are often shrouded in cloud and provide ideal growing conditions for species such as *Dracaena draco, Teline pallida* and the lichen *Roccella*.

Cliffs and ravines also function as refuges, because they tend to be inaccessible to people and introduced herbivores and predators, and so allow species to survive after they have been lost elsewhere. On the Canaries the most obvious animal examples are the large endemic lizards (*Gallotia* species) which probably survive on cliffs because they can be fairly safe there from introduced dogs and cats. Similarly, some plants that now have their main strongholds on cliffs probably had much wider distribution before humans and agile goats arrived. Some of these plants are typical of the high mountain zone, such as *Juniperus cedrus, Bencomia exstipulata, Rhamnus integrifolia* and *Senecio palmensis,* but others are dry woodland species

Corona de la reina *Gonospermum fruticosum*

Cruzadilla *Hypericum reflexum*

Palomera *Pericallis lanata*

Salvia broussonetii

Madama de Risco *Allagopappus canariensis*

such as *Dracaena draco, Juniperus turbinata, Olea cerasiformis, Maytenus canariensis, Sideroxylon canariense* and *Limonium arborescens*.

Life on bare rocks or cliffs, however, requires special adaptations, and the rock-dwelling or 'rupicolous' plant communities are often dominated by members of a small number of plant families, especially the Crassulaceae, Asteraceae and Lamiaceae, although many other groups are also represented.

The Crassulaceae include many cliff-adapted species including *Aeonium canariense* and *Greenovia aurea*, found on cliffs and steep dry slopes in many areas, and *Aeonium smithii* and *Greenovia aizoon,* which are also rupicolous but less widely distributed. On damper cliffs, usually north-facing, one can find the plate-like *Aeonium tabulaeforme*. Smaller species include the pendulous *Monanthes laxiflora* with silver-grey egg-shaped leaves, and other species of *Monanthes* that form succulent rosettes.

Among the Asteraceae (composites) a plant especially noticeable in Teno is *Vieria laevigata*, the only member of a genus endemic to Tenerife; it is a deep green shrub that can grow on dry rock faces bare of almost all other plants. There is also a group of closely related *Sonchus* species, *S. fauces-orci, S. gummifer, S. radicatus* and *S. tectifolius*, which live on dry cliffs. In addition, a Canary endemic genus *Atalanthus* has been established for a group of *Sonchus*-relatives including *Atalanthus capillaris, A. microcarpus* and *A. pinnatus*, which occur mainly on cliffs and in dry woodland; they have stiff, branched, finger-thick stems topped by feathery tufts of finely divided leaves and tall flowering stems with clusters of small yellow flowers.

4 Dry woodland remnants — 'bosque termófilo'

Sonchus cf *tectifolius*

Pelotilla escamosa *Monanthes laxiflora*

Capitana pegajosa *Phyllis viscosa*

Balillo *Atalanthus capillaris*

Bejeque *Aeonium canariense*

Siempreviva arbórea *Limonium arborescens* on Cuevas Negras cliff (note tall stems on right)

Siempreviva de Anaga *Limonium macrophyllum* near the Faro de Anaga

Sideritis kuegleriana. Photo: Rubén Barone

Other characteristic cliff composites are *Gonospermum fruticosum* and the scarce *Gonospermum* (ex *Lugoa*) *revolutum*, the Tenerife representatives of a genus endemic to the Canaries, while another pair are the widespread *Reichardia ligulata* and the coastal and local *R. crystallina*. Also scarce are the various rock-adapted knapweed relatives in the genus *Cheirolophus*, with pink or cream flowers, replacing each other in different parts of the island. One of the most attractive composites is the purple-flowered, silver leaved *Pericallis lanata*, common on banks and cliffs in the south. Much harder to find is the critically endangered *Hypochaeris oligocephala*, with shiny leaves and large, short-stalked yellow flowers, which is known only from a few cliff sites in Teno.

The Lamiaceae or labiates include the genus *Sideritis*, some species of which are highly visible on Tenerife cliffs because of their silver-felted stems and leaves. In general, *Sideritis dendro-chahorra* is the dry woodland species in the east and *S. cretica* takes its place in the west. One green but rare cliff-dwelling member of the group is *Sideritis infernalis*, named after Barranco del Infierno in which it occurs, while another green one is *S. kuegleriana*, which occurs on the north coast. Less conspicuous are the cliff-dwelling species of *Micromeria*, such as *M. teneriffae*, which occurs in the southeast between Anaga and Arico. Also scarce and local is *Silene lagunensis*, a Tenerife endemic campion (Caryophyllaceae) that often occupies cliff ledges. St. Johns worts (Hypericaceae) are represented on forest cliffs in the humid north by *Hypericum glandulosum*, but on dry cliffs one finds *Hypericum reflexum*, and it may be that the strange form of the leaves, whose bases wrap around the stems, is an adaptation for catching raindrops, helping the plant to survive in hot, dry conditions.

Cliff-dwelling members of other families include several species of *Crambe* (Cruciferae) that share with the composites a tendency to have tall inflorescences standing out rigidly from cliff faces. More robust shrubs, however, may simply get a good hold and develop a flattened bush-like form, as shown well by the buckthorn *Rhamnus integrifolia* (Rhamnaceae) and *Phyllis viscosa* (Rubiaceae). Ferns also flourish in barrancos; the most obvious is usually the richly coloured *Davallia canariensis*, but other species such as *Polypodium macaronesicum* and *Adiantum reniforme* are also often present.

In moist ravines there is a tendency for massive curtains of vegetation to form, since the availability of water allows luxuriant growth, often with *Bosea yervamora* and sometimes with scramblers such as *Smilax aspera*, *Bryonia verrucosa* and *Canarina canariensis*, but frequently dominated by the invasive South American vine *Delairea odorata*.

Animals of dry woodland

The extent of the dry woodland ecosystem on Tenerife is now so small that it is difficult to know what the original fauna was like. It is clear, however, that there would never have been a sharp distinction between the animal communities of the coastal and lowland shrubland and those of the dry woodland, and indeed these two ecosystems must always have graded into each other, with vertebrates especially likely to range over both. Lizards of the genus *Gallotia* are almost ubiquitous in Tenerife, and in the north of the island are represented by *Gallotia galloti* subspecies *eisentrauti*. Tenerife Wall Geckos *Tarentola delalandii* occur up to 1800-2000 m in warm areas, and West Canary Skinks *Chalcides viridanus* have been found even higher up, though they are more abundant lower down.

Molluscs are not conspicuous in dry woodland, or in the scrub lower down, mainly because these habitats are hostile to animals that prefer a moist environment. However, careful search in many of the sites reveals a wide variety of native molluscs, many of them endemic.

Pits made by larvae of ant lions, Myrmeleontidae, which lurk at the bottom to grab prey that fall in

Tenerife Lizards *Gallotia galloti* courting above Taganana

Oil beetle *Meloe flavicomus*, with golden hairs. Photo: Pedro Oromí

Male of *Orthetrum chrysostigma*, a dragonfly often conspicuous in open habitats

Several kinds of beetles can be seen in dry woodland habitats, one of the most easily recognised being the slow-moving oil beetle *Meloe flavicomus*, endemic to the Canaries and Madeira. Oil beetles have complex life cycles, laying enormous numbers of eggs hatching into tiny larvae which search for and eat the eggs of grasshoppers or bees. Mantids and dragonflies are also widespread, though the former are hard to see and the latter occur mainly in barrancos.

An example of the blurring of the dry woodland and coastal and lowland ecosystems is provided by the areas in the Teno massif where the dry woodland shrub *Retama rhodorhizoides* is abundant, but where there are also many *Euphorbia lamarckii* and Cardón *Euphorbia canariensis*, shrubs more typical of lowland areas. In 2001 there was an intriguing discovery relating to these transitional places. After two sightings of very large grasshoppers in remote parts of Teno, a systematic search led to the capture of a few individuals of *Acrostira tenerifae*, a new species of wingless grasshopper in the obscure Old World family Pamphagidae; a laboratory population was later established in La Laguna University. Species of *Acrostira* were known from other Canary islands and it is surprising that the Tenerife one had not been discovered earlier, in view of the large size (females up to 67 mm long). However, these grasshoppers are extremely hard to see, as well as being very scarce and apparently occurring only in remote parts of Teno. They feed primarily on *Euphorbia lamarckii*, but small nymphs may also use *Retama rhodorhizoides*. When disturbed, they immediately hide behind the branch, so that only the tips of the front two pairs of legs are visible (the third pair are used for jumping in emergency). In spite of this well developed anti-predator behaviour, they are eaten by the feral cats that are widespread in Teno, and this predation may be a reason for their current scarcity.

Barranco del Natero in Teno, with a mosaic of *Retama rhodorhizoides*, *Euphorbia lamarckii* and other shrubs. Photo: David Hernández-Teixidor

Pair of *Acrostira tenerifae*, a wingless grasshopper in which males are much smaller than females
Photo: Pedro Oromí

The White-tailed Laurel Pigeon restoration project on Gran Canaria

Bolle's Pigeon *Columba bollii* and White-tailed Laurel Pigeon *Columba junoniae* are two endemic birds from the Canary Islands that are often considered to be restricted to the laurel forest. While this is true for Bolle's Pigeon, there is strong evidence that the original habitat of the White-tailed Laurel Pigeon was the lowland thermophilous forest (here termed dry woodland) that was largely destroyed after the European Conquest of the archipelago in the 15th Century. Habitat destruction has now restricted this species to steep places and ravines, especially near the lower limit of the laurel forest; it may also be present in some areas of pine forest.

White-tailed Laurel Pigeon
Columba junoniae

Both species are now confined to the western Canary Islands after their extinction on Gran Canaria. However, there is some historical evidence from Fuerteventura, around the time of the Conquest, of large pigeons with white ends to their tails that may have been the White-tailed Laurel Pigeon.

In 2007 the Cabildo (island government) of Gran Canaria started a conservation project aiming to reintroduce the White-tailed Laurel Pigeon and restore part of the former laurel forest of the island. With the collaboration of the University of La Laguna and the Cabildo of La Palma, eggs and chicks of the pigeon from that island were fostered by domestic Barbary doves *Streptopelia risoria* to establish a small captive breeding population of eleven pairs. After initial difficulties the pigeons first bred in captivity in 2010.

Young White-tailed Laurel Pigeon being fed by fostering Barbary Dove *Streptopelia risoria*

Between March 2012 and December 2013, 57 laurel pigeons were released from an aviary at Barranco de la Virgen in the north of Gran Canaria. Two feeding platforms, proofed against rats and cats, were constructed near the aviary to offer supplementary food. First breeding in the wild took place in July 2013 and by January 2014 seven juveniles had been observed near the release site. In January 2014 at least 35% of the released birds were alive and 8 pigeons have survived more than one year. The pigeons move around in an area of 14 km² and some have been found at places 4 km from the release site.

In 2013 the project obtained European Union LIFE funding to support it until 2017. About half a million trees are expected to be planted during that period and the reintroduction of Bolle's pigeon is planned for the near future.

Text and photographs: Aurelio Martín
University of La Laguna

Sabinar de Afur

The most extensive surviving area of juniper scrub on Tenerife, with a few magnificent ancient specimens on the ridges.

The Sabinar de Afur occupies a large part of a dry amphitheatre sheltered by a high ridge from the moistening effect of the northeast trade winds. This rain shadow has inhibited development of laurel forest, and juniper woodland extends up to about 450 m. However, the upper fringe is mixed with Brezo *Erica arborea,* and there are scattered laurel forest trees (including *Morella faya, Ilex canariensis* and *Rhamnus glandulosa*) high up in the relatively damp gullies. Here we also found the scramblers *Canarina canariensis, Rubia peregrina* and brambles *Rubus* sp., along with the slender umbellifer *Cryptotaenia elegans* and bracken *Pteridium aquilinum*; lower down in the gullies there are Canary Palms *Phoenix canariensis*, as well as the reed *Arundo donax.*

On the open slopes the junipers are fairly widely scattered, with the space between them occupied by degraded vegetation with shrubs such as *Artemisia thuscula, Rubia fruticosa, Euphorbia lamarckii, Opuntia maxima,* along with with some *Globularia salicina, Hypericum canariensis, Jasminum odoratissimum* and *Carlina salicifolia.* Smaller plants on the slopes

4 Dry woodland remnants — 'bosque termófilo'

Pinillo Plantago arborescens

Aeonium canariense & *Gonospermum fruticosum*

Trébol basto Trifolium squarrosum

Helianthemum broussonetii

Gomereta Aeonium lindleyi

Monanthes laxiflora

Gamona Asphodelus ramosus

Phagnalon saxatile

Estrella Romulea columnae

Wahlenbergia lobelioides

Capitana Phyllis nobla dwarf form

Sabina *Juniperus turbinata*, on a ridge top near Afur, one of the largest specimens on Tenerife

include *Plantago arborescens, Aeonium lindleyi, Helianthemum broussonetii, Trifolium squarrosum, Aeonium canariense, Gonospermum fruticosum, Wahlenbergia lobelioides,* a species of Resedaceae and an unusual dwarf form of *Phyllis nobla*. Bulbs include *Asphodelus ramosus* and also *Romulea columnae*, a charming crocus that also occurs in the pine forest.

Gonospernum revolutum

Rhamnus crenulata, heavily browsed

Although the few bushes of *Rhamnus crenulata* have clearly been browsed, it may be that the remoteness from settlements has lessened the impact of people in this area; for example, collection of juniper wood for fuel would have been easier closer to homes.

The most remote part of the sabinar, high above the seaward end of the Barranco de Afur and the Playa de Tamadite, is also the most impressive, since a few of the junipers here are in such inaccessible places that they have been growing in their natural form for many centuries. Seeing the largest ones on the skyline from a kilometre away, we mistook them for Canary Pines, but the trek across several barrancos to examine them was rewarded by close views of some of the most remarkable trees on Tenerife.

The slopes on the seaward side of the Roque de Marrubial, overlooking the coastal track from Taganana to Chinamada, are steep but mainly accessible to goats; the vegetation has clearly suffered greatly, but retains some species that would have been important in the original dry woodland, including *Echium simplex, Limonium macrophyllum, Ceropegia dichotoma* and *Gonospermum revolutum*. Now that few goats are pastured here, this area may soon regain much of the richness of the seaward slopes beyond Chamorga.

Some other notable plants
Aeonium tabulaeforme, Andryala pinnatifida, Asparagus sp., *Bituminaria bituminosa, Centaurium* sp., *Convolvulus floridus, Cyperus* sp., *Davallia canariensis, Echium plantagineum, Euphorbia canariensis, Ilex canariensis, Lavandula canariensis, Micromeria varia, Oxalis pes-caprae, Parentucellia viscosa, Paronychia canariensis, Periploca laevigata, Rumex lunaria, Reseda scoparia, Salvia canariensis, Sideritis dendrochahorra* and *Sonchus acaulis*.

Aeonium canariense above Barranco de Afur

There are small signs of recovery of the vegetation on seaward slopes west of Taganana

Roque de los Pinos, Chinamada

Flourishing Canary Pines at their only site in Anaga, and a wide variety of typical dry woodland trees and shrubs.

This is a part of Anaga where moist laurel forest and vegetation moulded by agriculture are the two main surviving habitats. On Roque de los Pinos, however, Canary Pines are flourishing between about 450 m and near the summit at 548 m. Their presence here is probably due to a combination of factors: the steepness of the rock, which has limited the impact of people, and the rain-shadow provided by the high Chinamada ridge to the northeast, which leads to drier conditions than is normal at this altitude in Anaga, and inhibits full development of laurel forest. This provides a niche for the pines, but also for a wide range of dry woodland species. A speciality of the site is the Canary endemic shrub *Cistus chinamadensis*, with spectacular

Cistus chinamadensis, which on Tenerife occurs only in Anaga

pink flowers and densely hairy seed capsules. It grows mainly on the rock itself and is closely associated with the pines, thus taking the place of its close relative *Cistus symphytifolius*, which is a key member of the pine forest understorey elsewhere on the island.

Although goats were doubtless once common here, the steepness of the rock allowed many trees and shrubs to survive, and some of them are now expanding onto the less precipitous areas. As a result, Roque de los Pinos is one of the best places on Tenerife to see characteristic dry woodland trees mixing with those more typical of laurel forest at higher levels. One example of the latter is *Pleiomeris canariensis*, which grows high up on the rock, while others include *Apollonias barbujana* (regenerating freely on the access ridge), *Erica arborea*, *Ilex canariensis*, *Laurus novocanariensis*, *Morella faya* and *Rhamnus glandulosa*. Typical dry woodland trees that can also be seen here include Sabina *Juniperus turbinata*, *Olea cerasiformis*, *Pistacia atlantica*, *Visnea mocanera* and the rare *Sideroxylon canariense*.

This site also has a great diversity of shrubs. A few species such as *Euphorbia lamarckii*, *Kleinia neriifolia*, *Periploca laevigata* and *Rubia fruticosa* are routinely also present in lowland shrubland, but others are typical of dry woodland. One of the most attractive is *Bystropogon odoratissimus*, a tall, strong-smelling bush found only in part of Anaga, Teno and a few sites near Adeje in the southwest. Other shrubs include *Aeonium lindleyi*, *Atalanthus pinnatus*, *Bryonia verrucosa*, *Carlina salicifolia*, *Ceropegia dichotoma*, *Convolvulus floridus*, *Descurainia millefo-*

Canary Pine at summit of Roque de los Pinos

Marmolán *Sideroxylon canariense* with flower buds

lia, *Echium* cf. *virescens*, *Globularia salicina*, *Hypericum canariense*, *Jasminum odoratissimum*, *Marcetella moquiniana*, *Paronychia canariensis*, *Plantago arborescens*, *Sideritis dendro-chahorra*, *Smilax aspera*, *Solanum vespertilio* and *Tamus edulis*. The fern *Davallia canariensis* grows well on the shady sides of rocks. The introduced cactus *Opuntia maxima* is also present here.

Pajonera canaria *Descurainia millefolia*

Bystropogon odoratissimus

Delfino *Pleiomeris canariensis* with flowers

Barbusano *Apollonias barbujana* with flowers

Dry woodland on the seaward face of Roque de las Ánimas

Roque de las Ánimas, Taganana

A phonolitic rock towering nearly 400 m above the rapidly eroding shore east of Taganana, with an array of endemic plants growing on the rock faces and ledges out of reach of goats.

The name Roque de las Ánimas is said to relate to the souls or 'ánimas' of the 'orchilleros' who lost their lives while collecting 'orchilla', the *Roccella* lichens that flourish on these seaward-facing damp cliffs. Orchilla was the source of a purple dye exploited on an industrial scale in the Canaries over thousands of years, and the lichen populations became severely depleted at various times. The trade has now ceased and the lichens again grow luxuriantly on the rocks in many places where moisture and nutrients are brought by air currents from the sea.

The rock has ledges that provide space for diverse dry woodland. On the precipitous seaward face, where rising air leads to condensation, are several large Dragos and abundant yellow-flowered *Teline pallida,* a species occurring only near coasts in the north of Anaga and in Teno. We were reminded of the dynamic nature of the plant communities in these precarious situations on noticing that a substantial *Teline pallida* bush just to the left of the trunk of the

Drago in the main photograph was missing two years later, although we could just make out the dead trunk. The cliff habitat is also rich in members of the Crassulaceae, including *Aeonium lindleyi* and *A. urbicum,* along with *Monanthes polyphylla.* The related *Monanthes laxiflora* tolerates slightly dryer situations nearby, where one can also find *Aeonium tabulaeforme.*

The landward side of the rock and nearby slopes are dryer, and there is still active goat grazing, which clearly exerts a major influence on the vegetation. However, it is possible to find trees such as Sabina *Juniperus turbinata* and Canary Olive *Olea cerasiformis,* along with many Cardón *Euphorbia canariensis* and dry woodland shrubs such as *Artemisia thuscula, Ceropegia dichotoma, Convolvulus floridus, Jasminum odoratissimum, Lavandula* sp., *Lotus* sp., *Pancratium canariense, Sideritis dendro-chahorra, Sonchus acaulis* and the spectacular *Teucrium heterophyllum,* with silver-grey leaves and brilliant red flowers.

The rare yellow-flowered legume *Teline pallida*, with Cardón *Euphorbia canariensis* and other cliff plants

Gomereta *Aeonium lindleyi* is common on cliffs in Anaga

Pastel de risco *Aeonium tabulaeforme* and *Gonospermum revolutum*, with *Roccella* lichen

4 Dry woodland remnants — 'bosque termófilo'

View over Taganana with Sabina *Juniperus turbinata* (bottom centre) Acebuche *Olea cerasiformis* and Tabaiba amarga *Euphorbia lamarckii*

Nearby is an ancient specimen of Carob *Ceratonia siliqua* in such a remote and precipitous place that it is tempting to consider that this species might be native to Tenerife, but it is not accepted as such. Also probably introduced is a scabious, *Scabiosa atropurpurea,* which grows on the cliff tops here, close to where one can still find its rare endemic relative, *Pterocephalus virens,* with leaves in small rosettes. Another rare species recorded here is *Cheirolophus tagananensis,* with creamy white flowers.

Other species growing here or around Roque de Enmedio nearby include Canary Holly *Ilex canariensis, Rumex lunaria* and *Rubia fruticosa.* There are also clumps of *Cistus monspeliensis,* a species often associated with a history of intensive grazing. Another shrub that seems fairly resistant to grazing is *Globularia salicina,* with small blue flowers that become brown and brittle when they die, leaving the bushes cluttered with dead fragments. The Asteraceae are represented here by the rare *Gonospermum revolutum* and a coastal form of *Andryala pinnatifida,* a plant that can adapt to a wide range of situations, from sea level to Las Cañadas.

Chahorra *Sideritis dendro-chahorra*

Arrebol *Echium simplex* overlooking Roque de Fuera

ANAGA BEYOND CHAMORGA

A remote and beautiful area that provides a glimpse of the transition from laurel forest to the diverse dry woodland that once existed between it and the coastal shrubland on the slopes closer to the sea.

The high path from Chamorga to the Faro de Anaga, via the ruined houses of Tafada, climbs up to a 600 m ridge and then descends gradually across north-facing slopes. Several of the conspicuous endemic plants here are known only from this remote part of Tenerife, while others are typical of dry woodland in other parts of the island.

The first part of the walk up from Chamorga is through scrubby woodland with some laurel forest species such as *Erica arborea*, *Rhamnus glandulosa* and *Sideritis macrostachys*. Also present are various plants that grow in light gaps in the laurel forest but also flourish in the transition to dry woodland a little lower down; they include *Argyranthemum broussonetii*, *Teline canariensis*, *Isoplexis canariensis*, *Canarina canariensis* and the Arum relative *Dracunculus canariensis*, with leaves reminiscent of an extended hand. Other plants are associated with rocky banks and cliffs within laurel forest or dry woodland; they include *Aeonium cuneatum*, *Monanthes laxiflora* and *Arisarum simorrhinum*. *Sonchus acaulis* is common here, often with its rosette about half a metre above the ground on a thick trunk that is obscured by a 'skirt' of drooping dead leaves.

It is on starting to follow the path down from the top of the ridge that one encounters two of the little-known plant treasures of Tenerife. While all visitors know about *Echium wildpretii*, the red 'Taginaste rojo' of Las Cañadas that features on every tourist brochure, few people are aware of the spectacular 'Arrebol' *Echium simplex*, confined to a remote part of the island visited by fewer tourists. It has a ground-level rosette of large, silver green pointed leaves, from which springs a gracefully tapering column up to 2 m tall, densely clothed with white flowers. The second striking plant of this area is *Limonium macrophyllum,* an inland relative of the sea-lavenders of the coast; it is a large perennial with a rosette of bright green cabbage-like leaves, upright and slightly 'winged' flowering stems, and masses of 'everlasting' papery blue flowers highlighted by a scattering of white blooms.

The dead-leaf skirts of Cerrajón de monte *Sonchus acaulis* evoke a dance-floor scene

The presence nearby of a regenerating Drago *Dracaena draco* – with the Roque de Tierra population of this species also visible just offshore – raises the intriguing possibility that this area may once have had a covering of dry woodland very distinct from that elsewhere in the island. We think it was probably dominated by Dragos and Almácigos *Pistacia atlantica*, with sporadic occurrence of the tree-sized nightshade-relative *Withania aristata*. There was evidently also space for the big *Limonium* and the *Echium*, as well as for the several kinds of large shrubs that still persist in the area.

We were intrigued to notice that in the area where *Echium simplex* and *Limonium macrophyllum* were commonest, there were also Tabaibas *Euphorbia lamarckii* densely coated with lichens, something that we have not noticed on this species elsewhere; it is presumably caused by strong condensation on these steep north-facing slopes at about 500 m

Other shrubs in the area include the umbellifer *Bupleurum salicifolium*, common in dry situations high up in Teno, which is more luxuriant here in the moister conditions of Anaga. *Teline canariensis* is everywhere, and one can also find *Rhamnus crenulata, Jasminum odoratissimum* and the scarce endemic nightshade *Solanum vespertilio*. Also present is the dense bush *Paronychia canariensis,* with its elegant lanceolate fleshy leaves coming to a sharp point, and the large umbellifer *Todaroa aurea*, with yellow flowers but also foliage that turns a brilliant golden colour as it withers. Relevant smaller species include the vetch *Vicia cirrhosa*, with rows of large, dangling, off-white flowers; the yellow-flowered crucifer *Descurainia millefolia*, cousin to the Hierba pajonera *Descurainia bourgeauana* that turns swathes of Las Cañadas bright yellow in early summer. One can also see the curious endemic plantain *Plantago arborescens,* with tufts of narrow leaves that make it look superficially like a grass,

Pelotilla escamosa *Monanthes laxiflora*

Incienso *Artemisia thuscula*

Vicia cirrhosa

Siempreviva *Limonium macrophyllum*

Cardo monte *Carduus clavulatus*

Tacarontilla *Dracunculus canariensis*

Meloncillo *Bryonia verrucosa*

Nevadilla *Paronychia canariensis*

Rejalgadera *Solanum vespertilio*

and the campion *Silene lagunensis*, which often grows on cliffs and steep rocky slopes.

More noticeable in this area are the much branched thistle *Carduus clavulatus*, with broad basal leaves, which is endemic to the Canaries, and the shrub *Sideritis dendro-chahorra*, one of the largest members of its diverse genus, in which young leaves are covered with yellowish silver hairs. The scramblers *Bryonia verrucosa* and *Tamus edulis* are also typical dry woodland species found here, as is the large bulb *Asphodelus ramosus*.

Near here and just to the east of the remote hamlet of Las Palmas is Roque

Roque del Aderno (left) is a mainland extension of the Roques de Anaga, and both it and the steep slopes nearby have rich arrays of endemic plants

del Aderno (461 m), the northernmost high point on Tenerife. It represents the landward end of the series of intrusive rock masses that continues offshore with Roque de Tierra (also known as Roque de Dentro) and Roque de Fuera. Roque del Aderno is so precipitous that it has functioned as a refuge for dry woodland trees and shrubs eliminated by goats from gentler slopes, presumably including its namesake Aderno *Heberdenia excelsa*, but also with Sabina *Juniperus turbinata*, Acebuche *Olea cerasiformis*, Drago *Dracaena draco* and Peralillo *Maytenus canariensis*.

If one continues down to the Faro de Anaga (after which one can return by the track up the Barranco de Roque Bermejo) there is a chance to see a transition to the *Euphorbia* dominated lowland and coastal shrubland of this eastern point of Tenerife, with many additional plants.

Some other notable plants
Aeonium canariense, Aeonium lindleyi, Artemisia thuscula, Bituminaria bituminosa, Carlina salicifolia, Convolvulus floridus, Dittrichia viscosa, Echium strictum, Foeniculum vulgare, Lavatera acerifolia, Micromeria sp.*, Panacratium canariense, Pericallis appendiculata, Pericallis cruenta, Pimpinella anagodendron, Rubia fruticosa, Rubus* sp. *Rumex lunaria, Sonchus radicatus, Teucrium heterophyllum.*

Silene lagunensis is beautiful but sticky

Olives and Cardón on the slopes above Valle Brosque

Valle Brosque, Anaga

Easily accessible remnants of dry woodland in a traditional agricultural landscape in the dry south of Anaga.

The southern slopes of Anaga are dry habitats that have been subject to millennia of intensive grazing and cultivation, so it is difficult to visualise the original vegetation. There is still limited 'crofting style' agriculture in the valleys, with small cultivated plots, fruit trees, and many enclosures and small structures where goats, rabbits, poultry, dogs and even horses are kept. However, in some of the deep valleys that run northwards from the coast between Santa Cruz and Igueste de San Andrés into the heart of the massif, there are fascinating remnants of the dry woodland. We describe one of these, but there many places with similar vegetation.

From the small parking place in Valle Brosque (at the end of the road), two tracks are accessible. The left hand (western) one eventually reaches Las Casas de la Cumbre on the summit ridge of Anaga. The right hand track leads up, gently at first, to the old Casa Forestal that marks the top of Las Vueltas de Taganana, on the ridge road (TF-12); this track is thus one of the ancient routes from the south coast of Anaga to the north coast. On the way it passes through land that has clearly been in agricultural use for many centuries. The most conspicuous trees around the village are Carob, *Ceratonia siliqua*, with dark green, pinnate leaves with up to seven pairs of leaflets, and long green seed pods; this is a species native to the Mediterranean, which was probably introduced to Tenerife by humans long ago. Some way up the valley there are a few buildings and small cultivated areas, and nearby is a grape-crushing bowl carved out of the top of a volcanic rock, but there are now no significant vineyards nearby. The barranco here has large boulders with shaded north faces, on which there are fine communities of ferns and orchids.

However, the most interesting remnants of the original vegetation are high up on the less accessible slopes. Even here there are signs of grazing by goats, but around the crags at a height of about 500 m one can find diverse dry woodland dominated by Canary Olive *Olea cerasiformis*, juniper *Juniperus turbinata*, accompanied by species such as *Globularia salicina*, *Aeonium lindleyi*, Cardón *Euphorbia canariensis* and Tabaiba amarga *Euphorbia lamarckii*, as well as Brezo *Erica arborea*. Furthermore, on a steep intrusive rock mass just across the valley there is one of the best remnants of sabinar in Anaga (in 'The main types of dry woodland').

It seems likely, therefore, that the slopes of the series of barrancos such as Valle Brosque that run southwards from the east-west spine of Anaga to the coast may once have been largely covered by dry woodland. The laurel forest near the ridge would have given way to juniper-dominated woodland at around 600 m, with increasing presence of olives and various dry woodland shrubs on the slopes lower down, and then a transition to dominance of palms and willows in the valley bottoms. The shrubland dominated by euphorbias and drought-resistant shrubs such as *Kleinia neriifolia* and *Plocama pendula*, which now extends far inland, may originally have been more restricted to coastal areas.

Carob *Ceratonia siliqua*

Orchid *Habenaria tridactylites* with ferns *Polypodium macaronesicum* and *Davallia canariensis*

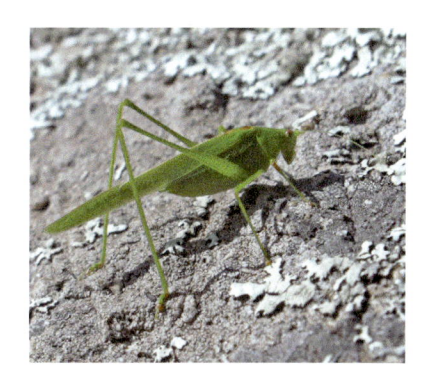

Katydid *Phaneroptera sparsa*, which is probably native

Some other notable plants

Aeonium canariense, Aeonium urbicum, Agave americana, Allium sp.*, Artemisia thuscula, Arundo donax, Asparagus umbellatus, Bituminaria bituminosa, Bosea yervamora, Carlina salicifolia, Ceropegia dichotoma, Foeniculum vulgare, Echium leucophaeum, Hypericum canariense, Hypericum reflexum, Jasminum odoratissimum, Lobularia canariensis, Micromeria* sp.*, Opuntia maxima, Paronychia canariensis, Pericallis cruenta, Periploca laevigata, Plantago arborescens, Polycarpaea latifolia, Punica granatum, Rhamnus crenulata, Rubus* sp.*, Salix canariensis, Salvia canariensis, Scilla latifolia, Sonchus tectifolius, Solanum vespertilio, Todaroa aurea*.

Ladera de Güímar in summer with *Pterocephalus dumetorus*,
Euphorbia atropurpurea, *Rumex lunaria* and *Euphorbia canariensis*

Ladera de Güímar

A long north-facing inland cliff at an altitude of about 500 m, often affected by low cloud. Home to a rich community of endemic trees and shrubs, including several that are hard to find elsewhere, now spreading back on to the less precipitous slopes.

The Ladera de Güímar is a dramatic north-facing inland cliff, created at the time of the Güímar collapse about 830,000 years ago, when a large part of the east coast of Tenerife slid into the sea (see Geology chapter). The cliff is more or less vertical in places, but there are also steep slopes with some soil. There are three possible access points near the seaward end of the cliff and even a short visit is worthwhile, but dedicated botanists may also wish to explore the less accessible parts of the cliff further inland.

Myrtle on the water channel, 1985

4 Dry woodland remnants — 'bosque termófilo'

The Ladera has long been famous for the many unusual plants that can be found there. The high diversity is due partly to the biological richness of the Güímar valley as a whole, partly to a tendency for cloud to settle along the top of the southern wall of the valley, and partly just because of the steepness of the cliffs, which allowed plants to survive the long onslaught of the goats and sheep. These plants are now spreading back over the somewhat gentler denuded slopes, giving a taste of the dry woodland that would once have covered the sides of the valley.

The small Sabinas *Juniperus turbinata* that we found here in the 1980s have now grown into substantial trees, and we suspect that they may have dominated the vegetation of these cliffs in the distant past; one can also find the typical dry woodland trees *Visnea mocanera* and *Olea cerasiformis*, and there are records of *Pistacia atlantica*. Shrubs include *Rhamnus crenulata*, the rare gymnosperm *Ephedra fragilis*, and the mallow *Lavatera acerifolia*, with delicate pale mauve, hibiscus-like flowers.

These species are accompanied by three of the large euphorbias – the cliff specialist Cardón *Euphorbia canariensis*, Tabaiba amarga *E. lamarckii* and also Tabaiba majorera *E. atropurpurea*, a beautiful Tenerife endemic species that occurs mainly in the southwest of the island but has an outpost here. A fourth, much rarer species, *Euphorbia bourgeana* (often spelled in other ways) has also been recorded here. A number of other endemic plants grow on the cliffs, many of them with very restricted distribution. One speciality is the pink-flowered scabious relative *Pterocephalus dumetorus*, which has a close relative in the high mountain zone; another is the crucifer *Crambe arborea*, with a tall spreading inflo-

Threadlike parasitic Dodder *Cuscuta* sp. envelops *Hypericum reflexum* and kills the leaves

Tinguarra cervariaefolia with celery-like leaves

Sonchus gummifer is well adapted to life on cliffs

rescence bearing large numbers of tiny white flowers; it is unusual in this genus in having finely divided leaves with linear lobes. Another rarity is the perennial white-flowered umbellifer *Tinguarra cervariaefolia*, sole member of a Canary endemic genus. In spring the brilliant purple flowers of *Pericallis lanata*, with its small silvery leaves, are particularly attractive, but in a dry summer this and many of the other smaller plants may become withered and hard to recognise. Another composite that is conspicuous here is one of the endemic sow-thistles, *Sonchus gummifer,* with blue-green leaves with reddish stems, which is capable of getting a grip in the most unpromising situations.

This is also a good place to see a species of *Atalanthus*, a genus closely related to *Sonchus* but with much-divided linear leaf lobes. The taxonomy is complex but the plants here are *Atalanthus microcarpus*, with really threadlike leaf lobes, while in plants from the north of the island (*A. pinnatus*) they are very fine but flattened. On shaded cliff faces you may find the fern *Adiantum reniforme,* with its kidney-shaped glossy green fronds, which occurs at relatively few widely scattered sites on Tenerife.

We were pleased to see a Barbary Falcon *Falco pelegrinoides* here recently, and have been struck by the general increase in this magnificent bird since our time in the island in the 1980s. Another notable animal recorded from here is the Long-tailed Blue *Lampides boeticus,* which is one of the less common butterflies on the island.

Some other notable plants
Aeonium arboreum, Allagopappus canariensis, Artemisia thuscula, Asparagus sp.*, Asteriscus aquaticus, Campylanthus salsoloides, Carlina salicifolia, Cistus monspeliensis, Davallia canariensis,*

Corregüelón *Convolulus canariensis*, widespread in the laurel forest, occurs here in regenerating dry woodland

A seedling Drago *Dracaena draco* symbolises the revival of dry woodland on the Ladera

4 Dry woodland remnants — 'bosque termófilo'

The dry woodland can spread back on to these gentler slopes now that the goats have gone

Argyranthemum frutescens ssp. *gracilescens*

Acebuche *Olea cerasiformis*

Palomera *Pericallis lanata* with one flower still surviving in midsummer

Visnea mocanera, a tree typical of the transition from dry woodland to laurel forest

Dracunculus canariensis, Echium strictum, Erica arborea, Erucastrum sp., *Gonospermum fruticosum, Globularia salicina, Helianthemum teneriffae* (not seen by us), *Hypericum canariense, Hypericum reflexum, Juniperus turbinata, Kleinia neriifolia, Lotus sessilifolius, Micromeria teneriffae, Monanthes brachycaulos, Opuntia maxima, Parietaria filamentosa, Periploca laevigata, Phagnalon* sp., *Roccella* sp., *Rubia fruticosa, Salvia canariensis, Silene* sp., *Sonchus acaulis, Tamus edulis.*

Crambe arborea is tall enough to compete with other dry woodland shrubs

Pterocephalus dumetorus

Aeonium urbicum drops its leaves as the inflorescence matures

Malva de risco *Lavatera acerifolia* is flourishing on the Ladera

4 Dry woodland remnants — 'bosque termófilo'

SIETE LOMAS AND THE BARRANCOS OF THE GÜÍMAR VALLEY

A relatively moist area of hills and ravines in the otherwise dry southeast, with many plants that do not occur elsewhere on the island.

High up behind the towns of Güímar, Arafo and Candelaria is a maze of 'lomas' (hills) and 'barrancos' (ravines) that run up to the top of the vast Güímar valley and the dorsal ridge of the island. On the other side of the ridge is the Orotava valley, which was formed – like the Güímar valley – by a giant landslide (see Geology chapter). Exploring this area is complicated and time-consuming; the rewards are breathtaking views down into the valley and a chance to see a fascinating array of native plants and associated animals.

The Güímar valley as a whole is home to a large number of species not found in other parts of Tenerife, perhaps partly because it is one of the few places in the dry south of the island where the high ground is often shrouded in cloud. Warming of the sheltered Güímar valley by the sun produces rising air, an onshore breeze and condensation above the back of the valley, so that cloud forms over the slopes behind Arafo and Güímar, even extending to the Ladera de Güímar. As a result of this moisture, plant communities on the high slopes and in the shaded barrancos include some species also found in

Bencomia caudata with seeds

Visnea mocanera, Arbutus canariensis, Picconia excelsa and *Pinus canariensis* in Las Coloradas

the laurel forests of the north, even though the main dorsal ridge of the island at the head of the valley is clothed with pine.

Several of the seven 'lomas' can be accessed by means of extremely steep but paved tracks that run improbably far up into the mountains at the back of the Güímar valley. We have explored some of those closest to the southern wall of the valley, following roads that originate in Güímar. There is agriculture along these ridges, but high up on them, between about

Salvia canaria *Salvia canariensis*

Cheirolophus metlesicsii

800 and 1000 m, there are habitat fragments between the cultivation below and the pine forest above, which show features of both laurel forest and dry woodland.

A particularly good example can be found on a ridge called Las Coloradas towards the west side of the Güímar valley. Conspicuous broadleaf trees at around 925 m here include Mocán *Visnea mocanera*, Palo blanco *Picconia excelsa* and Madroño *Arbutus canariensis*, all species typical of the lower limits of the laurel forest, but here mixed with pines that dominate the steep slopes running up the back of the Güímar valley. Other trees here are Acebiño *Ilex canariensis*, Faya *Morella faya*, Loro *Laurus novocanariensis*, Sanguino *Rhamnus glandulosa* and Follao *Viburnum rigidum*, as well as Brezo *Erica arborea*. Shrubs here and in the nearby Barranco de Badajoz include the Canary Olive *Olea cerasiformis*, the large St John's Wort *Hypericum canariense*, *Globularia salicina*, *Cistus monspeliensis*, *Asparagus umbellatus*, *Phyllis nobla*, *Tinguarra cervariaefolia*, *Carduus clavulatus*, *Convolvulus canariensis*, *Sonchus acaulis* and *Sonchus gummifer*.

Bystropogon canariensis, an aromatic plant in a genus endemic to the Canaries and Madeira

Beside the tracks up the ridges one can see other dry woodland plants characteristic of the Güímar valley, including *Bupleurum salicifolium*, and on shady banks the delicate green-flowered orchid *Habenaria tridactylites* and the vetch *Vicia cirrhosa*. Related to the last species are three rare endemic species in the genus *Dorycnium*, two of which have been recorded from this area, although we have not seen them. Also present in small numbers are the rare and beautiful dry woodland shrub *Anagyris latifolia* and the strange broom-like *Kunkeliella retamoides*, member of an endemic genus in the sandalwood family Santalaceae; care is needed to distinguish it from the rather similar shrubs *Ephedra fragilis* and *Retama rhodorhizoides*, both of which also occur in this area.

The ridges of Siete Lomas alternate with deep barrancos. Close to the Ladera de Güímar these include Barranco de Badajoz (Barranco del Rinconcito on some maps), Barranco del Agua and Barranco de las Ovejas, all of which have extremely interesting woodland fragments in their upper parts. However, two barrancos further north, the Barranco de las Saletas and the nearby Barranco de Añavingo, offer an easier way to see many of the most interesting plants.

Kunkeliella retamoides, a rare plant that has its main stronghold in Siete Lomas
Photo: Keith Emmerson

Barranco de las Saletas, with inset Aeonium arboreum

The narrow ravine of Barranco de Añavingo runs down from the main ridge and then opens out into a broader valley; here there is a fine stand of old Cork Oaks *Quercus suber* that are presumably planted, since this species is not considered native to the island. A little further up are several small specimens of the cherry *Prunus avium*, another species not recorded as growing wild on the island, which have probably also been planted.

Other tree species in the ravines are typical of the dry woodland and its transition to pine forest, and have doubtless arrived here naturally. Most striking is the Canary Strawberry Tree or Madroño *Arbutus canariensis*, with shiny leaves and richly coloured peeling bark, which is common on the bluffs along the sides of the barrancos, mixing with the lower Canary Pines *Pinus canariensis* that dominate the slopes higher up. A surprising but probably wild tree here is *Juniperus cedrus*, normally associated with the highest parts of Las Cañadas, but occurring in a few places down to below 1000 m. A conspicuous shrub in the bottom of the barrancos is *Pterocephalus dumetorus*, a local speciality, with pink scabious flowers. In Añavingo it is often found near to *Bencomia caudata*, another relatively scarce shrub usually found in the transition from dry woodland or laurel forest to pine forest. A rare plant that can be seen in bloom here

Descurainia lemsii

in midsummer is the knapweed-relative *Cheirolophus metlesicsii*, endemic to Tenerife and critically endangered, with large pink flowers. Also present here is the parsley-like *Pimpinella dendrotragium*, which usually grows in shady places.

Birdlife includes Hoopoe *Upupa epops*, Canary Blue Tit *Cyanistes teneriffae* (which we found nesting in a hole in the rock wall of a barranco) and Barbary Falcon *Falco pelegrinoides*. Bolle's Pigeon *Columba bollii* also occurs in the Güímar valley, in transitional habitats where the White-tailed Laurel Pigeon *Columba junoniae* might have been predicted; however, there are also records of the latter species in the area.

Some other notable plants

Adenocarpus foliolosus, Aichryson laxum, Allagopappus canariensis, Argyranthemum foeniculaceum, Argyranthemum frutescens, Artemisia thuscula, Bituminaria bituminosa, Bosea yervamora, Canarina canariensis, Carlina salicifolia, Ceballosia fruticosa, Centranthus calcitrapae, Cistus symphytifolius, Convolvulus floridus, Crambe strigosa, Daphne gnidium, Davallia canariensis, Descurainia lemsii, Dracunculus canariensis, Echium virescens, Euphorbia lamarckii, Foeniculum vulgare, Hypericum reflexum, Hypericum grandifolium, Isoplexis canariensis, Jasminum odoratissimum, Juniperus turbinata, Lavandula pinnata, Lobularia canariensis, Mercurialis sp., *Paronychia canariensis, Pericallis lanata, Phyllis nobla, Pterocephalus dumetorus, Pterocephalus lasiospermus, Retama rhodorhizoides, Rhamnus crenulata, Rubia fruticosa, Rubus* sp., *Rumex lunaria, Semele androgyna, Sideritis canariensis, Sideritis soluta, Silene berthelotiana* (not seen here by us), *Sonchus acaulis, Sonchus congestus, Sonchus gummifer, Tamus edulis, Tinguarra cervariaefolia.*

Echium strictum, with foraging endemic bumblebee *Bombus canariensis*

Clouded Yellow *Colias crocea* (above), with *Colias crocea* var. *helice* (below)
Photos: Juan José Bacallado

Barranco del Infierno with *Convolvulus floridus*

BARRANCO DEL INFIERNO

A magnificent gorge that runs from the top of the town of Adeje into the heart of the mountains of the ancient Conde massif. The vegetation here has some elements of the lowland shrubland, but most of the plants are more typical of dry woodland.

The Barranco del Infierno path (recently 'officially' closed) runs for almost three kilometres, starting at an altitude of about 400 m and terminating at a pool and waterfall among vertical cliffs at about 650 m. The first stretch of the track traverses along the side of the barranco, passing some beehives and giving fine views back down the barranco towards the sea, and then slowly converges with its floor. At first the vegetation is somewhat degraded, with plenty of Cardón *Euphorbia canariensis* and *Plocama pendula*, plants that resist grazing by goats. Some other species here are typical of lowland shrub areas, especially Tabaiba dulce *Euphorbia balsamifera*, *Neochamaelea pulverulenta* and *Launaea arborescens*, all unusually far inland. Others shrubs are typical of the transition between lowland shrubland and dry woodland; they include the widespread Tabaiba amarga *Euphorbia lamarckii* and several less well known endemic shrubs such as *Atalanthus capillaris*, a sow-thistle relative with threadlike leaves, *Justicia hyssopifolia* with its hooded white flowers, and *Allagopappus canariensis*, a member of a genus endemic to the Canaries, with narrow, toothed, sticky leaves and dense, flattish heads of yellow flowers.

4 Dry woodland remnants — 'bosque termófilo'

Left: Mantid *Blepharopsis mendica*. Photo: P Oromí

Right: Web made by larvae of the endemic moth *Yponomeuta gigas*, with adult (dead) inset. Photos: J J Bacallado

Although most of the other plants are also native to Tenerife, there are a few introduced ones; these include a prickly pear *Opuntia maxima*, Tree Tobacco *Nicotiana glauca* and occasional fruit trees such as Mulberry, Fig, Almond, Walnut and Pear. Some of the time you will be walking close to a water canal designed to take water from higher up the gorge down to the towns and agricultural areas below; this may be why the gorge is usually almost dry lower down. As you descend towards the damper bottom of the gorge the vegetation changes; there are thickets with brambles *Rubus* sp., Bracken *Pteridium aquilinum*, Maidenhair Fern *Adiantum capillus-veneris* and also the Canary Willow or 'Sauce' *Salix canariensis*. This willow, which can be found in damp places in many parts of Tenerife, sometimes provides an extraordinary entomological spectacle, since in early spring the trees are often infested by thousands of caterpillars of the endemic moth *Yponomeuta gigas*: the caterpillars weave a communal web that may entirely envelop the tree, thus giving them protection while they consume the leaves. A large predatory insect that you might see around here is the spotted African mantid *Blepharopsis mendica*, which is up to 6 cm long.

After this the barranco becomes quite narrow and the path stops where the cliff walls finally converge. Here there is a waterfall, which even in the middle of summer has water trickling down the rock face into the pool and on down the barranco, thus ensuring that

Echium aculeatum, a species of western Tenerife

Justicia hyssopifolia, another western species

Sideritis infernalis

Echium virescens

damp watery habitats remain available throughout the year. As a result, there are rich communities of plants adapted to wet places, and insects such as dragonflies, water beetles and water skaters.

If you look upward to the high cliffs along this stretch you will see wild Dragon Trees *Dracaena draco* clinging to the rocks. These are not the huge trees growing in sheltered conditions in town parks, but isolated scrawny specimens growing under harsh natural conditions. Also on these cliffs are Canary Pines *Pinus canariensis* at the lower limit of the pine forest. There are also many Sabinas *Juniperus turbinata* on steep rock faces and ledges, and the cliffs and steep banks along parts of the gorge provide refuges for other scarce trees and shrubs, including *Maytenus canariensis,* which seems to have no difficulty in establishing itself on almost vertical cliffs. We have also seen one specimen of the rare Marmolán *Sideroxylon canariense,* which looks much less happy on the cliff. Another uncommon tree that has probably survived here because of the cliffs is *Rhamnus integrifolia*, now confined to Las Cañadas and a few scattered places at much lower levels in the southwest. Other interesting shrubs include *Marcetella moquiniana* and *Lavatera acerifolia*, as well as the rare *Sideritis infernalis,* named after this gorge, and *Argyranthemum gracile*, a showy daisy with threadlike leaf lobes that is common right along the southwest side of the island. We have also found rosettes of *Reichardia ligulata* and of an unusual species of *Tolpis,* probably best treated as *Tolpis crassiuscula,* though it has been suggested that this little population may be a form of *Tolpis lagopoda*, which occurs in laurel forest in the east.

Duraznillo *Ceballosia fruticosa*

Tolpis cf *crassiuscula*

High up on some of the slopes, where they run up to the bottom of vertical cliffs, you may be able to see holes belonging to Cory's Shearwater 'Pardela Cenicienta' *Calonectris diomedea*. These birds come inland at dusk to their nests, sometimes in old rabbit holes, in barrancos throughout the island; their weird cries at night give rise to many a strange legend, and may even account for the name of this barranco.

Campylanthus salsoloides, Atalanthus capillaris and Cardón

Reichardia ligulata

The whole valley is a good place for birds; there are Sardinian Warblers *Sylvia melanocephala* and Blackcaps *Sylvia atricapilla*, singing tantalizingly hidden in the dense thickets of shrubs, Barbary Partridges *Alectoris barbara* calling from the slopes, Kestrels *Falco tinnunculus* hunting for the Canary Lizards and Buzzards *Buteo buteo* calling overhead.

Some other notable plants

Aeonium urbicum/pseudourbicum, Aeonium arboreum, Bencomia caudata, Bituminaria bituminosa, Bosea yervamora, Campylanthus salsoloides, Ceropegia fusca, Chamaecytisus proliferus, Cistus monspeliensis, Cistus symphytifolius, Convolvulus floridus, Crambe scaberrima, Descurainia millefolia, Echium aculeatum, Echium strictum, Echium virescens, Erysimum bicolor, Euphorbia atropurpurea, Euphorbia lamarckii, Euphorbia cf *peplus, Forsskaolea angustifolia, Globularia salicina, Helianthemum canariense* (not seen by us), *Hypericum reflexum, Jasminum odoratissimum, Kickxia scoparia, Kleinia neriifolia, Lavandula* sp, *Pancratium canariense, Periploca laevigata, Phoenix canariensis, Reseda scoparia, Rhamnus crenulata, Rubia fruticosa, Rumex lunaria, Ruta pinnata, Salvia canariensis.*

Sonchus fauces-orci

Argyranthemum gracile

Degollada de Cherfe and the Masca road

Mountain pass at the threshold of a geologically ancient part of the island, with spectacular eroded ridges and barrancos. Vegetation in many parts is 'retamar blanco', a form of dry woodland dominated by Retama blanca *Retama rhodorhizoides*, and with many other plants characteristic of the original dry woodland of the Teno massif.

The Degollada de Cherfe is the pass at over 1000 m on the road between Santiago del Teide and Masca, in the Teno massif; it is the highest of the dry woodland localities that we discuss. From the parking place it is easy to walk out along the ridge for a little way and explore a nearby gully. We also mention here some additional plants from similar areas a few miles further along the road to the villages of Masca and Los Carrizales (the latter considerably lower down) and from the pass where the road leaves the Masca area and descends towards Buenavista.

In the Teno massif, most of the land that is not precipitous has been used for agriculture for many centuries. Broad stairways of terraced plots have been cultivated, and goats and sheep have grazed almost everywhere else. In recent decades, however, cultivated areas have been progressively abandoned, and the number of goats has diminished; as a result, signs of recovery of the dry woodland are easy to find. However, as in many denuded areas around the world, many plants have been entirely eliminated or have become so rare that regeneration is hampered by the lack of seed sources. Near Teno Alto in the north of the massif, the island government has purchased a finca (farm) in order to restore a large area of dry woodland, as described earlier.

The dominant plant in this part of Teno is the white-flowered broom Retama blanca *Retama rhodorhizoides*. It can cover large areas, sometimes almost as a monoculture, perhaps as a result of intensive grazing in the past. However, this pressure is now reduced, and even near the road there are some areas with high plant diversity; many of the shrubs are endemic and only to be found in this kind of dry woodland.

At the altitude of the pass the most conspicuous associate of the Retama is Tabaiba majorera *Euphorbia atropurpurea*, with spectacular red-purple flowers, which has its main stronghold in this area; it tends to grow with a single straight, golden brown trunk that then splits to form about four thick branches.

In some places there are fine clumps of *Sonchus canariensis*, the largest species of its genus on Tenerife. It is a heavily branched shrub that can reach 3 m in height. Also present is the largest of the unbranched species in this genus, *Sonchus acaulis*, with rosettes near ground level that may be over a metre across, and *Atalanthus capillaris*, previously included in the genus *Sonchus*, which has very fine threadlike leaf lobes clustered at the tips of stiff branches. *Argyranthemum*

Retama rhodorhizoides & *Euphorbia atropurpurea*

Retamón canario *Teline canariensis*

Aeonium pseudourbicum

Sonchus canariensis, sometimes referred to as the Tree Dandelion, in regenerating dry woodland

Waterfall on a rainy day near Masca

foeniculaceum and *Sideritis brevicaulis*, both of which also occur in this area, are members of diverse groups with local representatives in various parts of the island, in which taxonomic revisions – and thus name changes – are frequent.

Of special interest here is *Phyllis viscosa*, a small shrub with sticky, bright green, long-stalked and pointed leaves, which sometimes forms dense hanging clumps on cliffs. It is one of the two Tenerife species of *Phyllis*, a genus found on the Canaries and Madeira, but with its closest relatives on the other side of the equator in southernmost Africa. More widely distributed dry woodland shrubs here include the spreading yellow-flowered umbellifer *Bupleurum salicifolium*, the yellow thistle *Carlina salicifolia*, Escobón *Chamaecytisus proliferus*, *Teline canariensis*, *Echium virescens* and *Echium aculeatum*, as well as *Globularia salicina*. As in most dry areas below the forests in Tenerife, there are also some invasive alien species, especially *Agave americana*, which is common in most parts of Teno, and a prickly pear *Opuntia maxima*.

Anís de risco *Bupleurum salicifolium*, one of the plants capable of growing out of solid volcanic rock

4 Dry woodland remnants — 'bosque termófilo'

Bejequillo tinerfeño *Aeonium haworthii*

Chahorra de Teno *Sideritis brevicaulis*

Parentucelllia viscosa

Bea tinerfeña
Greenovia dodrentalis

Lavandula minutollii

Colderrisco escabrosa *Crambe scaberrima*

Pelotilla pálida *Monanthes pallens*

At Los Carrizales vegetables are still grown near the village, but an outlying farm and cultivation terraces lie abandoned, as well as an old cave dwelling (inset)

It is worth keeping a sharp eye on the birds here since this is one of the few areas where you might conceivably see the Rock Sparrow *Petronia petronia*, a bird that has declined drastically in Tenerife in the last half century, but which may be hanging on in at least one part of Teno.

We tend to think of Teno as a dry place, so it was startling on our most recent visit to find an impressive waterfall only a few hundred yards from from the village of Masca, at about 600 m, with Canary Willow *Salix canariensis* and generally lush vegetation in a wet landscape that also included Canary Palms *Phoenix canariensis*. The floor of the barranco here has stands of Caña *Arundo donax*, an introduced giant reed occurring in many wet gullies on the island.

At lower levels in this area, around the hamlet of Los Carrizales at about 400 m, many additional endemic plants become more common, especially on the sides of the gorge that runs down to Playa de Carrizal 1.5 km below. These include Cardón *Euphorbia canariensis* and *Euphorbia lamarckii*, along with *Artemisia thuscula* and the beautiful *Aeonium canariense*. Most of these species, however, are more typical of the coastal and lowland shrubland rather than the dry woodland.

Some other notable plants

Aeonium arboreum, Aeonium spathulatum, Bituminaria bituminosa, Cistus monspeliensis, Dicheranthus plocamoides, Davallia canariensis, Erica arborea, Foeniculum vulgare, Forsskaolea angustifolia, Gonospermum fruticosum, Hypericum canariense, Hypericum reflexum, Kleinia neriifolia, Morella faya, Paronychia canariensis, Periploca laevigata, Plocama pendula, Rubia fruticosa, Rumex lunaria, Salvia canariensis, Todaroa aurea, Vieria laevigata.

Montaña de la Mulata (Camino del Risco de Teno Alto)

An ancient mule track leading through one of the richest areas for endemic plants in Tenerife, which also offers fine views out over the Isla Baja, one of the few large low-lying coastal areas on the north side of the island.

This site is largely within the Teno Natural Park and the area referred to by some botanists as El Fraile. On the south side of the road approaching the tunnel on the way to the Punta de Teno there is a cobbled track (much eroded in places) that leads behind the water tanks and up the right-hand gully to the steep slopes above the cultivated zone. The lower part of the path has some typical lowland scrub species including *Euphorbia lamarckii*, *E. balsamifera* and *E. canariensis*, along with a few Prickly Pear *Opuntia maxima* that sometimes have infestations of the Cochineal Bug *Dactylopius coccus*. This open sunny area is an excellent place to watch warblers and other birds, jumping spiders and a good range of butterflies.

The gully and the slopes higher up would once have had a rich dry woodland habitat, occupying a zone that has been almost entirely taken over by agriculture in most of the island. Here, although the goats have taken their toll – and still do so at the higher levels – many of the endemic shrubs can still

Malva de risco *Lavatera acerifolia*

Looking down on Isla Baja, with Buenavista, the northwest coast and lots of bananas

be found. Characteristic species include the buckthorn *Rhamnus crenulata* and *Jasminum odoratissimum*, the latter with glossy green pinnate leaves, elongate yellow flowers and black-brown oval fruit. You will also see the straggly shrub *Lavatera acerifolia*, an endemic relative of the hibiscus, with delicate white and pink flowers. Other plants include an endemic arum *Dracunculus canariensis* and the small endemic St John's Wort *Hypericum reflexum*. Also present is *Teucrium heterophyllum,* an attractive red-flowered sage relative with oval, sometimes toothed leaves that are rich green above but white felted below; it has curious seed pods that open at the top.

An attractive shrub that is conspicuous here is the composite *Atalanthus pinnatus*, with thin shoots springing from the top of stubby woody stems and carrying pinnate leaves with threadlike leaflets, and sprays of groundsel-like flowers. The *Echium* here is *E. aculeatum*, notable for its very spiny leaves and compact clusters of white flowers. On this and other shrubs you may see a strange tangle of threads that is parasitic dodder *Cuscuta* sp., which attaches itself to a plant of another species and forms connections with its vascular system, after which the dodder's own roots die and it depends on the host for its nutrition.

Uncommon trees here include the Canary Olive *Olea cerasiformis*, which grows in some of the driest places. This is also one

Parasitic Dodder Cuscuta *sp. spreading over* Echium aculeatum

of the relatively few sites on the island where there is a surviving population of Almácigo *Pistacia atlantica*, a deciduous tree with thin, spreading branches that can achieve a hemispherical form; its pinnate leaves are somewhat reminiscent of Rowan *Sorbus aucuparia*. A commoner species here is the purple flowered *Euphorbia atropurpurea*, more typical of dry woodland further south in Teno; here it coexists not only with *Euphorbia lamarckii* and *E. canariensis*, but also with a high altitude population of the normally coastal *Euphorbia balsamifera*, which grows well in steep rocky gullies, and also with the euphorbia-like *Kleinia neriifolia*.

Only a few struggling shrubs survive in the mist in this severely eroded area close to Teno Alto

On some cliff faces near the head of the gully are beautiful growths of lichens, including *Roccella*, but also a variety of cliff-dwelling shrubs including *Vieria laevigata*, *Phyllis viscosa*, *Ceropegia dichotoma*, rosettes of *Monanthes pallens*, and *Monanthes laxiflora* with its leaves like bunches of pallid grapes. High up on the track to Teno Alto one passes through land where goats still have an obvious impact and some weedy plants are noticeable, and close to the village much of the land is devastated by erosion, probably starting many centuries ago.

Some other notable plants

Aeonium cuneatum, Aeonium tabulaeforme, Agave americana, Andryala pinnatifida, Argyranthemum frutescens, Artemisia thuscula, Arundo donax, Asparagus sp., *Asphodelus ramosus, Bituminaria bituminosa, Ceballosia fruticosa, Convolvulus floridus, Echium plantagineum, Gladiolus italicus, Globularia salicina, Gonospermum fruticosum, Habenaria tridactylites, Hypericum canariense, Ilex canariensis, Justicia hyssopifolia, Lavandula canariensis, Lavandula buchii, Micromeria varia, Pancratium canariense, Periploca laevigata, Phoenix canariensis, Rubia fruticosa, Rubus* sp., *Rumex lunaria, Scilla haemorrhoidalis, Scilla latifolia, Sideritis cretica, Sonchus acaulis, Sonchus radicatus.*

Gonepteryx cleobule on *Jasminum odoratissimum*

Atalanthus pinnatus

Degraded cliff vegetation above Los Silos, with Sabina (centre) now regenerating

Barranco de Cuevas Negras, Los Silos

A fine barranco, cutting into the ancient cliffs of northwest Tenerife, which is famous for its special endemic plants.

This barranco, eventually rising to over 600 m, has an excellent track (PR TF-53) that soon begins a steady ascent, through vegetation with many trees and shrubs typical of dry woodland, including the green-stemmed *Bosea yervamora*, *Marcetella moquiniana* and *Rhamnus crenulata*. The track eventually reaches an area moist enough to support laurel forest, where trees such as *Laurus novocanariensis*, *Morella faya*, *Visnea mocanera*, *Picconia excelsa*, *Heberdenia excelsa*, *Apollonias barbujana* and *Viburnum rigidum* can still be found, in spite of the major impact of centuries of agriculture and the presence of much prickly pear *Opuntia maxima*. Some of the laurel trees here have curious growths of the fungus *Laurobasidium lauri*.

One can also see here the small tree *Maytenus canariensis*, but it seems especially at home on cliffs where it manages to grow in places apparently devoid of soil, where there is less competition from larger trees. The basalt cliffs at this site are especially important, since they have allowed the survival of many rare shrubs indigenous to the Teno massif, which have been eliminated by human activity from less precipitous areas. Many of the cliffs are inaccessible, and we have obtained only a distant view of one of the specialities of the area, the rare giant *Limonium arborescens*, which produces a tight bunch of brilliant blue flowers at a height of up to 1.8 m above the ground (see photo on page 103).

4 Dry woodland remnants — 'bosque termófilo'

Cliff at Cuevas Negras with Tedera *Bituminaria bituminosa*, *Sonchus congestus*, *Asparagus scoparius* and *Ceropegia dichotoma*

There are, however, many places where one can see interesting cliff plants close-up, including members of the Crassulaceae such as *Aeonium tabulaeforme* and two species of *Monanthes* with succulent and almost globular leaves. The omnipresent *Bituminaria bituminosa* here shows its ability to grow opportunistically hanging down on cliffs, while in the case of *Ceropegia dichotoma*, a succulent in the family Asclepiadaceae, some stems droop initially but recover and grow firmly upwards to form an erect clump. In contrast, *Asparagus scoparius* is committed to upright growth, though often straying from the vertical eventually. Another plant capable of growing in a pendulous form is *Sonchus congestus,* with thin, straggly branches that reach up to the light in laurel forest but dangle down cliffs when that is more appropriate; its cousin *Sonchus radicatus* has a flat rosette close to the cliff face. The crucifer *Descurainia millefolia* can colonise tiny ledges, while the

Monanthes polyphylla with *Atalanthus* cf *pinnatus*

widespread fern *Davallia canariensis* has creeping rhizomes that cling to the rocks, putting out fronds at intervals.

An important feature of this site is that the less precipitous sides of the barranco appear to be recovering ecologically as a result of reduced agricultural activity. Both Dragos *Dracaena draco* and Sabinas *Juniperus turbinata* have survived here on inaccessible cliffs exposed to moisture-laden updraughts, but the Sabina at least is now spreading onto steep slopes nearby where species of *Euphorbia* are currently abundant. These dry slopes also have one of the few Tenerife populations of the hemi-parasitic shrub *Osyris lanceolata*, a species that we have seen only at a different site. Two other unusual species recorded here are the very rare mallow *Navaea phoenicea* and the leguminous *Dorycnium spectabile*.

Peralillo *Maytenus canariensis* with ripe fruit (inset)

Chahorra de Daute *Sideritis cretica*

Alhelí de mediania *Erysimum bicolor*

4 Dry woodland remnants — 'bosque termófilo'

Some other notable plants

Adiantum reniforme, Aeonium ciliatum, Aeonium urbicum, Andryala pinnatifida, Artemisia thuscula, Atalanthus pinnatus, Arundo donax, Canarina canariensis, Carlina salicifolia, Convolvulus floridus, Dorycnium spectabile, Erica arborea, Erysimum bicolor, Euphorbia canariensis, Euphorbia lamarckii, Globularia salicina, Gonospermum fruticosum, Hypericum reflexum, Hypericum canariense, Ilex canariensis, Jasminum odoratissimum, Lobularia canariensis, Monanthes laxiflora, Monanthes polyphylla, Opuntia maxima, Paronychia canariensis, Phoenix canariensis, Pericallis echinata, Rubia fruticosa, Rubus sp., Rumex lunaria, Semele androgyna, Sideritis cretica, Solanum vespertilio, Sonchus acaulis, Tamus edulis, Todaroa aurea.

The rare endemic *Navaea phoenicea* growing near El Palmar

Taginaste gigante *Echium giganteum*

Paniqueso *Lobularia canariensis*

Dry woodland with *Marcetella moquiniana* (bottom left)
and Almácigo *Pistacia atlantica* (centre left)

Interián cliffs, Los Silos

A rich remnant of dry woodland of a type found only near the western end of the precipitous north coast of Tenerife. For anyone who is short of time but wishes to see the dry woodland ecosystem in a relatively intact state, this is an excellent place to visit. Access is easy, with only a short level walk needed to find many unusual and beautiful plants.

The minor road from Los Silos to La Tierra del Trigo climbs steeply up the cliffs of Interián, just to the east of the Barranco de Cuevas Negras (the previous site). The area is a protected site (Sitio de Interés Científico de Interián) and forms part of the protected landscape of the Acantilado de La Culata. At a hairpin bend with a small pull-off the road crosses an aqueduct known as the Canal de Las Carvas, at a height of a little over 200 m. By walking along this to the east or the west one can gain easy access to a fascinating stretch of dry woodland growing on very steep ground, with a remarkably high proportion of the trees and shrubs characteristic of this habitat. Furthermore, the cliff is so steep that it shows in a compressed form, hints of the transitions between different original woodland types suited to different altitudes. The many vertical rock faces and unstable slopes must always have prevented development of a closed canopy forest covering the whole cliff. Tall trees can grow in places, and in future these may increase under reduced pressure from domestic herbivores, but there will always be room

4 Dry woodland remnants — 'bosque termófilo'

Palo de Sangre *Marcetella moquiniana*

Dicheranthus plocamoides

Mataprieta *Justicia hyssopifolia*

Vieria laevigata

Saquitero *Heberdenia excelsa* with fruit

Allagopappus canariensis

Canary Palms and *Marcetella moquiniana* in a ravine at Interián

for tall shrubs and for plants living on rock faces. This is doubtless the reason for the great diversity of the woodland here.

In ravines and on the low ground at the foot of the cliffs Canary Palms *Phoenix canariensis* are the most noticeable trees, and here they may once have formed a closed canopy, perhaps along with Dragon Trees *Dracaena draco*, a few of which are still present, but this area is now much modified by humans. Around the transition from cultivated ground up into semi-natural woodland there is much *Artemisia thuscula*, perhaps reflecting an ability of this shrub to colonise disturbed ground. On the steep slopes above this, Almácigo *Pistacia atlantica* is abundant, and easily recognized by the broad hemispherical shape of the mature trees, the pinnate leaves and the bright red fruits; this area is now one of the main strongholds of the species on Tenerife. The other common large tree here is an unusual form of *Heberdenia excelsa*, locally known as 'Saquitero'; these are tall trees, often with slanted dark crowns, which stand out from the more bushy trees and shrubs nearby. It is only in this region of Tenerife that *Pistacia* and *Heberdenia* occur together in large numbers. There are also some oaks *Quercus* cf *ilex* at the foot of the slope, but these are probably planted; other non-native plants here are *Agave americana* and the prickly pear *Opuntia maxima*, but fortunately the latter species shows no sign of becoming abundant in this area.

Other trees occurring here include *Laurus novocanariensis, Visnea mocanera, Maytenus canariensis* and *Withania aristata*, while a few other species such as *Bosea yervamora, Pleiomeris canariensis, Sideroxylon canariense* and *Spartocytisus filipes* occur in this general area, though we have not seen them here. Conspicuous shrubs on the slopes above and below the 'canal' include *Asparagus scoparius*, with tall, arching stems and foliage that was turning to brilliant orange yellow when we made a midsummer visit to the site. *Hypericum canariense* also often turns yellow or orange, as does *Marcetella moquiniana*, an endemic species that is common in parts of the Interián cliffs but rather scarce elsewhere in Tenerife; vigorous young stems are sometimes brilliant crimson, giving this shrub its local name of Palo de Sangre. Other shubs here include *Atalanthus pinnatus, Ceballosia fruticosa, Globularia salicina, Jasminum odoratissimum, Justicia hyssopifolia, Kleinia neriifolia, Lavatera acerifolia, Periploca laevigata, Rhamnus crenulata, Rubia fruticosa* and *Ruta pinnata*, as well as a few *Teucrium heterophyllum*.

4 Dry woodland remnants — 'bosque termófilo'

The Cardón *Euphorbia canariensis* is mainly on craggy places higher up, but *Euphorbia lamarckii* occurs on the slopes, along with a few *Euphorbia balsamifera*, unusually far from the coast.

Following the aqueduct to the east, one eventually reaches a rock face with a fascinating array of cliff plants including *Aeonium tabulaeforme, Phyllis viscosa, Allagopappus canariensis, Sideritis cretica, Dicheranthus plocamoides, Monanthes laxiflora, Paronychia canariensis, Reichardia ligulata* and *Sonchus acaulis*. On the rocks there are beautiful orange 'crustose' lichens, as well as the well known *Roccella* lichens with long fronds reaching out to catch moisture from humid air forced upwards by the cliffs. In one part of the cliff there is a spectacular display of the composite shrub *Vieria laevigata*, the sole member of a genus that is endemic to the island of Tenerife. In steep places above, a few Dragon Trees *Dracaena draco* cling to the cliff, and there are also Canary Olives *Olea cerasiformis* and some Sabina *Juniperus turbinata*.

Walking from the road towards the west and negotiating a change of level in the water channel, there are interesting views down to plantations at the foot of the cliff, and eventually also into a steep ravine with a dense mass of vegetation. Species here include the introduced reed *Arundo donax* and much *Convolvulus floridus, Rumex lunaria*, brambles *Rubus fruticosa, Salix canariensis* and the introduced and invasive composite *Ageratina adenophora*. On the edge of the ravine is a fine specimen of Barbusano *Apollonias barbujana*, a laurel forest tree that also grows in the downward transition from laurel forest to dry woodland. Other species rather typical of the transition zone are *Echium giganteum*, a lanky shrub found only in the west of the north coast, and *Canarina canariensis*, though we did not see this here.

Saquitero, the dry woodland form of *Heberdenia excelsa*

This rock face on the Interián cliffs has a diverse plant community, with *Lavatera acerifolia* below

Dry woodland on the cliff below Genovés, with *Juniperus turbinata* (top centre) and *Spartocytisus filipes* (top right)

GENOVÉS AND EL GUINCHO

A remarkable site retaining one of the richest fragments of dry woodland to be found anywhere on Tenerife, with great variety of characteristic trees and shrubs.

The village of Genovés lies half way between Icod de los Vinos and Garachico on the TF82 road to Santiago del Teide, which roughly follows the division between the ancient Teno massif and the rest of the island. The area around Genovés, known as the Acantilado de La Culata, is part of the extraordinarily steep north face of Tenerife, created at a time in the distant past – long before the giant landslide that formed the Orotava valley – when a vast chunk of the island slipped away into the sea. Just below the village, on cliff-top land previously used for agriculture at a height of about 350 m, is a wonderfully diverse patch of regenerating dry woodland (see also photograph at start of this chapter).

The shrubs here reflect the fact that the site has been subject to the pressures of agriculture for a long time. It is likely that the pristine vegetation would have been 'sabinar', dominated by Sabina *Juniperus turbinata*, and a few specimens survive on the most precipitous parts of the site. With them are several

Zarzaparrilla *Smilax aspera*

bushes of the broom *Spartocytisus filipes,* which is perhaps an ecological analogue of the related *Retama rhodorhizoides,* which dominates the dry woodland in parts of Teno. *Spartocytisus filipes* has been shown (on La Palma) to be highly susceptible to the effects of introduced herbivores such as rabbits *Oryctolagus cuniculus,* and it is likely that it was much more abundant on Tenerife in the past.

A speciality of this site is the rare *Osyris lanceolata,* a member of the family Santalaceae, which parasitizes nearby shrubs via their root systems, but also photosynthesizes using its own chlorophyll. Another unusual plant here is *Ephedra fragilis,* found in only a few sites on

Asparagus umbellatus on *Euphorbia lamarckii*

Tenerife, which is a gymnosperm and thus related to the conifers rather than to the angiosperms (flowering plants); it has masses of thin branches that superficially resemble those of broom.

More widely distributed dry woodland shrubs here include *Echium giganteum* and *Ruta pinnata,* both species tall enough to hold their own in dense shrubby vegetation, and *Aeonium arboreum, Todaroa aurea* and *Paronychia canariensis,* all of which thrive in rocky places. The two species of *Asparagus* that occur here, *A. scoparius* and *A. umbellatus,* are both characteristic of dry woodland, but have different life strategies. *A. scoparius* (which also often grows on cliffs) produces a cluster of tall erect stems that can reach the light above surrounding shrubs, while *A. umbellatus* sprawls over adjacent shrubs, using them as support. Another plant using other shrubs as support is *Smilax aspera,* which shows that it is capable of producing a large crop of fruit on the rare occasions when it is able to grow in strong light.

We are so close to the sea here that it is no surprise to find some shrubs which are also common in the lowland and coastal shrubland, such as *Artemisia thuscula, Rubia fruti-*

Scilla haemorrhoidalis

Crassula tillaea

cosa and *Euphorbia lamarckii*. However, the presence of *Pleiomeris canariensis, Apollonias barbujana* and *Laurus novocanariensis*, all laurel forest trees that turn up in a number of dry woodland sites, suggests that Genovés is at about the height at which there would once have been a gradual transition from dry woodland to the laurel forest above it. This conclusion is reinforced by a walk along a track to the site of an old bath house just to the west of Genovés, where a break in the line of cliffs and a steep-sided ravine offers a less exposed and moister environment. Here – in addition to the species just mentioned – there are other trees characteristic of the transition, including *Ilex canariensis, Sideroxylon canariense, Withania aristata* and the dry woodland form of *Heberdenia excelsa*.

Sabina *Juniperus turbinata* (lower left) near sea level with *Euphorbia balsamifera* at El Guincho

On the coast just below Genovés is a site which demonstrates that in some parts of the north coast of Tenerife, the dry woodland once extended right down to the sea, almost eliminating the coastal scrub that is so extensive in many other parts of the island. Punta de la Sabina (Juniper Point) is a promontory below the village of El Guincho (the Osprey). More level areas a short distance inland have banana plantations, but the rugged basaltic coastal strip has dry woodland remnants as well as elements of coastal scrub. *Juniperus turbinata* flourishes and produces seedlings within 50 m of the shore, and the Canary Olive *Olea cerasiformis* also grows there, though both show some signs of damage by salt spray. In an adjacent but even more exposed area there is a small area of coastal scrub dominated by *Euphorbia balsamifera* and *Schizogyne sericea*, with a few other halophytic plants. Plants with bulbs or corms are conspicuous here in early spring, and the delicate *Scilla haemorrhoidalis* was particularly beautiful during a recent January visit, while the ubiquitous *Asphodelus ramosus* was already in bloom. The commonest of the Tenerife orchids, *Habenaria tridactylites*, also occurs here, even very close to the sea.

Some other notable plants around Genovés and El Guincho

Agave americana, Allium sp.*, Argyranthemum frutescens, Atalanthus pinnatus, Bituminaria bituminosa, Bosea yervamora, Davallia canariensis, Erica arborea, Erysimum bicolor, Globularia salicina, Jasminum odoratissimum, Kleinia neriifolia, Lotus tenellus, Opuntia maxima, Periploca laevigata, Phagnalon* sp.*, Polypodium macaronesicum, Reichardia ligulata, Rhamnus crenulata, Roccella* sp.*, Rumex lunaria, Salsola divaricata, Sideritis kuegleriana, Sonchus acaulis.*

Laurel forest on the Vueltas de Taganana

5

Laurel forest – 'monteverde'

The laurel forest ecosystem

The evergreen laurel forest is one of the most remarkable features of the western Canary Islands. Ecologists consider that it is a relict of a type of subtropical forest that covered large parts of southern Europe and the Mediterranean area for a long period prior to the middle of the Miocene epoch around 15 million years ago, but which largely disappeared during the subsequent cooling that culminated in the Pleistocene Ice Ages. Similar relict forests are present in Madeira and the Azores, and also in more distant places including southeastern USA, China and Japan, and the southernmost parts of South America, Africa and Australasia. On the Canaries the most extensive surviving area is in the Parque Nacional de Garajonay on the island of La Gomera, but on the steep northern slopes of Tenerife there are still impressive tracts of relatively undisturbed laurel forest, especially in the Anaga peninsula and in the northwest.

In these tracts of forest one can still walk in the dim light beneath the dense canopy formed by enormous laurels and other trees of similar form – many of them heavily draped

Píjara *Woodwardia radicans* is a giant fern that forms dense undergrowth in the laurisilva of Anaga

with lianas, lichens and mosses – and get a feel for what the laurel forest of Tenerife was like in its pristine state. Although one may be tempted to wait for a rare clear day and walk in patchy sunlight, the wet days are much more evocative of the true nature of this habitat; even when it is not actually raining the forest is frequently shrouded in cloud.

The prevalence of the clouds produces an altitudinal zone where temperatures are relatively mild (12–16°C) and frosts do not occur. Humidity is almost continuously high and annual rainfall measured by conventional techniques is in the range of about 500 to 900 mm per year. However, these rainfall figures represent only part of the real situation, because of the profound effect of the trees themselves in augmenting the capture of moisture, as discussed more fully in Chapter 1. Moisture from the clouds condenses on the leaves and drips to the ground, an effect that is especially important for the trees and other plants during the summer months (June, July and August) when there is virtually no direct precipitation.

Over millions of years, these conditions led to development of a wide band of laurel forest along the north-facing (windward) slopes and cliffs of Tenerife, from Anaga in the east to the edge of the Teno massif in the west. In Anaga the laurel forest on the northern slopes reached the tops of the ridges and extended a short way down on the other (southern) side. Further to the west – where the mountains are higher – the laurel forest extended upwards as far as the moisture allowed. During the last 200,000 years, however, the forests towards the western end of the north coast have been repeatedly devastated by eruptions from the Teide volcanoes.

The laurel forest can exist only within the altitudinal band influenced by the sea of clouds produced by the north-easterly trade winds, which keeps it moist even during the dry summer period. In winter the cloud sea is higher and influences a zone of transition from laurel forest to humid pine forest at about 1200 metres above sea level; this is locally termed 'pinar mixto' and has a canopy of pine but an understorey mainly of *Erica arborea, Morella faya, Laurus novocanariensis* and *Ilex canariensis*. The lower limit of the laurel forest on the northern slopes is related to the bottom of the cloud sea and is now generally around 500 m, but may have been as low as 200 m in a few places in the past; below it there was once a narrow band of dry woodland, with small areas of coastal shrubland near the shore.

On the southern slopes of Tenerife the laurel forest may never have existed in a fully developed form. However, some of the less moisture-demanding species in the forest community are still present at around 1000 m in the Güímar valley, where cloud is often present, suggesting that in this area a type of laurel forest referred to as 'laurisilva termófila' or 'laurisilva xérica' once occurred more widely in the damper places between the dry woodland below and the pine forest above.

This idea is reinforced by a recent study of fossil remains of snails, although it relates to a much earlier time. The cinder cone of Montaña Negra (121 m) in the dry plains 4 km west of Tenerife Sur airport, has deposits about 300,000 years old, sandwiched between layers of volcanic rocks. These contain shells of several species of land snails; most of them are typical of fairly dry habitats, but two belong to the semislug genus '*Plutonia*' (see comment in 'Animals of the laurel forest'). These snails have reduced shells and cannot retract into them, so that they can only live in damp habitats, and are now found in the laurel forest in the north of the island. It seems likely, therefore, that a form of humid woodland was once present near the south coast of Tenerife. However, this woodland would have had a chequered history in view of the dramatic volcanism that has affected this part of the island during the last two million years.

The soils of the laurel forest are acid, rich in humus and fertile, with effective recycling of nutrients. Light levels are high in the canopy but very low beneath it, so where the canopy is complete it inhibits development of a shrubby understorey, and only seedlings of laurel forest trees can be found, with ferns in damper places. However, breaks in the canopy enable an assortment of climbing plants ('lianas') to reach up towards the light, and allow establishment of a variety of shrubs, herbs and ferns. The forest floor has a substantial layer of dead plant material, especially in hollows, and is rich in mosses, fungi and lichens, along with an array of associated invertebrates such as snails, harvestmen, centipedes, spiders and beetles.

In favoured places on Tenerife a diverse closed-canopy high forest termed 'laurisilva' survives, but in other areas there are several forms of shrub-forest known locally as 'fayal-brezal'. In some places a form of this is natural, but in many others it is the result of destruction of mature forest by agricultural or forestry activity, followed by the development of more or less degraded habitats. Together, laurisilva and fayal-brezal are referred to in Spanish as 'monteverde', and we use the term laurel forest in the same sense.

Before the arrival of the Guanche people from Africa, 2–3000 years ago, the laurel forest probably had a total extent of 300–600 km^2 (~15–30% of the area of the island). A recent assessment indicates that Tenerife now has just over 100 km^2 of laurel forest, of which only about one fifth is true laurisilva and four-fifths are fayal-brezal, so it is clear that the island has lost a very large proportion of its original evergreen broadleaf forest. Recovery is now under way in many areas, but the return of fully developed laurisilva will take several centuries.

In the canopy of the laurel forest, where many kinds of trees are flowering and fruiting, scramblers like Corregüelón de monte *Convolvulus canariensis* (centre) can also reach the light

Laurisilva

The best surviving areas of relatively intact laurel forest or 'laurisilva' are in the highest parts of the Anaga peninsula. There is a more restricted area in the east of the Teno massif, and small patches in deep ravines along the north coast near Santa Úrsula and Los Realejos, and in the Güímar valley in the south. In its undisturbed state, the laurisilva has a dense canopy 20–30 m above the ground and can include up to 17 species of broad-leaved evergreen trees and tall shrubs, plus two others that occur only on the fringes. Not all of these trees are normally found in the same area, some being associated with the drier parts of the forest and some flourishing mainly in particularly humid ravines, where the canopy reaches its greatest height.

Four characteristic trees of the evergreen forest belong to the laurel family (Lauraceae) and dominate the canopy in favourable places, so it is natural that they have given their name to the laurel forest. However, while laurels are the most significant group, the other species that typically occur in the forest are extremely diverse taxonomically, belonging to a dozen different plant families. Since all are adapted to life in the same damp environment, it is not surprising that many of them have a similar growth form and share with the laurels certain characteristics such as waxy elongate leaves that shed rainwater.

The similarity in the appearance of the trees gives the forest a feeling of uniformity, but also makes identification difficult, especially from ground level. Most of the species are either confined to the Canary Islands alone or to the archipelagoes of Macaronesia. The common species can be learnt fairly easily, and being able to identify them adds greatly to the interest of exploring the forested parts of the island. We therefore provide a key to the leaves and a photo gallery with accompanying text which we hope will help. In using this information one needs to remember that leaves of young trees are often very different from those of adults.

As well as the broad-leaved trees, two members of the heather family – Brezo *Erica arborea* and Tejo *Erica platycodon* – are of great importance. They have needle-like leaves and are easily recognisable as heaths, except for the fact that they are the size of small trees or very tall shrubs (>5 m), in contrast to the low-growing *Erica* species of the moorlands in northern Europe. Brezo is the more tolerant of dry conditions, flourishing in steep rocky places and colonising open areas that once held laurisilva, but also penetrates the less arid parts of the pine forest. Tejo, however, prefers extremely moist situations. It has bark that tends to peel off in strips (unlike Brezo with its hard finely grooved bark) and the spaces under the loose strips provide good refuges for endemic beetles such as the black weevil *Laparocerus undatus* and the tenebrionid *Nesotes conformis*.

The four members of the Lauraceae are Loro *Laurus novo-canariensis,* Barbusano *Apollonias barbujana,* Viñátigo *Persea indica*

Contrasting foliage of Brezo (left) and Tejo

and Til *Ocotea foetens*. All are capable of growing to more than 20 m and thus tend to exclude trees with smaller stature. They often have multiple trunks, with young shoots springing up from the base and eventually replacing the main tree, so making the trees extremely long-lived (and also hard to kill). *Laurus* is the commonest, dominating the forest in many places and readily colonising vacant ground. *Apollonias* is locally frequent within the forest and also extends its range down onto steep rocky places below the main forest zone. *Persea* normally grows in ravines and other places where surface water is available, while *Ocotea* is characteristic of deep and humid ravines, where it can reach a height of over 30 m, doubtless living for several centuries.

The holly family (Aquifoliaceae) has two members on Tenerife, both of them capable of growing to more than 15 m. Acebiño *Ilex canariensis* is very common in the laurel forest as well as occurring in the transition to pine forest. Naranjero salvaje *Ilex perado* ssp. *platyphylla* is moderately common in the laurel forest from Anaga westwards as far as Santa Úrsula.

Til *Ocotea foetens*, a member of the Lauraceae that produces abundant suckers

The forest also includes tall trees in several other families. Hija *Prunus lusitanica* (Rosaceae) is common near the ridge road and in other parts of Anaga, and occurs very locally along the north coast as far as the Orotava valley, but is absent from the northwest. Palo Blanco *Picconia excelsa* represents the olive family (Oleaceae); it is typical of middle altitudes in the forest, but requires rich soil and moisture. Two species characteristic of the drier parts of the forest are Mocán *Visnea mocanera* (in the tea family Theaceae) and Madroño canario *Arbutus canariensis* (a broad-leaved member of the Ericaceae); both can also be found in transitional areas between laurel forest and dry woodland or pine forest, in the north and also in parts of the south such as the Güímar valley.

Faya *Morella faya* (previously *Myrica faya*) in the family Myricaceae, is a relative of Bog Myrtle, a fragrant shrub of wet heaths in western Europe. It is a pioneer, quickly colonising gaps in the canopy, along with the two species of *Erica*, and is thus preadapted to exploiting areas disturbed by people. It now typically occurs in the form of small trees in secondary formations of fayal-brezal, but can grow much taller in well preserved fragments of laurisilva.

Two laurel forest trees are in the little known family Myrsinaceae, and both have become scarce: Delfino *Pleiomeris canariensis* rarely reaches a height of 15 m, but has the largest leaves of all the forest trees. *Heberdenia excelsa* is similar but of somewhat smaller stature and is curious in that it exists in two forms that may be on the way to becoming separate species, though they differ very little genetically. The form known as Aderno occurs in the laurel forest, while the Saquitero or Sacatero has smaller leaves and is found in the dry woodland.

Three additional smaller trees have patchy distribution in the forest. Sanguino *Rhamnus glandulosa* is the laurel forest member of the Rhamnaceae, with close relatives in other habitats on the island; it is most frequent in ravines and near forest edges. The willow Sauce *Salix canariensis* (Salicaceae) requires moisture and may not be able to compete with the taller *Persea* and *Ocotea* in undisturbed ravines, but is now frequently found in the laurel forest as well as damp places elsewhere. Marmolán *Sideroxylon canariense*, in the family Sapotaceae, is now rare but was probably always more typical of dry woodland than laurel forest.

The two Tenerife members of the elder family Sambucaceae are both tall shrubs or small trees of the laurel forest. Follao *Viburnum rigidum* is the most abundant understorey species in some areas, especially

One of the surviving specimens of Tabaiba de monteverde *Euphorbia mellifera*

in the west (where a potential competitor *Prunus lusitanica* is absent). Saúco *Sambucus nigra* ssp. *palmensis* is a rare and endangered tree, occurring in the heart of the forest in a few places in the north, and westwards to beyond Los Realejos. Two other tall shrubs occur in the laurel forest and may also grow to 5 m, but neither is easy to find. Peralillo *Maytenus canariensis*, in the spindle family Celastraceae, sometimes occurs in the lower parts of the evergreen forests but is primarily a species of the dry woodland and now survives mainly on cliffs. Ortigón de monte *Gesnouinia arborea*, a member of the nettle family Urticaceae, is a scarce shrub found mainly on cliffs in the north.

One additional tree species, Tabaiba de monteverde or Adelfa de monte *Euphorbia mellifera*, is omitted from our key and photo gallery because it is so rare; it is a slender upright tree which can reach 15 m, with leaves clustered towards the tips of the branches and the flowers in a terminal cluster. Like other species of *Euphorbia*, it contains milky latex sap.

Fayal-brezal

Fayal-brezal is a name used for three reasonably distinct types of shrub-forest that occur in the evergreen forest zone but lack a full canopy of broad-leaved trees and include relatively few tree species. The term fayal-brezal refers to the fact that in this vegetation type the

Tree-heathers dominate the fayal-brezal on the tops of the ridges in Anaga

dominant species are normally the laurel-like *Morella faya* (locally known as Faya) and one or both of the giant heathers Brezo *Erica arborea* and Tejo *E. platycodon*. However, there is often also a scattering of other broadleaved trees, especially the holly *Ilex canariensis* and *Laurus novocanariensis*. Unlike most of the other trees of the laurel forest, Brezo is widely distributed outside the Canaries, occurring elsewhere in Macaronesia, southwest Europe and also in Africa, both in the north and on mountains in the tropics.

In all three types of fayal-brezal the shrub-forest normally reaches a height of 5–10 m. The first type, which is probably a natural 'climax' community and which we call 'ridge-top fayal-brezal' occurs on the most exposed, steepest and highest ridges that the cloud can normally cross, especially along the crest of the Anaga peninsula at around 800–950 m. Here Tejo *Erica platycodon* is often the dominant species, apparently thriving in these windy situations, where condensation from the swirling mists is very evident. However, Brezo *Erica arborea* and Faya *Morella faya* are normally also present, and there are often other species such as *Ilex canariensis*, *Laurus novocanariensis*, *Heberdenia excelsa* and *Prunus lusitanica*. This cloud forest is of relatively short stature, and sunlight can often get down to ground level, so there is a rich flora of small shrubs, herbs and ferns.

A second and more or less natural type of fayal-brezal occurs in the transition between laurel forest and pine forest, in situations too dry for the most moisture-demanding laurel forest species. Here Brezo *Erica arborea* is the normal representative of the tree-heaths; it tolerates relatively dry conditions and occurs alongside *Morella faya*, *Ilex canariensis* and sometimes also *Laurus novocanariensis*. Members of this group of species often also occur in the pine forest and

Fayal-brezal develops where there was once laurisilva, after unsustainable exploitation

may be found as high at 1700 m in a few places. Additional species sometimes occurring in the transition zone are *Arbutus canariensis*, *Rhamnus glandulosa* and *Visnea mocanera*.

The third and most extensive type of fayal-brezal, sometimes referred to as woody heath, is a plant community in degraded areas that were once laurisilva. It results from unsustainable exploitation such as felling of large trees, persistent cutting of saplings or collection of firewood, grazing by domestic stock (or intensive collection of fodder) and sometimes burning. In this type of shrub-forest Brezo is almost always the dominant species, but *Morella faya*, *Ilex canariensis* and sometimes also *Laurus novocanariensis* are also typically present. The shrub *Daphne gnidium*, more typical of pine forest, sometimes makes its appearance in fayal-brezal, as do several shrubs more typical of dry woodland such as *Hypericum canariense*, *Jasminum odoratissimum*, *Globularia salicina* and *Gonospermum fruticosum*.

OTHER PLANTS OF THE LAUREL FOREST

Lianas and other scrambling plants play a significant role in the laurel forest. A few species flourish in real laurisilva, reaching up into the canopy. They probably need to get started in a light gap where they can gather enough solar energy to climb tens of metres up into the trees. If they can get up there, they have the opportunity to spread over the canopy. *Convolvulus canariensis* sometimes manages this, and then develops massive liana stems of a type generally associated with forests in the tropics; it also forms dense curtains in larger light gaps, where it sometimes produces masses of pale blue or mauve flowers.

Saúco *Sambucus nigra ssp. palmensis*

A rare small tree related to Elder of Europe. The opposite leaves are pinnate, with 7 leaflets - 3 pairs and one at the tip. (The smaller shrub *Bencomia* has pinnate but alternate leaves.)

Naranjero salvaje *Ilex perado*

A tree with large, broad leaves with a sharp prickle at the tip and usually small forward-pointing prickles on the edges (prickles are rare in *I. canariensis*).

Follao *Viburnum rigidum*

A vigorous small tree, common in the understorey, with opposite leaves that are broad and rough, hairy above and below.

Sanguino *Rhamnus glandulosa*

A medium-sized tree with broad leaves that have toothed edges and conspicuous swollen glands near the base. The fruits resemble those of the hollies (*Ilex*).

Palo blanco *Picconia excelsa*

A tree with opposite leaves that are smooth and leathery; they can be serrated when young. The bark of twigs and stems has characteristic white warts.

5 Laurel forest – 'monteverde'

Hija *Prunus lusitanica*
A tree with leaves well spread along the twigs, with crinkled, notched edges. They are lance-shaped, long (>10 cm) and broad at the base, with leaf stalks sharply separated off and normally pink.

Madroño *Arbutus canariensis*
A tree with long, fairly narrow, leathery leaves mainly at the tips of the branches. The bark is velvety golden green, peeling in rich brown or purple flakes.

Faya *Morella faya*
A tree with fairly small, lance-shaped leaves, usually with wavy or irregular edges (not the regular serrations that are usual in *Visnea*). Confusion with *Laurus* is possible, but its leaves have glands.

Sauce *Salix canariensis*
A tree with the leaves slightly scalloped or smooth-edged; they are up to 6 cm long, narrow and rather limp, with white hairs on the back and stalk.

Mocán *Visnea mocanera*
A tree with fairly small leaves (<6 cm), glossy and leathery, lance-shaped, broadest in the middle and tapering rather gradually into the leaf-stalks. Twigs may have warts.

Peralillo *Maytenus canariensis*

A small tree with leaves that are oval, conspicuously broad and blunt-tipped, the edges often with slight and widely spaced irregularities. Not a laurel forest species but confusable with some of them.

Marmolán *Sideroxylon canariense*

A rare tree with long, blunt-ended leaves, boat-shaped and long-stalked. Not typically in the laurel forest.

Aderno *Heberdenia excelsa*

A tree with large, lance-shaped leaves tapering back to the leaf-stalk and usually also with a pointed tip (cf *Pleiomeris*); they are stiff, leathery and often down-turned at the edges, which may be yellowish.

Acebiño *Ilex canariensis*

A tree with leaves that are rather small, oval, leathery and stiff, with somewhat thickened edges; sometimes with a small prickle at the tip and a few elsewhere.

Delfino *Pleiomeris canariensis*

A tree with very large (up to 15 by 5 cm) dark green leaves that normally taper to a blunt rounded tip. The flowers have 5 petals fused at the base forming a tube (in *Heberdenia* the 5 petals are separate.)

5 Laurel forest – 'monteverde'

Loro *Laurus novocanariensis*
A tall tree. The leaves have small glands along their whole length at the bases of the veins, providing a distinction from *Apollonias*, *Persea* and *Morella faya*, which all lack glands.

Til *Ocotea foetens*
A tall tree with leaves that have a few swollen glands near the base (rather like those on the serrated leaves of *Rhamnus glandulosa*). The fruit has a deep cup.

Barbusano *Apollonias barbujana*
A tall tree with bright green rather glossy lance-shaped leaves. Some of the leaves on almost any tree have conspicuous, randomly distributed, lumpy galls.

Viñátigo *Persea indica*
A tall tree with large leaves without glands. Usually there are some brilliant gold or red dying leaves and the red fallen leaves often indicate that a tree is nearby.

Ortigón de monte *Gesnouinia arborea*
A shrub or small and slender tree, related to nettles. Leaves lance-shaped, deep-veined and hairy, often crowded near tips of branches.

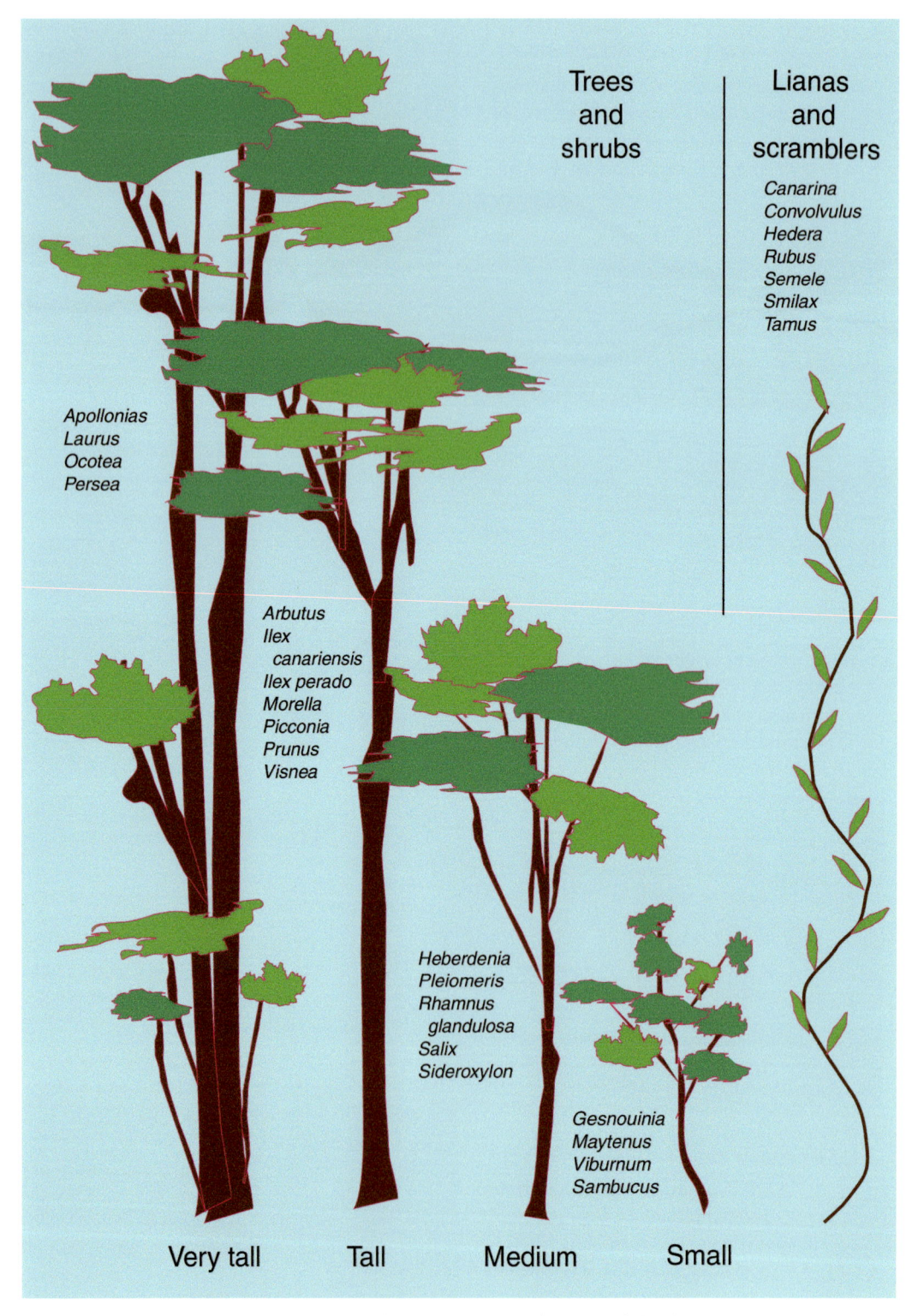

Diagram indicating the relative sizes of trees and shrubs in the laurel forest, although these are extremely variable

5 Laurel forest – 'monteverde'

OPPOSITE LEAVES

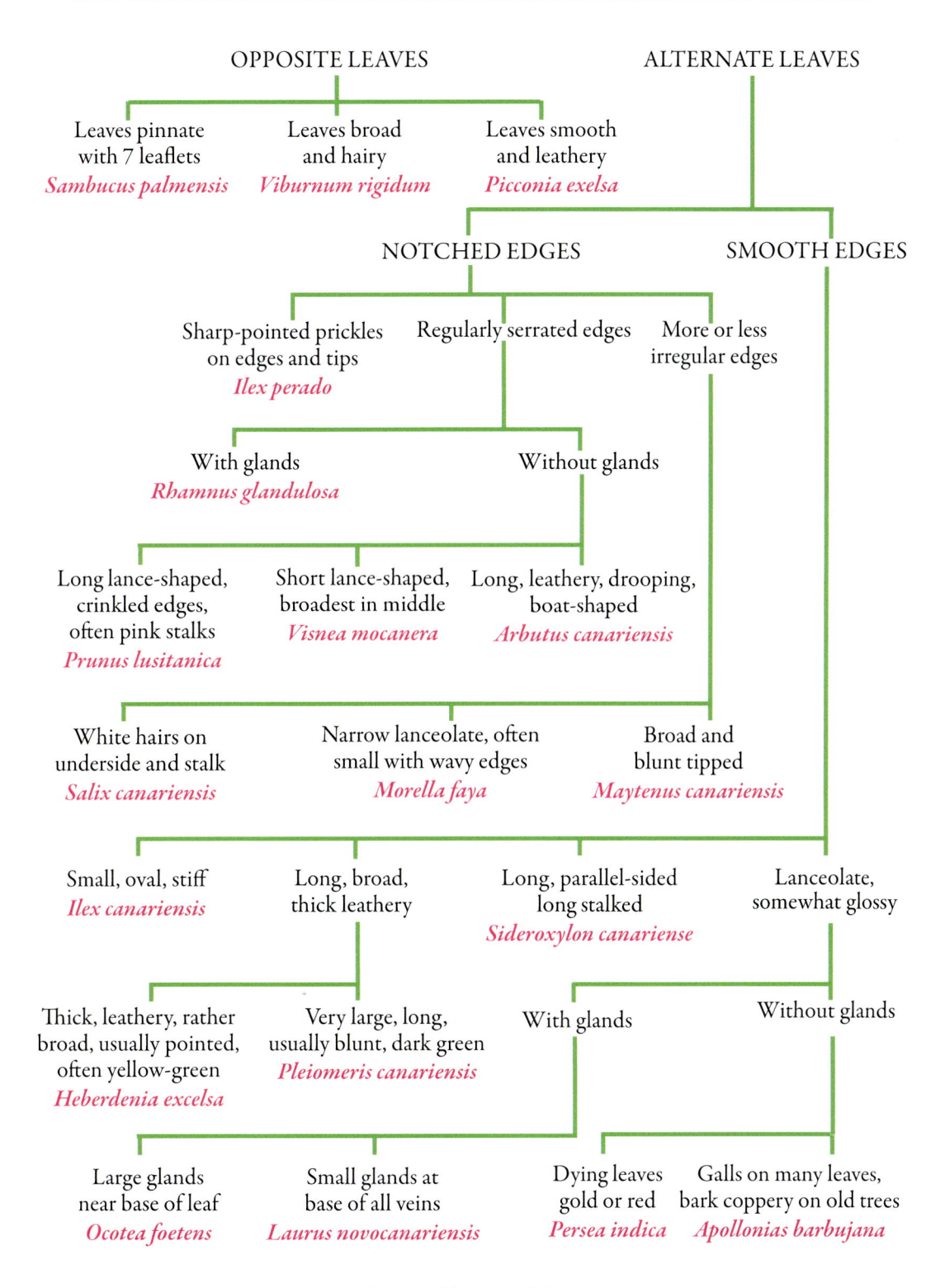

- **Leaves pinnate with 7 leaflets** — *Sambucus palmensis*
- **Leaves broad and hairy** — *Viburnum rigidum*
- **Leaves smooth and leathery** — *Picconia exelsa*

ALTERNATE LEAVES

NOTCHED EDGES

- **Sharp-pointed prickles on edges and tips** — *Ilex perado*
- **Regularly serrated edges**
  - **With glands** — *Rhamnus glandulosa*
  - **Without glands**
    - Long lance-shaped, crinkled edges, often pink stalks — *Prunus lusitanica*
    - Short lance-shaped, broadest in middle — *Visnea mocanera*
    - Long, leathery, drooping, boat-shaped — *Arbutus canariensis*
- **More or less irregular edges**
  - White hairs on underside and stalk — *Salix canariensis*
  - Narrow lanceolate, often small with wavy edges — *Morella faya*
  - Broad and blunt tipped — *Maytenus canariensis*

SMOOTH EDGES

- Small, oval, stiff — *Ilex canariensis*
- Long, broad, thick leathery
  - Thick, leathery, rather broad, usually pointed, often yellow-green — *Heberdenia excelsa*
  - Very large, long, usually blunt, dark green — *Pleiomeris canariensis*
- Long, parallel-sided long stalked — *Sideroxylon canariense*
- Lanceolate, somewhat glossy
  - **With glands**
    - Large glands near base of leaf — *Ocotea foetens*
    - Small glands at base of all veins — *Laurus novocanariensis*
  - **Without glands**
    - Dying leaves gold or red — *Persea indica*
    - Galls on many leaves, bark coppery on old trees — *Apollonias barbujana*

Key to leaves of the Laurel Forest trees

Gibalbera *Semele androgyna* is an easily recognised climber

Zarzaparilla *Smilax canariensis* can become rampant

Three species of climbers are 'monocotyledons', members of the major group of plants that includes the grasses, palms, lilies and orchids. *Semele androgyna*, with curious shiny pseudo-leaves (cladodes) bearing flowers and fruit on their edges, is often found hanging on trees near ground level in forest of fairly low stature, but also occurs on cliffs. The other two species can be confusing: *Smilax canariensis* is a perennial, woody, wiry plant using tendrils to climb, and sometimes forming massive canopies on forest trees. *Tamus edulis* is more delicate, having a large, perennial, underground tuber and annual, thin, twining stems with which it forms tangled masses of foliage on other plants; it was probably once widely distributed in dry woodland.

The ivy *Hedera canariensis* is abundant in a few places in the laurel forest; it is reminiscent of the European *Hedera helix* (which has been introduced to the island) but differs in the shape of its leaves, which are broader than long, slightly heart-shaped or almost circular. Several other species are more scramblers than climbers. The most spectacular is Bicacarera *Canarina canariensis*, a giant campanula whose closest relatives are found on mountains in East Africa, with bell-shaped, six-parted, orange flowers, large globular fruits and opposite, long, strongly-toothed and triangular leaves. It is a vigorous vine that puts out shoots across the forest floor, scrambling on other plants but rarely getting above head height; it dies back to a tuberous root each year. If you shake the flowers of the vine, drops of sweet nectar will fall.

Brambles *Rubus* spp. are also familiar though taxonomically difficult; there are thought to be three native species on Tenerife, including the endemic *R. palmensis*, but they are not easy to tell apart; all can occur in the laurel forest and one also infests degraded areas.

Other scramblers sometimes found in the forest include species of Rubiaceae, and also the extremely rare or possibly extinct *Normania* (or *Solanum*) *nava*, a member of the nightshade family (Solanaceae) with glandular-hairy stems and leaves, and orange fruit.

Apart from the large shrubs already mentioned, which can grow to a height of 5 m, several smaller shrubs can be found in the laurel forest. The most intriguing is *Bencomia caudata*, with alternate, pinnate leaves, which reaches 2-3 m and is now found in remote sites in both the north and south of Tenerife. Its original habitat was probably rocky places in the transition zones between dry woodland and laurel or pine

Corregüelón *Convolvulus canariensis* rarely flowers unless it can reach the light

Norza *Tamus edulis* has delicate foliage but also a massive underground edible tuber

The endemic ivy Hiedra de monte *Hedera canariensis* has noticeably broad leaves

Bicácaro *Canarina canariensis*, deservedly honoured with the name of the archipelago

forest. Other shrubs occurring in the transition zones are *Hypericum canariense, Globularia salicina, Jasminum odoratissimum* and *Daphne gnidium*.

Many of the smaller plants of the laurel forest are largely confined to natural light gaps that let in sunlight for part of the day, or to the edges of roads and tracks. Conspicuous along the summit road in Anaga in spring is *Teline canariensis*, a bush that can grow to 3 m, with small trifoliate leaves, large yellow pea-flowers and silky stems and seed pods. Its relative *Adenocarpus foliolosus* also occurs in laurel forest, fayal-brezal and pine forest and is confusingly similar, but the stems are not silky and the pods are sticky, with glandular papillae.

Along tracks in summer and autumn conspicuous species include *Ixanthus viscosus*, a generic endemic gentian-relative, locally called Reina del Monte (queen of the mountain); it has almost stalkless opposite leaves with a few parallel veins, and pale yellow flowers. More striking is Cresta de gallo *Isoplexis canariensis*, with brilliant orange flowering spikes; this genus is a fairly close relative of the foxgloves and is found only in Madeira and the Canaries. Also related to *Isoplexis* is the figwort *Scrophularia smithii*, with a tall inflorescence but tiny whitish green flowers.

Another common roadside plant is *Sonchus congestus*, a dandelion-relative with leaf rosettes at the ends of thick wiry stems, inflorescences covered with velvety white hairs and bunches of bright yellow flowers followed by globular seed heads; it also occurs deep in the forest, although it is much less luxuriant there. In the same family Asteraceae are several endemic members of the genus *Pericallis*. The widespread *P. appendiculata* is a small shrub with numerous white flowers and lobed leaves, green above and white-felted below. *P. cruenta* is similar but with purple flowers and purple undersides to the leaves; it occurs mainly in the centre of the north coast. *P. tussilaginis* is adapted to the dampest situations and is a relatively delicate herbaceous perennial with small groups of mauve flowers; it occurs in Anaga and westwards almost to the Orotava valley. Other conspicuous herbaceous plants include the tall forest buttercup *Ranunculus cortusifolius*, while the deep purple cranesbill *Geranium reuteri* makes a change from the predominant yellow of the majority of the forest flowers.

In Tenerife one is rarely far from a species of *Hypericum* (St John's Worts), but they can be confusing. In the laurel forest the typical species is Malfurada *H. grandifolium*, which occurs in laurisilva and in fayal-brezal, but *H. glandulosum* is also present, mainly on cliffs and at the lower edge of the forest. A third species, Granadillo *H. canariense*, is primarily a species of the dry woodland. Another native shrub that is not really a plant of the forest is the giant dock *Rumex lunaria*, which occurs in disturbed places and along roads; it is probably dispersed by vehicles.

Reina del monte *Ixanthus viscosus* is common along forest tracks

Cresta de gallo *Isoplexis canariensis* flourishes in light gaps

5 Laurel forest – 'monteverde'

Retamón *Teline canariensis*, with silky foliage, is conspicuous along roadsides in spring

Cerraja de monte *Sonchus congestus* varies in its growth form, often hanging on cliffs

Bejeque puntero *Aeonium urbicum* exploits roofs as well as dry cliffs within the laurel forest

Tusilago morado *Pericallis cruenta* is found mainly in the laurel forest near the north coast

Palomera *Pericallis appendiculata* dominates parts of the laurel forest floor

Bugallón *Pericallis tussilaginis* is widespread in laurel forest and damp lowland areas

Natural History of Tenerife

Morgallana *Ranunculus cortusifolius*

Lluvia *Crambe strigosa*

Pata de gallo *Geranium reuteri*

Capitana *Phyllis nobla*

Malfurada *Hypericum grandifolium*

Algaritofe *Cedronella canariensis*

The cabbage family (Brassicaceae) is represented by the striking though scarce Lluvia *Crambe strigosa*, with rough, toothed leaves and tall, slender flowering stems that carry up the tiny flowers to form a cloud of white specks high above the plant. Another plant often conspicuous along trails is Capitana *Phyllis nobla* in the family Rubiaceae, a member of a group endemic to the Macaronesian islands; it is an upright shrublet with shiny but hairy leaves in whorls, and loose clusters of whitish flowers. Scarce and less conspicuous is the umbellifer *Cryptotaenia elegans*, with slender stems and white flowers.

The family Lamiaceae is well represented in the laurel forest. Algaritofe *Cedronella canariensis*, with wiry stems and dense spikes of mauve flowers, seems very patchy in its occurrence, often occurring in shady ravines and areas of fayal-brezal on the edge of pine forest. The strongly scented shrubs in the genus *Bystropogon* are also in the family Lamiacae and two of them are frequent in the forests of Tenerife. Poleo de monte *B. canariensis* is the normal species in the laurel forest along the east and central parts of the north coast and Poleo *B. origanifolius* is mainly in the pine forest in the west and southwest. The genus is of special interest since taxonomists once thought that as well as the endemic species in the Canaries and Madeira, there were also – improbably – representatives in South America. The latter are now placed in a separate genus, but recent DNA studies still suggest that *Bystropogon* of Macaronesia provides a link between the mints of the New World and those of the Old World. The related *Origanum vulgare* is also present, with flowers spread out along the stems (in contrast to the terminal clusters of *Bystropgon*), as well as a couple of species of *Micromeria*, though these are not really laurel forest plants. In contrast, shrubs in the large genus *Sideritis*, characterised by their felted and usually whitish hairs, occur widely in the laurel forest; the usual species are *S. canariensis*, in which the tall inflorescence has well-separated whorls of flowers, and *S. macrostachys*, in which the inflorescence is very compact and cylindrical or slim-conical.

The genera *Myosotis* (forgetmenots) and *Urtica* (stinging nettles) both have several native species on the island, often occurring in deeply shaded and damp places. However, *Urtica* species need to be carefully distinguished from *Mercurialis canariensis* (notable for its

Cryptotaenia elegans, a scarce plant of the laurel forest

Chahorra de monte *Sideritis macrostachys*

Gladiolo silvestre *Gladiolus italicus*. Photo: Manisha Pittelkow

Aeonium ciliatum occurs in laurel forest and lower down in Anaga

Violeta de Anaga *Viola anagae* is endemic to the Anaga peninsula

Asparagus fallax is a rare shrub of the laurel forest

stems which are square in cross section), a newly discovered endemic in the *Euphorbia* family, and *Mercurialis annua*, more often in dry woodland. The Boraginaceae, so well represented in other habitats on Tenerife, play only a small role in the laurel forest, the inconspicuous broad-leaved *Echium strictum* being the only species normally present. Several native violets are also present, the rarest being the Tenerife endemic *Viola anagae*, which lives only in a few places in remote parts of the forests of Anaga. A number of other groups of herbaceous plants are represented, including widespread but probably native species of *Galium* (Rubiaceae) and weedy introduced composites (Asteraceae) such as *Bidens* species and *Ageratina adenophora*.

Apart from the three climbers mentioned above, monocotyledonous plants are not conspicuous in the laurel forest except at ground level. However, the shrub *Asparagus fallax*, with very dense foliage and small scarce flowers, can be found in heavily shaded places, and sometimes also its relative *A. umbellatus*, with a scrambling habit and somewhat larger flowers. The woodrushes *Luzula* have complex taxonomy, but at least two species are present, including the endemic *L. canariensis*. The iris *Gladiolus italicus* also occurs in the forest, though it flowers more freely in relatively open areas.

Many species of ferns (Pteridophyta) grow abundantly on some parts of the floor of the laurel forest. Most impressive is *Woodwardia radicans*, with fronds up to 2 m in length, which is abundant in some core areas. It is one of rather few kinds of ferns with the capacity to produce new individuals at the tips of the fronds. Bracken *Pteridium aquilinum*, now sometimes split into several species but traditionally considered one of the most widely distributed of all plants, grows in open spaces, mainly in fayal-brezal. Other ferns are found mainly in the most shady and moist parts of the laurisilva, and this is where one can find the only endemic species of fern on Tenerife, the buckler-fern *Dryopteris oligodonta*. Another significant fern is *Culcita macrocarpa*, which in the Canaries is found only in a small part of the Anaga peninsula; it is related to the tree ferns (*Dicksonia*) a group that has representatives in humid tropical and subtropical forests in many parts of the world. *C. macrocarpa* does not have a trunk like a typical tree fern but forms small groves of vertical fronds that may reach more than 2 m in height. Other ferns are included in the Box overleaf.

5 Laurel forest – 'monteverde'

Bejeque de monte *Aeonium cuneatum* flourishes near ridge tops

Mercurialis canariensis, a newly described endemic in the laurel forest

Estornudera *Andryala pinnatifida*, which is remarkably varied in its growth form

Urtica morifolia, found in damp and shady places in the laurel forest

Estrellita *Aichryson laxum*

Aichryson pachycaulon ssp. *immaculatum*

Asplenium hemionitis

Dryopteris guanchica

Diplazium caudatum

Blechnum spicant

Woodwardia radicans

FERNS OF THE LAUREL FORESTS OF TENERIFE

Ferns are among the most interesting plants from the biogeographical, ecological and conservationist points of view, and laurel forest is undoubtedly the best ecosystem for ferns and fern-allies in the Macaronesian or Mid-Atlantic archipelagos: Azores, Madeira and the Canary Islands. More than 30% of the species of Pteridophyta recorded in the Canaries live in the more or less humid types of laurel forest, and the abundance and biomass of fern species here is certainly higher than in any other ecosystem represented in the archipelago.

In the Canary Islands about 30 species of ferns and fern-allies are found in the laurel forests, but not all of them are common in this environment. In the case of Tenerife, we can consider that there are 20 species typical of the laurel forest, plus another eight which are found inside it occasionally. The more humid laurel forest holds very interesting and even endangered fern species like *Culcita macrocarpa*, *Vandenboschia speciosa* (=*Trichomanes speciosum*), *Diplazium caudatum*, *Hymenophyllum tunbrigense* and *Blechnum spicant*. In the subhumid and humid types of this habitat we can also find *Woodwardia radicans*, *Pteris incompleta*, *Polystichum setiferum*, *Dryopteris oligodonta*, *Dryopteris guanchica*, *Cystopteris fragilis* and *Selaginella denticulata*. On the other hand, in the dry or thermophilous type of laurel forest *Adiantum reniforme*, *Davallia canariensis* and *Polypodium macaronesicum* are locally common. Other ferns such as *Asplenium hemionitis* and *Asplenium onopteris* are usually abundant everywhere in the interior of such forests, while Bracken *Pteridium*

Vandenboschia speciosa

Pteris incompleta

Adiantum reniforme

Hymenophyllum tunbrigense

Polystichum setiferum

aquilinum is mainly present around the tree line, in old cultivations in open fields, and along the edges of tracks and paths. Two additional species, *Athyrium filix-femina* and *Dryopteris affinis* ssp. *affinis*, are among the rarer and poorly-known ferns in Tenerife and the Canary Islands in general; they prefer the more humid sites.

In transition areas with the pine forest, and in rocky habitats, one is likely to find species such as *Asplenium aureum* and *A. trichomanes,* but they can also appear locally in sites with pure laurel and tree-heath forests (laurisilva and fayal-brezal). Fortunately, the introduced and invasive fern *Cyrtomium falcatum* is very rare in the laurel forest on Tenerife, in contrast with the situation in the Azores, where it is spreading in many wooded localities. The situation is similar with *Christella dentata,* a very scarce and local fern in Tenerife, which is much more common in La Gomera and La Palma and seems to be introduced to the archipelago. Finally, *Notholaena marantae* ssp. *subcordata* prefers open, sunny areas, and *Adiantum capillus-veneris*, *Anogramma leptophylla* and *Pteris vittata* are very scarce or very local in the laurel forest.

In Tenerife the best sites to find the fern species mentioned above are the Anaga peninsula, the forest of Agua García, the mountains of Santa Úrsula and La Orotava, the massif of Tigaiga and the forest of Monte del Agua in the northwest.

Text and photographs: **Rubén Barone**

Shady and moist habitats are of course also ideal for the growth of mosses and liverworts (Bryophyta), lichens and fungi, and all these groups are well represented, especially in those areas where the forest floor is not subject to disturbance. Fungi are especially common on dead wood, but there is one large and conspicuous species, *Laurobasidium lauri*, which is parasitic on living laurels *Laurus novocanariensis*.

Less steep rocky places are colonized by such plants as the endemic *Sonchus congestus* and the variable composite *Andryala pinnatifida*, with its slightly white-velvety leaves, as well as *Aeonium canariense* with broad and slightly hairy leaves. The similar *Aeonium cuneatum* is distinguished by its more straplike hairless leaves; it seems to thrive in the damp conditions near the ridge and gives a dramatic effect when viewed through swirling mist, since large specimens have leafy flowering stems up to a metre high, ending in spectacular cones of yellow flowers.

Cliffs and banks within the forest zone often let in some sunlight and tend to have a distinct flora. In some places one can see the branched *Aeonium ciliatum*, which also occurs in more open places lower down. Shady and very moist cliffs, including some along the Anaga ridge road, offer niches for the plate-like *Aeonium tabulaeforme* and for three delicate *Aeonium* relatives, the hairy *Aichryson laxum* and two of its more or less hairless relatives: *Aichryson punctatum*, distinguished by the purple or black spots and wrinkles on the leaves, and the unblemished but much less common *A. pachycaulon* ssp. *immaculatum*.

Light penetrates the forest where there are cliffs or steep banks

Culcita macrocarpa, a rare tree-fern relative in the laurel forest

Animals of the laurel forest

The laurel forest (and especially the relatively intact laurisilva) is rich in invertebrates, and the easiest way to find them is to go to shady places where there are deep accumulations of moist leaf litter. Here one can always find a diverse array of animals including snails, harvestmen, centipedes, spiders and beetles, a high proportion of them being endemic. One example is the amphipod crustacean *Orchestia guancha*, a Tenerife endemic that thrives in the humid forest; it belongs to the family Talitridae, a group more familiar as sandhoppers on the sea shore.

Snails also flourish in the laurel forest. A common species is *Hemicycla bidentalis* in the family Helicidae, which also occurs on the isolated Roque de Fuera off the north coast of Anaga, where it has evolved to form a distinct subspecies. If you search the leaves of saplings you may find a 'semislug' of the family Vitrinidae, a group of snails in which the external shell is much reduced, so that they look much like slugs. These animals live in continuously humid environments and so do not need to be able to withdraw into their shells to survive dry periods. Many endemic species occur in the Canaries and other Macaronesian islands, with six species on Tenerife, but there has evidently been much convergent and parallel evolution, so that relationships among them are still partly unresolved and scientific names such as the genera *Insulivitrina* (ex *Plutonia*), '*Guerrina*' and '*Canarivitrina*' are in flux.

Another important group of snails is the genus *Canariella* (in the leaf snail family Hygromiidae) with nine species (plus some subspecies) currently recognised on Tenerife, all of them endemic to the island. Even more diverse is the genus *Napaeus*, endemic to the Canaries and with 21 species endemic to Tenerife. *Napaeus variatus*, which occurs in the laurel forest along the north coast, normally has its shell covered with a complete layer of soil that makes it very hard to spot and so presumably reduces the chance of it being eaten by birds or lizards.

Beetles have also diversified freely in the Canaries. Analysis of the ground beetle family Carabidae shows that almost four-fifths of the species and subspecies occurring naturally in the laurisilva of Tenerife are endemic to the Canaries. Some stocks of these beetles have undergone remarkable evolutionary radiation; for instance, ten species of the genus *Calathus* are found only in the Tenerife laurel forest. The ground beetles are mainly nocturnal hunters, and

Semislug *Insulivitrina* (*ex Plutonia*) sp. in the family Vitrinidae

Earwig *Guanchia cabrerae* male
Photo: Pedro Oromí

Painted lady *Vanessa cardui*
Photo: Juan José Bacallado

the best way of finding them is by turning over loose rocks or logs (but one should of course replace these carefully afterwards). However, even in daylight there is a chance of seeing the large *Carabus faustus*, which is black with brilliant green highlights (see photo page 343).

The weevil genus *Laparocerus* also has many representatives in the laurel forest, but this is another nocturnally active group, so that these plant-feeding beetles are best found by searching foliage in the forest at night.

The bugs (Heteroptera and Homoptera) have also been studied in some detail, and they present a picture similar to the beetles, with a very high proportion of species in the laurisilva being endemic. The cultivated areas nearby, on the other hand, have very few endemic bugs but a large number of widespread species.

One of the special insects of the forest is the endemic earwig *Guanchia cabrerae,* while another is a delicate endemic cockroach *Phyllodromica brullei,* which has winged males but wingless females. The flies (Diptera) in the forest also include species with limited flying ability: craneflies (family Tipulidae) are all weak flyers but *Tipula macquarti* is a large endemic species that has entirely abandoned the use of its tiny wings.

A more impressive insect with much reduced wings is the large (up to 2.5 cm) green bush-cricket *Calliphona konigi* (family Tettigoniidae) which is known only from the northern side of Tenerife, although it has relatives with full wings on other islands in Macaronesia. Members of this genus have abandoned the normal vegetarian diet of bush-crickets and become predators on other invertebrates. *Calliphona konigi* does not fly and is cryptic and hard to see, but its black and red hind wings can be exposed suddenly to startle potential enemies; immatures can be found in flowers such as buttercups while the adults are usually in trees. Another bush-cricket that can be found on the leaves of the laurisilva trees is *Canariola nubigena*. It is much smaller, wingless, and with antennae four times its body length; the females are brown but the males have marbled grey bodies with patches of other colours.

One of the characteristic butterflies of the laurel forest is the familiar looking Canary Speckled Wood *Pararge xiphioides* (sometimes treated as a subspecies of the ordinary Speckled Wood *P. aegeria*) which can often be seen on patrol or resting in patches of sunlight. Also frequent in the forest are the large yellow endemic Canary Brimstone *Gonepteryx cleobule* and the Canary Red Admiral *Vanessa vulcania*, which is more common in this habitat than the ordinary Red Admiral *Vanessa atalanta*. Several other butterflies occur in the laurel forest as well as in other habitats, including the Cardinal

Canary Speckled Wood
Pararge xiphioides

Argynnis pandora (whose larvae feed on violets) and the Canary endemic Lulworth Skipper *Thymelicus christi*. The Queen of Spain Fritillary *Issoria lathonia* has turned up in laurel forest of the northwest, but is now thought to be a vagrant. Among the spiders of the laurel forest is the large clubionid *Cheiracanthium pelasgicum*, a member of a genus in which many species are poisonous.

Lizards and geckos do not often live in the tall laurisilva, which offers little sun for basking, but lizards are found on emerging rocks and in open spaces in fayal-brezal; we do not know the distribution of skinks in the laurel forest. Frogs are mainly to be found in ravines and other places with permanent water, but tree frogs can be heard croaking in winter and spring in areas well away from water.

Canary Red Admiral
Vanessa vulcania

It is generally quiet in the forest, but if you suddenly hear a noisy bird flying off from the trees above you, it is likely to be one of the two endemic fruit-eating pigeons: Bolle's Pigeon *Columba bollii* with a dark tail or the White-tailed Laurel Pigeon *Columba junoniae* with a largely white tail. The first of these can be found in forest in most parts of the north coast, but the second is less common and has a more patchy distribution. Both pigeons feed largely on the fruits of the laurel forest trees, but *C. bollii* also eats some invertebrates and *C. junoniae* seems more prone to add fruit of cultivated trees and cereal crops to its diet. Both species undoubtedly disperse the seeds of the laurels and other species such as hollies, and *C. bollii* may be especially important in the dispersal of Palo Blanco *Picconia excelsa* and other trees including *Rhamnus glandulosa*.

Many other kinds of birds can be seen in or over the laurel forest, and we mention only the most characteristic ones. Buzzards *Buteo buteo* can often be seen soaring overhead, and sometimes also Kestrels *Falco tinnunculus*, while Sparrowhawks *Accipiter nisus* are routinely present. Long-eared Owls *Asio otus* are present but are easier to hear than to see, while Woodcock *Scolopax rusticola* are also inconspicuous, although widespread in both laurisilva and fayal-brezal. Among songbirds the Common Chaffinch *Fringilla coelebs* is a prominent species, while other normal residents in this habitat include the endemic Canary Chiffchaff *Phylloscopus canariensis*, Goldcrest *Regulus regulus*, Tenerife Blue Tit *Cyanistes teneriffae*, Blackbird *Turdus merula* and European Robin *Erithacus rubecula*. However, most of these species are more easily seen in fayal-brezal, forest edges and along tracks than in high laurisilva, and fayal-brezal is also used by additional species such

Red Admiral *Vanessa atalanta*
Photo: Juan José Bacallado

Columba bollii, the commonest pigeon in the laurel forest. Photo: Aurelio Martín

Barbastella barbastellus guanchae, found in forests and lowland areas in the north
Photo: Rubén Barone

as Sardinian Warbler *Sylvia melanocephala*, Blackcap *Sylvia atricapilla*, Canary *Serinus canaria* and Linnet *Carduelis cannabina*.

Mammals are not easily seen in the laurel forest, but several species of bats occur there, including *Pipistrellus maderensis* and the recently discovered *Barbastella barbastellus guanchae*, a subspecies endemic to Tenerife and La Gomera. The latter occurs in both laurel and pine forest, but also in lowland areas in the north of the island.

One also often comes across signs of activity by the introduced Roof (or Black) Rat *Rattus rattus*. The rats bite off leafy twigs from the trees in such numbers that they are obvious when walking along forest trails. It is not clear how much of the twig is actually eaten, but the leaves are often intact and provide convenient clues as to which tree species one is walking under. Trees subject to the attacks include *Laurus*, *Heberdenia*, *Pleiomeris*, *Persea*, *Picconia* and doubtless others, and we have also found damage to the small shrub *Phyllis nobla*. Intriguingly, there is a hint from the anatomy of the extinct Tenerife Giant Rat *Canariomys bravoi* that it may have fed partly on the leaves of trees (and presumably also the fruits), so it is possible that the trees of the laurel forest have been subject to this type of attack for many millions of years.

Heberdenia excelsa foliage pruned off trees by rats

Some laurel forest trees can produce 'suckers' far above the ground, even from old trunks

LAUREL FOREST OF ANAGA

The largest surviving area of laurel forest on Tenerife, within the Anaga Rural Park, and with examples of both laurisilva and fayal-brezal. The laurisilva here is the most diverse on the island and several rare or endangered species occur in the more remote places, some of which are under special protection.

The Anaga peninsula that forms the northeastern tip of Tenerife is an ancient volcanic massif worn down to its roots by millions of years of erosion. During the early part of its history Anaga was more or less isolated from the rest of the island and evolution proceeded somewhat independently. As a result, Anaga is home to an extraordinary variety of living things and it includes – at Cruz del Carmen – the place that has the highest recorded biological diversity to be found anywhere within the European Union. A recent submission to UNESCO by the island government resulted in June 2015 in the official designation of the peninsula as the 'Reserva de Biosfera del Macizo de Anaga' in recognition of its unique cultural and biological interest. Permission is needed for visits to certain special areas.

Before the arrival of the Guanche people two to three thousand years ago, the central ridge of Anaga and the northern slopes down to 200–450 m were probably almost entirely covered by evergreen laurel forest. The forest would have extended lower on slopes fully exposed to the northeast trade winds than on slopes in the rain shadows created by lateral ridges such as those near Afur, Taborno and Chinamada.

On the south side of Anaga and on its eastern tip the lower limit of the laurel forest was probably close to 600 m, since the clouds streaming over the central ridge tend to evaporate at about that height, and there would have been a broad band of dry woodland below it.

The arrival of people brought major changes to Anaga, with felling and burning of trees in the more accessible areas and herding of goats over all of the lower slopes, with negative impacts on the native vegetation. Cultivation presumably extended gradually, perhaps reaching its greatest development around the end of the 19th Century, and declined rapidly during the second half of the 20th Century. Now there are many slopes covered by abandoned terraces with scrubby unnatural vegetation, often grazed by goats (and also some sheep).

Laurel forest would once have extended much further down on the northern slopes of Anaga

On the north side of the peninsula these processes led to the virtually complete elimination of the original laurel forest from the slopes below about 500–600 m, but leaving the areas above that relatively intact. On the south side of the ridge there was anyhow little forest below that level, leading to the current situation in which the forest extends fairly symmetrically along the spine of the peninsula, reaching its lower limit around 600 m on either side. However, the most luxuriant forest has probably always been on the northern slopes and the crest of the summit ridge, and it is in remote parts of these areas that magnificent forest can still be found.

The road running eastwards along the jagged mountainous spine of Anaga is usually shrouded in mist, providing a taste of the conditions that have permitted development of the laurel forest. The clouds – or rain – whip across the ridge from the north, and even when it has not been raining the road is often wet below the trees. On brighter days it is possible to look down on the canopy of the forest in some places. It is in the canopy – where there is sunlight or at least somewhat brighter light – that the leaves of the trees can capture the energy needed for reproduction, so abundant flowers and fruit are often visible.

Cruz del Carmen and Pico del Inglés

Although the ridge road is narrow there are places where one can pull off and get easy access to several trails – some of them well signposted – from which the forest can be explored. The first place to stop at is the easy-to-miss information centre below the car park at Cruz del

Carmen. It is a good starting point for learning about the native trees and also for access to a couple of trails (and a restaurant).

A couple of kilometres further east there is a 'mirador' or viewpoint called Pico del Inglés on a spur road overlooking laurel forest. Several of the typical trees grow close to the parking area, including *Laurus novocanariensis, Pleiomeris canariensis, Rhamnus glandulosa* and *Viburnum rigidum*. Both this and Cruz del Carmen are good places to compare the two tree-heaths *Erica arborea* and *E. platycodon*, and to see other native plants including *Phyllis nobla* (a member of a Macaronesian endemic genus), the yellow-flowered shrub *Teline canariensis*, the buttercup *Ranunculus cortusifolius*, the thistle *Carlina salicifolia* and the nettle *Urtica morifolia*. The wide view out over forest makes this one of the easiest spots in the east of the island to see the endemic Bolle's Pigeon *Columba bollii*. The morning – just as the sun reaches the treetops and before crowds of visitors arrive at this popular viewpoint – is the best time to watch for pigeons as they fly above the canopy far below you, on the lookout for trees with ripe fruit. The other endemic pigeon – the White-tailed Laurel Pigeon *Columba junoniae* – is less common and is not so easy to see at this end of the island, except in a few sites.

If this is your only visit to the laurel forest it is worth walking down into part of the forest from Pico del Inglés. There is a marked path starting from the car park that goes right down to the south coast just east of Santa Cruz, but not much of it is in forest. There is also an obscure and unmarked path starting immediately opposite an abandoned restaurant about half a kilometre back along the dead-end road to Pico del Inglés, but permission is needed to follow this path. At first it is rather steep and slippery, but it soon levels out and winds through forest that still retains many of the typical trees, shrubs and vines (including *Smilax canariensis* and *Tamus edulis*), although it has clearly been cut over in the past. The two tree-heaths are both conspicuous along this path, but in rather different situations. In the lower, less steep areas are fine specimens of Brezo *Erica arborea* with hard, finely-grooved bark; higher up, and on the steeper and more exposed slopes, you will see more of the smaller Tejo *E. platycodon*, characterized by its more geometric

Trees capture moisture from mist, which then drips to the ground

Ortigón de monte *Gesnouinia arborea*, a scarce understorey shrub

leaf arrangement and the tendency for its bark to peel off in strips. If you go further down the track you will also see the barred entrance to a 'galería' for collecting water, and nearby there used to be a water drip where one could see forest birds drinking. In this damp spot – characteristically – there are specimens of the laurel-relative *Persea indica*, and you will probably notice the orange-red dead leaves on the ground.

Vueltas de Taganana

After returning to the main ridge road another few minutes drive eastwards brings one to an access point for an ancient track named Vueltas de Taganana, which gives excellent opportunities to see many laurel forest trees and other plants. The old Casa Forestal by the roadside has a few planted endemic shrubs around it, including *Gesnouinia arborea*, growing here much more luxuriantly than it usually does in the forest. Also here is the rare elder *Sambucus nigra* ssp. *palmensis* with opposite pinnate leaves, each with 7 (or fewer) leaflets, and also *Bencomia caudata* with alternate pinnate leaves with 7-11 leaflets, and with very different, drooping flowering spikes. Smaller species include the crucifer *Crambe strigosa*, the gentian-relative *Ixanthus viscosus* and the forest buttercup *Ranunculus cortusifolius*.

Even a short downhill walk from here is worthwhile, with trees such as *Ocotea foetens*, *Laurus novocanariensis*, *Apollonias barbujana*, *Rhamnus glandulosa*, *Visnea mocanera*, *Ilex perado* and *Prunus lusitanica*, shrubs including *Hypericum grandifolium* and *Teline canariensis*, the scramblers *Semele androgyna* and *Hedera canariensis*, and many kinds of ferns including *Diplazium caudatum* and *Woodwardia radicans*.

It is also easy to walk up the Vueltas from the village of Taganana far below, climbing initially past vineyards and partially abandoned agricultural land that would once have held dry woodland, and then reaching a gradual transition to laurel forest, which here comes lower than in almost any other part of Anaga. There are fine displays of lianas belonging to *Convolvulus canariensis*, *Semele androgyna* and *Smilax canariensis*. There are also vines of *Canarina canariensis*, scrambling *Rubia peregrina*, brambles *Rubus* sp. and specimens of the rare *Asparagus fallax*, member of a group typical of more open habitats. A ravine at about 500 m is particularly impressive, with accumulations of leaf litter rich in invertebrate animals in spite of the apparent dryness of the ground at this low altitude. Here and there are ancient specimens of Til *Ocotea foetens*, with suckering multiple trunks, and also the relatively rare *Pleiomeris canariensis* with

Climbers on climbers: *Semele androgyna* on an ancient liana of *Convolvulus canariensis*

its enormous leaves. Between this point and the top of the Vueltas at the Casa Forestal one can see almost all of the characteristic trees of the laurel forest.

El Bailadero and further east

Further east along the ridge road lies the conspicuous abandoned restaurant El Bailadero (referring either to a traditional meeting place for dances, or to a place where the sheep used to meet) close to where the road from San Andrés on the south coast to Taganana in the north crosses under the main ridge by a tunnel. There is a government-run hostel about a kilometre beyond the old restaurant. If you stay the night there – and especially if it is misty – it is worth looking out for forest insects attracted to the lights, some of which stay on the adjacent walls during the day. You may find a variety of species, including the very large Convolvulus Hawkmoth *Agrius convolvuli*, with pink bands round its abdomen.

Beyond El Bailadero lies a relatively undisturbed area of laurel forest, centred on the peak of Chinobre (910 m). Trails here are generally unmarked, since the authorities do not want to encourage heavy use of such a sensitive area, but there are some pull-offs from which one can get a good feel for the habitat. The ridge road finishes at Chamorga and the last patches of forest are close by, since the remote easternmost tip of Tenerife beyond this (the 'Reserva Natural Integral de Ijuana') has only fragments of forest in the uppermost parts of the ravines.

There is a well marked trail up the ravine running northwest from Chamorga to the ridge. The area near the village has been used by people for many centuries and the slopes where there is no longer agriculture are covered by scrubby artificial habitats with elements of dry woodland and of fayal-brezal, such as *Hypericum canariense* and *Bupleurum salicifolium*. In relatively light places there is much regeneration of *Picconia excelsa*, along with slender specimens of *Gesnouinia arborea* and large ferns. Also here is *Canarina canariensis* sprawling on the forest floor or scrambling up into low branches. Near the watershed there is good ridge-top fayal-brezal with many *Rhamnus glandulosa* and *Picconia*. From the ridge itself there are fine views down to the coast towards Punta del Hidalgo and the remote settlement of El Draguillo, only recently given a road connection to Taganana.

Trunk of large Brezo *Erica arborea*, with inset flowers, above an ancient specimen of Barbujano *Apollonias barbujana* with flaky bark

Chanajiga

For visitors staying in the north of the island and without an opportunity to visit Anaga, a visit to the more limited laurel forest of Chanajiga is a worthwhile alternative.

This area is frequently under cloud, so it's best to choose a clear day. Chanajiga is a recreation site perched on the side of the Risco de la Zarza (cliff of the bramble) that forms part of the western wall of the Orotava valley, at an altitude of 1200 m above Los Realejos. The cliffs were created at the same time as the valley, when a giant landslide took an enormous bite out of the northern side of Tenerife and millions of tons of rock disappeared into the sea below.

On these steep slopes some forestry planting of non-native Monterey Pine *Pinus radiata* and other trees has been carried out in the past, but there are also remnants of the laurel forest that once extended along the whole of the north side of the island. A forest track goes along the contour of the incredibly steep slope and eventually reaches the Barranco de Ruiz and several other recreation sites further to the west. The steepness of the terrain makes it hard to walk into the forest, but many of the trees can be seen from the track and there is a good chance of seeing one or both of the endemic fruit pigeons, the dark-tailed Bolle's Pigeon *Columba bollii* and the White-tailed Laurel Pigeon *Columba junoniae*.

Chanajiga is also a good place to see other native birds, as well as butterflies, which are abundant in open spaces along the track where flowering plants can flourish; there are also damp areas on the steep rock faces that add to the variety of habitats.

Some notable plants
Adenocarpus foliolosus, Aeonium sp.*, Bencomia caudata, Carlina salicifolia, Cedronella canariensis, Chamaecytisus proliferus, Cistus monspeliensis, C. symphytifolius, Echium giganteum, Erica arborea, Geranium reuteri, Hypericum grandifolium, Hypericum reflexum, Ilex canariensis, Laurus novocanariensis, Morella faya, Myosotis latifolia, Pericallis cruenta, Persea indica, Phyllis nobla, Pteridium aquilinum, Rubus bollei, Scrophularia smithii, Sideritis canariensis, Smilax canariensis, Viburnum rigidum.*

Butterflies you might see
Cardinal *Argynnis pandora*, Southern Brown Argus *Aricia cramera*, Clouded Yellow *Colias crocea*, Canary Blue *Cyclyrius webbianus*, Canary Brimstone *Gonepteryx cleobule*, Canary Grayling *Hipparchia wyssii*, Small Copper *Lycaena phlaeas*, Canary Speckled Wood *Pararge xiphioides*, Small White *Pieris rapae*, Bath White *Pontia daplidice*, Lulworth Skipper *Thymelicus christi*, Meadow Brown *Maniola jurtina*, Red Admiral *Vanessa atalanta*, Painted Lady *Vanessa cardui*, Canary Red Admiral *Vanessa vulcania*.

Barranco de Ruiz

The Barranco de Ruiz is a dramatic gorge cutting into the north side of Tenerife, where there is a fairly small surviving area of laurel forest. The area is a protected Site of Scientific Interest, but there is easy access on foot. Most of the trees typical of this habitat can be seen here, as well as both of the laurel pigeons and other endemic birds, and many kinds of butterflies.

The easiest approach to this site is from Icod el Alto. The track leads down towards a place where there is a major 'step' in the gorge: the relatively shallow ravine above the step, originating at the eastern end of La Fortaleza on the edge of Las Cañadas, suddenly becomes a landscape-scale gorge, with impressive vertical cliffs. Over millions of years the water rushing down the ravine in times of flood has cut back from the coast for two and a half kilometres into the layers of volcanic deposits that form the extraordinarily steep north coast of Tenerife. On its eastern side the step in the gorge is not vertical, and a steeply sloping area at a height of about 400–500 m provides conditions suitable for laurel forest. The track down is along a cliff with spectacular geology: near the top is a massive basaltic lava flow overlying a bright red layer of 'burnt soil', while further down there are deep deposits of whitish pyroclastic fragments.

Along this track, which is fully exposed to the western sun, butterflies are notably abundant. We have seen *Vanessa vulcania, Gonepteryx cleobule, Pieris rapae, Pieris cheiranthi* and *Pararge xiphioides*, but other species are doubtless often present.

The woods along the rim of the gorge are much modified but include numerous Canary Pines. Within the canyon, where humidity is doubtless higher and human access more restricted, a significant area of laurel forest has developed – and survived – on the slopes below a major rock wall. At the bottom of the forested slope there has been agricultural activity in the past on a shelf above the floor of the main gorge, but there is now little more than a ruin and a large patch of bracken *Pteridium aquilinum*. In the forest itself many of the main tree and shrub species are still present, although the forest was probably much used in the past and even during one of our visits we met a group of men digging up a large clump of fern, which they said they were collecting as a house-plant.

In this patch of forest Palo blanco *Picconia excelsa* seems to be the dominant tree. Bark of *Picconia* is rough, tending to form a pattern of small raised squares; although it is dark in colour, it seems to encourage growth of pale lichen, which covers the bark of old trees, justifying the Spanish name ('white pole'). Other trees here include *Laurus novocanariensis, Morella faya, Ocotea foetens, Persea indica, Pleiomeris canariensis* and a few *Viburnum rigidum, Salix canariensis* and *Arbutus canariensis*. The sun also gets in along the track that plunges down through the forest, enabling the many bushes of *Gesnouinia arborea* to produce abundant flowers. We have not yet managed to find the rare knapweed-relative *Cheirolophus webbianus*, with cream-coloured flowers, which also occurs in this area.

Other plants in the forest here include *Pericallis appendiculata*, a much more robust shrub than most members of this genus, with large leaves that are velvety silver on the underside; its flowers are white, but seem not to be produced in deep shade. Many of the other smaller plants typical of laurel forest are also present, and are supplemented by cliff-dwelling plants and also by open ground species – including weedy aliens – associated with the degraded areas above and below the small patch of forest.

Palo blanco *Picconia excelsa* dominates this patch of laurel forest; its bark often has white patches of lichen

Seedlings of Palo blanco *Picconia excelsa* are abundant in parts of the barranco floor

Taginaste gigante *Echium giganteum* in flower

Mould formed when liquid lava engulfed a tree trunk

There are a couple of other opportunities to explore the laurel forest in this area. One is by a small track beside the abandoned restaurant El Bosque on the TF342. This allows one to get through a mass of invasive plants (including *Ageratina adenophora*, *Oxalis pes-caprae* and *Delairea odorata*) and into the ravine, after which one can walk upstream a little and then scramble up along the eastern side. The bed of the ravine here is at about 700 m and along the sides there is a strip of laurel forest that snakes its way up into the mountains. This strip does not extend far up the sides of the ravine and there is a rapid transition to degraded pine forest.

The ravine itself, as it continues upwards, contains many species characteristic of disturbed laurel forest. The Palo blanco has produced many seedlings on the forest floor, and one can also see *Laurus novocanariensis*, *Morella faya*, *Viburnum rigidum* and *Erica arborea*, with large *Salix canariensis* in the bed of the ravine. Other plants include *Daphne gnidium*, *Hypericum canariense*, *Arisarum simorrhinum*, *Rubus bollei*, *Phyllis nobla*, *Urtica morifolia*, *Smilax aspera*, *Rubia peregrina* and bracken *Pteridium aquilinum*. There is abundant leaf litter, with many snails and other invertebrates, and we came across a highly cryptic grasshopper on a rock face near the road.

About 1 km further west on the TF342 there is a parking place on the west side of the gorge. Here there are steps down to the edge of a sheer cliff plunging into the ravine, and these provide a quick and spectacular view. There are fine specimens of *Apollonias barbujana* on the left and of *Laurus novocanariensis* on the right, clinging to the top of the cliff, and *Echium giganteum* is growing beside the steps. In the lava wall on the right are several moulds of trees overwhelmed by the liquid lava, the largest of which is about 20 cm in diameter.

Monte del Agua

Although Anaga is the most extensive laurel forest habitat on Tenerife, Monte del Agua may be more easily accessible for visitors staying in the west. It provides one of the best opportunities for seeing both Bolle's Pigeon and the White-tailed Laurel Pigeon.

This is a forested mountainous region at the head of a huge valley – Barranco de los Cochinos – which cuts deep into the north of the Teno massif. The approach is from Erjos del Tanque along an unpaved road that passes through fayal-brezal for about 1 km, with *Erica arborea, Morella faya, Adenocarpus foliolosus* and brambles *Rubus* sp.

Whereas in Anaga there are several forest trails that are easy to follow, here there are very few, but one can drive along the forest road as far as a conspicuous pylon. Park here and continue on foot to get excellent access to this fascinating area. The road soon enters an area of laurisilva with steep slopes densely covered with many species of laurel forest trees. In some of the gullies and dells there are fine large specimens, often with massive growths of lichens on their trunks. Common species here are *Laurus novocanariensis, Morella faya, Erica arborea, Ilex canariensis, Persea indica* and *Arbutus canariensis*, with *Viburnum rigidum* forming the main understorey. Interesting smaller shrubs include *Isoplexis canariensis, Geranium reuteri* and *Phyllis* cf *viscosa*; there are also still a few surviving specimens of a highly endangered laurel forest tree, Tabaiba de monteverde *Euphorbia mellifera*.

The road runs in and out of the side valleys along the side of the main ravine, and after about 4.5 km reaches a track junction beside a tall bare rock overlooking the forest. There are wide views from around here, offering a good chance to see both Bolle's Pigeon and the White-tailed Laurel Pigeon – especially early in the morning – flying over the tops of the trees in the valley below. Beside the track here are exposed sunny areas with several plants typical of the transition from dry woodland to laurel forest, including the tree *Visnea mocanera* and shrubs including *Globularia salicina*, *Carlina salicifolia* and *Echium aculeatum*. These open spaces are also good for viewing the many species of butterflies present in the area.

When we returned to this site in November 2012 we found that a major fire had passed through during the previous summer, and we were told that the inhabitants of the village of Erjos del Tanque had been evacuated for a while. The impact on the forest had been curiously patchy and it was hard to understand the route taken by the fire. In the distance were large patches of forest turned from green to brown (but not completely blackened) while along the track we passed through some areas where the undergrowth was browned but still present, and others where the undergrowth was almost absent and the bases of the trees (small in this eastern part of the site) were charred.

Some other notable plants
Apollonias barbujana,
Bystropogon canariensis,
Cedronella canariensis,
Cistus symphytifolius,
Crambe strigosa,
Foeniculum vulgare,
Heberdenia excelsa,
Hypericum grandifolium,
Ixanthus viscosus, Ocotea foetens, Picconia excelsa, Ranunculus cortusifolius, Rumex lunaria.

The fire in 2012 burnt only parts of the forest at Monte del Agua

Butterflies you might see
Cardinal *Argynnis pandora*, Clouded Yellow *Colias crocea*, Canary Blue *Cyclirius webbianus*, Canary Brimstone *Gonepteryx cleobule*, Small Copper *Lycaena phlaeas*, Canary Speckled Wood *Pararge xiphioides*, Small White *Pieris rapae*, Bath White *Pontia daplidice*, Lulworth Skipper *Thymelicus christi*, Meadow Brown *Maniola jurtina*, Red Admiral *Vanessa atalanta*, Painted Lady *Vanessa cardui*, American Painted Lady *Vanessa virginiensis*, Canary Red Admiral *Vanessa vulcania*.

Dry pine forest near Vilaflor

6

Pine forest – 'pinar'

The pine forest ecosystem

The pine forest of Tenerife in its simplest natural form consists almost entirely of Canary Pine *Pinus canariensis* – the only species of pine native to the island – with a sparse understorey. Although pines cover a substantial part of Tenerife, it is now hard to find areas of 'natural' pine forest. Human impact doubtless increased gradually after the arrival of the Guanche people from north Africa 2000–2500 years ago, and much of the more accessible native pine forest was destroyed or grossly modified after the Castilian conquest in the 15th century. Felling and replanting, removal of undergrowth, and harvesting of the 'pinocha' (the thick cushion of decaying pine needles) have been part of the island's economy for centuries. Maps dating from the middle of the 20th century show the natural pine forests that still

Pine needles (pinocha) collected in the forest and transported by mules (1985)

Canary Pines colonising 18th century volcano

Road cutting through open pine forest exposes pine roots far from the nearest trees

survived at the nadir, which covered about half the area of the original forests.

A major restoration project was started in 1946 and as a result of this 'repoblación' the pine forest now covers just over 400 km², which is one fifth of the island and probably about three quarters of the natural extent (see Box). Restored areas, which are extensive in both the north and the south, are sometimes difficult to distinguish from natural stands, but usually consist of closely spaced, even-aged and fairly young trees.

A few groups of very old Canary Pines still remain on Tenerife, and it is well worth making the effort to see some of these magnificent trees. There are fine groups in the north of the island around the eastern rim of the Orotava valley, on ridges that can be viewed either from tracks near Montaña de Joco or at Roque del Topo near Aguamansa. In the south, splendid trees can be seen by the roadside just north of Vilaflor, and impressive groups of old trees are present at about 1500 m near Madre del Agua and along the Barranco del Río a little further east. In the latter areas there are partially dead old trees and large pieces of timber on the ground, showing that these are patches of ancient forest in which natural ecological processes of death, decay and recycling of nutrients have been operating for many years.

The Canary Pine is also a pioneer species, well adapted to recently created volcanic terrain. This can be seen near Volcán de Samara and in the area above Garachico where pines are colonising 18th century lava flows and cinder fields. Other species that are quick to colonise include Brezo *Erica arborea* and some smaller shrubs such as *Rumex maderensis* and *Scrophularia glabrata* (see Chinyero site below).

6 Pine forest – 'pinar'

DRY AND MOIST PINE FOREST

The pine forest of Tenerife takes different forms on the leeward (southern) and windward (northern) slopes, leading local ecologists to treat them as separate ecosystems. In the south there is dry pine forest or 'pinar seco', while in the north there is moist pine forest or 'pinar húmedo'. On both windward and leeward slopes the pine forest occupies a very broad

Mature Canary Pine above Vilaflor

altitudinal range, generally extending upwards from around 700 m on the dry southern slopes and from about 1200 m on the moister northern slopes. On both aspects the upper limit is at 2200–2300 m around the rim of Las Cañadas. This pine forest zone has annual rainfall within the range of 400–1000 mm and mean annual temperature 10–15°C.

The dry pine forest on the high slopes in the south of the island once graded downwards into the dry woodland on the foothills of the mountains, now largely destroyed. In some places, however, a dry form of laurel forest existed between the pines and the dry woodland, and remnants of it survive today, especially in the Güímar valley (see Chapter 4).

Now that agricultural pressure on the land is reduced, a natural downward spread of pines seems to be under way, for instance around Chimiche and El Río at about 450 m, but at the same time some dry woodland trees such as junipers are also making a comeback.

Natural dry pine forest tends to be fairly open, individual trees having widespread root systems that extract almost all the moisture from the soil around them. The mature pines may be so widely spaced that a full canopy is not formed and there are spaces between them with hardly any plants. In contrast, areas of restoration are normally planted densely, although some self-thinning occurs as the trees mature.

On the northern slopes of the island the northeast trade winds bring moisture, allowing development of a band of laurel forest (now fragmented) above the dry woodland, in the zone where the 'cloud sea' ensures almost continously moist conditions. The moist pine forest occupies the zone above this that is is still influenced by the cloud sea. In general the laurel forest extends up to about 1200 m and the moist pine forest is developed above this, with a transition zone between them known as 'pinar mixto'. However, there are some wild pines as low as 450 m at one unusually dry site in Anaga.

Escobón *Chamaecytisus proliferus* thrives in dry parts of the pine forest where the canopy is not too dense

Ancient dry pine forest in the upper reaches of Barranco del Río

In the lower parts of the northern forests, where the pines are frequently within the cloud zone, they may be draped in lichens (*Usnea* sp.) and some kinds of broad-leaved trees can compete with the pines, so that there is a smooth transition to the laurel forest below. The trees that most often occur with pine are Faya *Morella faya*, Acebiño *Ilex canariensis* and Brezo *Erica arborea*. These are primarily trees of the laurel forest and are favoured by the moisture produced by the cloud sea, but all of them tolerate somewhat drier conditions and are often abundant in the lower parts of the pine forest and transitional areas (as well as in degraded 'fayal-brezal').

In the few places where the northern pine forest survives in a mature state, the trees grow fairly densely, giving a complete canopy at a height of about 30 m and almost complete shade on the ground. Here associated species are largely confined to rocky outcrops and gaps in the canopy that let in sunlight. In the moist north these 'light gaps' also offer the best opportunities for natural regeneration of the pines, since seedlings can sometimes find enough light to grow up into the canopy, eventually replacing the parent trees. In the dry south, however, shade and moisture rather than light are the limiting resource for seedlings, and they are normally killed by the sun unless they germinate in places shaded by the parent trees in the heat of the day.

The dry and moist forms of pine forest have rather different communities of associated shrubs, and there are also changes with altitude. In the dry pine forest the understorey may be almost absent. Away from the shade of the trees the intense sunlight and the dessicating effect of the spreading pine roots create hostile conditions, while under the trees the dead needles decay only very slowly, producing a smothering litter layer in which few plants can grow. This layer usually overlies rocks and thin acid soils, so that the plants are short of

On some dry southern slopes regenerating pines and small shrubs can survive only in areas shaded for part of the day

Moist replanted pine forest with understorey of *Adenocarpus foliolosus* and pine seedlings

Foresters planting trees below La Fortaleza in about 1949, on barren ground where the natural vegetation has been removed by goats. Photo from "Vegetatión y Flora Forestal de las Canarias Occidentales" by Luis Ceballos & Francisco Ortuño.

Reconstruction of the pine forests of Tenerife – the 'repoblación'

Planting of pines in Tenerife goes back a century or more, but the year 1946 saw the start of a major project to restore pine forest to areas from which it had disappeared. Intriguingly, it appears that wood production was not the primary reason for this initiative, though it must have been in the minds of the protagonists as a beneficial feature of the work. In their classic book on the "Vegetación y Flora Forestal de las Canarias Occidentales" (1st edn. 1951, 2nd edn. 1976) Luis Ceballos and Francisco Ortuño explained clearly (2nd edn. pp. 205-6) that the principal objective of the replanting programme was to augment the usable supply of water by increasing 'precipitaciones horizontales' involving condensation on the trees, and thus to increase the wealth of the region. It is clear that in modern terminology, this was a pioneering initiative aiming to restore the ecosystem services (benefits for people) conferred by natural vegetation; the benefits were specified in terms of economic potential and the beauty of the landscape.

However, when the authors mention the species of pine that were used, it becomes clear that they viewed the programme as a forestry project rather than a more purist attempt to recreate a fully natural system. At the time of the planting in the mid 20th century foresters tended simply to choose tree species that grew well and produced useful timber. The Canary Pine was used for almost all of the work in the south, probably because of its fire resistance, but large areas in the north – from Teno to the western part of Anaga – were planted with Monterey Pine *Pinus radiata* (= *P. insignis*) from California, with some Aleppo Pine *P. halepensis* and Maritime Pin*e P. pinaster* from the Mediterranean. The largest areas were on the lower slopes of the dorsal ridge of the island between La Esperanza and Los Realejos, and on hillsides south of El Tanque and Buenavista. As a result, one twentieth of the Tenerife pine forest is of introduced species. More recent forestry practice has been to replant entirely with the Canary Pine, in accord with modern conservation principles, which place emphasis on native species and natural ecosystems. The use of the native pine brings benefits in terms of vulnerability to gales, since the Canary Pine tends to snap off near the top and stay alive, while Monterey Pine is more often uprooted. It also provides benefits in terms of fire resistance. Forest fires have affected vast areas in the Canary Islands in recent years, and the pine forests are more flammable than the other types of vegetation. The introduced Monterey Pine does not survive intense fires, but Canary Pines are rarely killed by fire (see Box on Fire Ecology).

The early planting in the reforestation programme extended outside the original range of the Canary Pine, causing damage to the natural vegetation. The greatest extensions of pine forest downwards into areas where the original vegetation would have been laurel forest were on the slopes above Garachico, Icod de los Vinos and La Guancha, and around La Esperanza; pine was also planted on land once occupied by dry woodland between Santiago del Teide and Guía de Isora. Extensions of the upper edge of the pine forest were less extensive, but Canary Pines were planted in mountain shrubland habitat just below the southern rim of Las Cañadas, and in large areas below La Fortaleza.

In addition, there was experimental planting of Monterey Pine actually within the caldera, for instance in Cañada Blanca at 2200 m near the Roques de García, and in Cañada de la Mareta just north of Guajara. These plantations of introduced species were felled when they came to be viewed as inappropriate within the National Park, though the remains of stumps can still be seen.

Even when the Canary Pine was used, the restoration programme usually involved planting of large areas over a short period, and removal of lower branches, often resulting in the development of fairly dense, even-aged stands of trees with narrow pointed crowns, reminiscent of conifer forests in the Alps. Although these replanted forests are still very different from the ancient natural forest, it is worth noting that half a century later they have become suitable habitat for native animals such as the Great Spotted Woodpecker.

A nest hole in the dead tree at the top of the picture shows that this replanted pine forest is now a suitable habitat for Great Spotted Woodpeckers

water for much of the year, while the sporadic passage of forest fires eliminates any species that lack protective adaptations (see Box on Fire Ecology). In general, the undergrowth is sparse, with only a few kinds of shrubs and smaller plants.

In most of the pine forest in the south the typical shrub in the undergrowth is Jara or Amagante *Cistus symphytifolius*, with showy pink flowers. The ecology of this plant is closely linked to that of the pines. Like them, it can grow in dry, poor soils and withstand prolonged drought. It is also well adapted to fire, since although the shrubs themselves may be killed, the seeds germinate readily after fire has passed through, and the seedlings benefit from the nutrients in the ash and the extra light that reaches ground level when the branches of the pines have been burnt. *Cistus symphytifolius* is also common in the pine forest of the northern slopes, mainly at higher levels. It is sometimes associated with its white-flowered relative Jaguarzo *Cistus monspeliensis*, which is particularly successful in degraded areas where forest was once present. These two species of *Cistus* (both sometimes known by the name Jara) also benefit from the fact that they are not normally eaten by goats.

Old pine trees draped with lichens are almost obscured by Faya, Brezo and Jara in a transitional area of 'pinar mixto' that is often within the cloud sea on the north slope of the island

Another key species related to the dry pine forest is Escobón *Chamaecytisus proliferus*. This large, loose, white-flowered leguminous shrub is well adapted to the conditions in some dry areas that are also suitable for pine, especially at high altitudes. There are a few places outside the pine forest where it is the dominant plant in a special shrub-forest community or 'escobonal'. However, Escobón cannot survive under a dense canopy of pine, and conversely, attempts to establish pines within escobonales usually end in failure, so foresters often systematically destroy Escobón. Under natural conditions, competition between the two species sometimes leads to development of patches with pure Escobón and others dominated by pine, as well as intimately mixed patches.

Two other leguminous shrubs, Codeso *Adenocarpus foliolosus* and Codeso de cumbre *Adenocarpus viscosus,* are associated with the pine forest. The former is typical of middle levels on the northern slopes and on the dorsal ridge above La Esperanza, where it occurs among the pines up to about 1600 m. The latter is primarily a plant of Las Cañadas, but also grows freely on the surrounding slopes, in association with Escobón and Canary Pine. On the dorsal ridge it occurs down to 1000 m, and seems to hybridise with *A. foliolosus*. A related shrub, *Teline osyrioides*, occurs sparingly in pine forest high up in the southeast and south.

In some parts of the pine forest, especially replanted areas, the floor of the forest has numerous cushiony clumps of another yellow-flowered legume, Corazoncillo del pinar *Lotus campylocladus*, often partly covered by fallen pine needles; this is another plant that thrives after the passage of fire.

Experiments on pine forest legumes in La Palma by ecologists from La Laguna University, indicate that the scarcity of understorey plants in many of the pine forest areas is largely a result of the presence of alien herbivores for many centuries. Four species were studied: *Chamaecytisus proliferus*, *Teline stenopetala*, *Spartocytisus filipes* (now rare on Tenerife, in dry woodland remnants) and the vetch *Cicer canariense* (now extremely rare on Tenerife). All of these have seeds that germinate after fire, but unlike species of *Cistus*, all are known to be eaten by goats. At the start of the project seeds were sown both in open plots and inside enclosures designed to exclude the local (introduced) herbivores: feral goats, Rabbits *Oryctolagus cuniculus* and Barbary Sheep *Ammotragus lervia* (not present on Tenerife). The plants were counted at intervals over four years, with dramatic results. Seedlings were initially abundant in both open and enclosed plots, and four years later many medium and large plants of all four species had developed in the enclosures, but there were only tiny numbers in the open plots. Clearly, the pine forests may have been very different before these herbivores arrived.

Several aromatic herbs in the Lamiaceae are frequent in the pine forest, though not all are restricted to this habitat. The small white-flowered Tomillo *Micromeria hyssopifolia* occurs mainly in somewhat open places where the forest and the soil are degraded, while the Poleos *Bystropogon*

The pink-flowered Jara *Cistus symphytifolius* and white-flowered *Cistus monspeliensis* are typical pine forest plants

Codeso *Adenocarpus foliolosus* is frequent in both pine forest and laurel forest in the north

Corazoncillo *Lotus campylocladus* is a typical plant of the pine forest floor, including the highest areas

origanifolius and *B. plumosus* are somewhat larger and smell of mint; the former is traditionally used for making a herbal tea. Algaritofe *Cedronella canariensis*, a sprawling shrub with compact spikes of pink flowers, is typical of the laurel forest but also occurs in the transition to pine. The white-felted woody herbs in the genus *Sideritis* are represented in the pine forest most commonly by *S. oroteneriffae*, a Tenerife endemic typical of rocky places, and by *S. soluta* which occurs mainly in the south and southeast.

The Scrophulariaceae are represented in the dry pine forest rather surprisingly by a figwort *Scrophularia glabrata*, member of a group normally associated with damp habitats but here acting as a pioneer. Another typical but patchily distributed plant of the pine forest is the white-flowered, red-berried *Daphne gnidium* (Thymelaceae). The scabious genus *Pterocephalus* (Dipsacaceae) is also represented in the pine forest; *P. dumetorus* is the species found in the Güímar valley, but in the north we have seen plants resembling the high mountain species *P. lasiospermus*. The Boraginaceae are usually represented here by *Echium virescens*, which is widespread in forest zones, especially on rock faces and cliffs. It is a large, spreading bush producing many erect inflorescences, ususally with pink flowers.

Several other plants occur mainly on cliffs, both in the pine forest and elsewhere. Cliff-dwelling species that favour the pine forest include *Asparagus plocamoides*, a tall bush with groups of upright stems up to 3 m high and very fine, needle-like but quite short cladodes (<30 mm); it also occurs in dry woodland habitat.

Succulents in the family Crassulaceae are not abundant in the pine forest, but in rocky light gaps one can find the yellow-flowered, orange-stalked *Aeonium spathulatum*, while the wonderfully geometric cup-shaped rosettes of *Greenovia aurea* are also likely to catch the eye, even when the spectacular flowers are absent. Two other species typical of rocks and cliffs are the campion *Silene berthelotiana*, found mainly in deep ravines in the south, and *Pimpinella dendrotragium*, which occurs in the Güímar valley. Additional cliff-dwelling plants are mentioned in relation to the Boca del Valle site and in Chapter 4 on Dry Woodland Remnants.

Tomillo de pinar *Micromeria hyssopifolia* is the usual member of its genus in the pine forest

Orquídea canaria *Orchis canariensis* (= *O. patens*)

Echium virescens is also frequent in rocky places in the pine forest

Animals of the pine forest

It is in the pine forest that you will be able to see the only species of songbird on Tenerife that has always been recognised as a distinct endemic species; this is the Blue Chaffinch *Fringilla teydea*, which has distinct subspecies on Tenerife and Gran Canaria. Any chaffinch that you see in most of the south of Tenerife will be the Blue Chaffinch; it is restricted to pine forest, favouring especially areas that also have the leguminous shrubs Escobón and Codeso. However, in the north and in the Güímar valley the situation is more complex, since the local subspecies of the common Chaffinch, *Fringilla coelebs canariensis*, occupies the laurel forest, but also areas that are transitional between this and the pine forest. These include Aguamansa, El Lagar, La Montañeta and along the dorsal ridge between La Esperanza and the Mirador de Ortuño. The Blue Chaffinch also occupies these transitional areas and the two species can occur together. Since the upperparts of the male local Chaffinch are noticeably bluish, confusion is possible. However, the Blue Chaffinch lacks the double pure white wing bars noticeable at rest in both sexes of the ordinary Chaffinch.

In general the various parts of the pine forest areas have similar sets of birds, but with some differences between the north and south of the island, and between areas with dense trees or with more open ground. In any part of the pine forest you have a chance of seeing the local leaf warbler, the Canary Chiffchaff *Phylloscopus canariensis*. This is now considered to be an endemic species rather than a subspecies of the Chiffchaff *P. collybita,* and it occurs all over Tenerife including the pine forest. The Great Spotted Woodpecker, previously restricted to the mature pinewoods of the west and south, can now be found in most parts of the pine forest, since the majority of the pines planted in the restoration programme are now around half a century old and provide suitable nesting sites. Other species present in all parts of the pine forest are Buzzard, Kestrel, Sparrowhawk, Plain Swift (mainly summer) and Tenerife Blue Tit (which has been helped by provision of nest-boxes in some areas replanted with pines).

Species more common in the north are Long-eared Owl, Blackbird, Robin and Goldcrest, the latter favouring areas with Brezo, while the Grey Wagtail is mainly associated with ravines containing water. The Canary is rather scarce in the pine forest, although it is probably the commonest bird on the island, while the Greenfinch occurs only around the edges of the pine forest, mainly in winter. Woodcock and Bolle's Pigeon are primarily birds of laurel

Blue Chaffinch *Fringilla teydea*
Photo: Rubén Barone

Common Chaffinch *Fringilla coelebs*. Photo: Rubén Barone

forest and occur in pine forest almost exclusively in the north, mainly in areas of transition to laurel forest.

The Barbary Partridge, Rock Dove, Berthelot's Pipit, Canary and Turtle Dove (only in spring and summer) tend to avoid dense pine stands and are therefore more frequently encountered in open forest areas, especially in the east, south and southwest of the island. Certain other species such as the Hoopoe, Southern Grey Shrike, Spectacled Warbler and Greenfinch, which prefer other habitats, may occasionally be observed in the pine forest. The Raven is now rare on Tenerife and the only parts of the pine forest where it is likely to be seen is in the areas around Chinyero, Chío and Vilaflor.

Great Spotted Woodpecker pair
Photo: Rubén Barone

Mammals are not conspicuous in the pine forest. Cats are not often seen, but the prevalence of scats shows that they are widespread. They are often associated with houses but in some cases are probably feral (living as wild); many have coloration and physical aspect similar to Wildcats. Hedgehogs and at least five species of bats (*Barbastella barbastellus, Nyctalus leisleri, Pipistrellus maderensis, Plecotus teneriffae* and *Tadarida teniotis*) can also be seen in the pine forest. The House Mouse is present there, as are Rabbits, at least in open areas. Reptiles are represented mainly by the Tenerife Lizard *Gallotia galloti,* which is common in the drier and more open areas. The Canary Skink *Chalcides viridanus* and Canary Gecko *Tarentola delalandii* are inconspicuous, spending most of the daylight hours under rocks, but the gecko can sometimes be found hiding in cracks in the bark of Canary Pines.

The invertebrate life in the pine forest is less rich than in the laurel forest, and there are few species restricted to this zone. However, turning over rocks one may find a number of interesting animals, one of the most impressive being the poisonous centipede *Scolopendra valida*, which is probably native. Spiders of the genus *Dysdera* are also to be found under rocks, in silk cells; they are easily recognized by their orange legs, brown to purplish carapace and silky grey abdomen; a large number of species of this group have evolved on the archipelago and have occupied a wide variety of habitats, from the slopes of Mt. Teide to near sea level; there are also many species in caves.

The old pines provide special microhabitats, and some invertebrates are well adapted to them. The large endemic crab spider *Olios canariensis*, which has a flattened body, can be found under the wide pads of flaky bark that clothe the oldest pines; it is also common in the laurel forest. A number of other unusual spiders have been found in the masses of dead pine needles that sometimes accumulate on the lower branches of the pines.

Specialised insects in the pine forest include the brown-black heteropteran bug *Aradus canariensis*, a flat species adapted to life under the bark of dead trees. In some years there are very high numbers of the large endemic black beetle *Buprestis bertheloti,* but it is inconspicuous since it flies mainly near the tops of the trees.

Butterflies are less noticeable in the pine forest than in some other habitats on Tenerife, since nectar producing flowers are relatively scarce. However, in good weather one can see a variety of species, especially along tracks and in open spaces. Pine forest is the preferred habitat of one of the most handsome butterflies of the island, the endemic Tenerife Grayling *Hipparchia wyssii,* which is most common in the higher areas, especially in the south; the adults emerge in late spring and the larvae feed on grasses in summer and then hibernate.

Olios cf *canariensis* on pine trunk
Photo: Pedro Oromí

The leguminous plants so numerous among the pines (*Chamaecytisus, Adenocarpus, Teline* and *Lotus*) provide food for larvae of three other butterfly species widespread in pine forest: the ubiquitous endemic Canary Blue *Cyclyrius webbianus,* Long-tailed Blue *Lampides boeticus* and Clouded Yellow *Colias crocea.* Other species widespread in this habitat are Red Admiral *Vanessa atalanta,* Canary Speckled Wood *Pararge xiphioides* (more common in laurel forest) and Small White *Pieris rapae,* along with two species that occur mainly in open places: Small Copper *Lycaena phlaeas* and Bath White *Pontia daplidice.* Butterfflies more patchily distributed or uncommon in pinewoods include Southern Brown Argus *Aricia cramera* (whose larvae feed on legumes but also on members of the family Cistaceae), Canary Red Admiral *Vanessa vulcania* and Meadow Brown *Maniola jurtina* (the latter two found mainly in the north). The Green-striped White *Euchloe belemia* occurs in open areas at high altitude in the pine forest and also in Las Cañadas, where larvae have been found feeding on *Descurainia bourgeauana.* Finally, four species that are more typical of other habitats can also be found in areas of transition to pine; these are Large White *Pieris cheiranthi,* Canary Brimstone *Gonepteryx cleobule,* the large and richly coloured Cardinal *Argynnis pandora* and Lulworth Skipper *Thymelicus christi.*

Another ecologically important species is the endemic moth *Calliteara fortunata* in the family Lymantriidae. The hairy caterpillars of this species, which sport a series of purple tufts down their backs, are enormously abundant on the pines in some years and cause serious defoliation of both Canary Pines and Monterey Pines. These caterpillars can also be found on bushes of the broom Retama del Teide *Spartocytisus supranubius* in some parts of the high mountain zone.

Fire ecology on Tenerife

Fires have been important in the development of the forests of Tenerife. They can be started by thunderstorms, but these are relatively rare (around 2–3 per year) and occur mainly in winter. However, sporadic volcanic eruptions on all the oceanic islands of the Atlantic must have generated fires, so it is no surprise to find that the Canary Pine is one of the most fire-resistant of all pines: fires have evidently moulded the evolution of the species over millions of years.

The arrival of people increased the frequency of fires, and in the modern era fire resistance is of special relevance, since climate change is increasing the prevalence of intensely hot periods that heighten the risk of natural fires, as well as making the forests especially vulnerable to arson. In August 2012 a major fire engulfed much of the pine forest to the southwest of Las Cañadas. This zone already contained large areas – for example between Vilaflor and Las Lajas, and at Arenas Negras – with trees showing severely charred bark from previous fires, but many of them had already put out new shoots. After the 2012 fire, the Tenerife pine forest will recover much more quickly than the laurel forest of La Gomera, which suffered an even larger fire at the same time, but the setback to forestry will be severe.

Some of the Canary Pine's adaptations to fire are easy to see. It has thick, heat-insulating bark and is unusual among pines in its ability to produce new shoots from the trunk (or along branches) after being burnt. It can generate a bud and then a shoot directly on the trunk, from cells that are normally programmed to produce an increase in trunk diameter. It can also spring up from the base after destruction of the trunk, which not only enables it to regenerate after severe fire (or felling) but may also help it to establish itself in high places where occasional winter ground frosts might kill all the foliage of young saplings. Several other Tenerife plants can also sprout from the base after most of the branches are killed by fire. *Erica arborea* and *Morella faya* can grow shoots to one metre in two years after a burn. We have also been struck by the fire resistance of *Cheirolophus teydis*, the beautiful yellow-flowered knapweed-relative that grows so well near Boca Tauce, which recovered quickly after the 2012 fire. In contrast, Retama del Teide *Spartocytisus supranubius* showed no sign of life after having been charred by the fire.

Even badly charred old pines produce new shoots after fire

Cheirolophus teydis sprouts luxuriantly after a fire has passed, but Retama del Teide (back right) is killed

The wildfire of 2012 devastated a large area of pine forest south of Boca Tauce

Fire passing through an area increases the amount of light reaching the forest floor, and often also creates an excellent seed-bed, enriched with available nutrients from the ash. This favours seedlings of Canary Pine and many typical shrubs and other plants of the pine forest in which the seeds survive fires and germinate quickly afterwards. These include *Cistus* and *Echium*, legumes such as *Adenocarpus*, *Chamaecytisus*, *Lotus* and *Bituminaria*, as well as grasses. Other plants such as *Asphodelus* and the orchid *Habenaria*, which are adapted to survive through hot dry summers by remaining dormant underground in the form of bulbs or tubers, are usually unharmed by fire. We have also noted herbaceous perennials such as *Scrophularia glabrata*, *Nepeta teydea*, *Sonchus acaulis* and *Pimpinella dentrotragium* sprouting healthily from the roots after a fire has passed.

Fires in the Canary Islands are generally of low intensity, with intense ones mainly affecting steep slopes and ridges. This is crucial for the invertebrate animals of the forest floor, since low intensity fires heat only the surface layers and many animals can take refuge by going deeper. As a result, studies of invertebrate communities in forests affected by fire have shown less serious effects than expected. Furthermore, since the organic matter in the upper soil layers is usually not totally destroyed, absorption of water is not much affected and runoff leading to erosion is not much increased. However, the effects of fire generally last longer in dry parts of the island, since regeneration of the plants of the forest floor is dependent on a supply of moisture. It seems clear, therefore, that the scarcity of undergrowth in the pine forest is partly a result of fire.

Burnt Canary Pine, shown here in red evening light, with many new shoots

Moist pine forest at Los Órganos, Aguamansa

Aguamansa

Moist Canary Pine forest on the eastern cliffs and slopes of the Orotava valley at about 1200 m, with some trees and shrubs associated with a transition from the laurel forest that used to be present below this level. Access to mature pines at El Topo.

The forest above Aguamansa is an impressive example of the northern-aspect moist pine forest or 'pinar húmedo' on the ridges and cliffs forming the eastern wall of the Orotava valley. Part of the area comprises the strict reserve of Pinoleris, where access from above (along miles of serpentine forest roads) is restricted. However, there are many well marked walks starting from the car park a short distance up the road from the village of Aguamansa. One of the tracks leads to Los Órganos, impressive cliffs of volcanic rubble eroded by water into columns reminiscent of organ pipes.

If you take the track signed to El Topo you will see a fine group of ancient Canary Pines. Some of these are very large and draped with grey lichens, which can also be found on the shrubs in the undergrowth, reflecting the fact that Aguamansa is frequently shrouded in damp cloud for days on end, a situation more typical of laurel forest than pine forest. It seems probable that in the past the laurel forest extended up to about the height of Aguamansa. It has now been largely removed, and the presence of the non-native Sweet Chestnut *Castanea sativa* and Tasmanian Blue Gum *Eucalyptus globulus* testifies to past planting in the area,

6 Pine forest – 'pinar'

The succulent Bea dorada *Greenovia aurea* with concave leaf rosettes flourishes on rock outcrops in the pine forest

while the diverse understorey accompanying the pines indicates a broad zone of transition from laurel forest to the pine forest above.

Brezo *Erica arborea* and Faya *Morella faya* are dominant in the understorey here, along with the laurel forest trees *Laurus novocanariensis*, *Viburnum rigidum*, *Ilex canariensis* and a few *Ilex perado*. Other species seem to be related to the transition, including *Arbutus canariensis*, *Rhamnus glandulosa*, *Bencomia caudata* and *Daphne gnidium*. Leguminous shrubs include Escobón *Chamaecytisus proliferus* and Codeso *Adenocarpus foliolosus*, but also the tall yellow-flowered shrub *Teline stenopetala*, a species with very restricted distribution on Tenerife. The sticky-leaved, white-flowered shrub *Cistus monspeliensis* is often associated with fayal-brezal, but the related pink-flowered Jara

Southern Brown Argus *Aricia cramera* on Torvisco *Daphne gnidium*
Photo: Juan José Bacallado

Cistus symphytifolius, which is also present here, is typical of dry pine forest and becomes more abundant at higher levels. The *Cistus* species are food plants for larvae of a spiny brown leaf beetle, *Dicladispa occator*, an endemic species very typical of the pine forest.

The White-tailed Laurel Pigeon occurs in this area, but you will have a better chance of seeing it in some of the laurel forest sites. There is also a rich array of invertebrate animals. If the sun shines you are likely to see several species of butterflies, and if you are lucky you may come across the well known Praying Mantis *Mantis religiosa*, which disguises itself as part of a plant while waiting for its prey: it is, however, commoner in agricultural areas. Under rocks you may find the small grey-brown wingless cricket *Gryllomorpha canariensis* and also the two endemic earwigs in the genus *Guanchia*. In the summer you are likely to see one of the most impressive insects on Tenerife, the giant endemic asilid fly *Promachus vexator*, up to 30 mm long and a voracious predator on other insects; a specially intense burst of buzzing often indicates a mating attempt.

Only a few parts of the pine forest are moist enough for lichens *Usnea* sp. to grow on the trees

The leaf beetle *Dicladispa occator* feeds on *Cistus symphytifolius*. Photo: Pedro Oromí

Predatory fly *Promachus vexator* with prey

Moist replanted pine forest at El Lagar

EL LAGAR

A good place for visitors staying in the north of the island to see plantation pines with shrubby undergrowth, and with luck also the Blue Chaffinch and other island birds.

The Área Recreativa El Lagar is an area of reafforestation at a height of 1000 m, with fairly open undergrowth, mainly of *Morella faya*, *Erica arborea* and *Cistus symphytifolius*, as well as *Lotus campylocladus* partly covered by the accumulated pine needles. A fire has been through recently, but does not seem to have been severe. A few decades ago water taps were always left dripping here to attract thirsty birds. Recently, however, water-saving initiatives have led to the taps being turned off; this makes it much harder to see the local birds, and may even have reduced their populations. However, water is still sometimes available for the birds in the warden's compound, and there is a real chance of seeing the Blue Chaffinch, in the same area as the striking local subspecies of the ordinary Chaffinch. You may also see a Sparrowhawk or Great Spotted Woodpecker or the very dark-headed Tenerife Blue Tit: both the latter species can sometimes be seen pecking vigorously and tearing at the flaky bark of the Canary Pines, which harbours many insects and spiders. When we visited the site in October a few years ago we were intrigued to see a female Pied Flycatcher looking very much at home, hawking for flies from a perch on a pine tree: this species, however, does not breed on the island.

Aeonium spathulatum and *Sideritis oroteneriffae* in a rocky light-gap

Montaña de Joco

Fairly dense pine forest with shrub understorey at about 1900 m on the north side of the dorsal ridge of the island. One of the few easily accessible areas with very old Canary Pines; also Blue Chaffinches, and rock faces with beautiful endemic plants.

This forest area is less humid than Aguamansa 500 m below, although the cloud sea does sometimes reach up to this level. Some of the ridges on the right, below you, have the remains of ancient pine forest, although the higher slopes along the track are planted. The shrub understorey varies; in some places it is mainly Brezo *Erica arborea* and in others it is predominantly the legumes *Adenocarpus viscosus* and *Chamaecytisus proliferus*, with some *Cistus symphytifolius*, plus a wide variety of smaller shrubs in rocky places.

The forest track winds in and out round the heads of several ravines, giving opportunities for easy access to rock faces with communities of endemic plants that provide spectacular floral displays in summer. Most conspicuous is the brilliantly yellow-flowered, red-stemmed *Aeonium spathulatum*, contrasting with the velvety white *Sideritis oroteneriffae*. Close by are displays of the strange succulent *Greenovia aurea* with its cup-shaped leaf-rosettes, and a species of *Aeonium* that may be *Aeonium canariense* at an unusually high altitude.

6 Pine forest – 'pinar'

Around this altitude (1900 m) several species of mountain plants are approaching their lower limit, below which they are replaced by less specialised relatives. Examples include *Adenocarpus viscosus*, the composite *Tolpis webbii* and the scabious *Pterocephalus lasiospermus*, all of which have their main strongholds in Las Cañadas. In the case of *Adenocarpus* (and perhaps some other plants) there are signs of past hybridization between species typical of the high mountain zone and those found lower down: the opportunity for this may have arisen because of man-made changes in the extent of the pine forest. In at least two cases both the low and high altitude members of a pair have been found close to this site: one is the thistle *Carlina salicifolia* and its high altitude replacement *Carlina xeranthemoides*, and the other is *Micromeria hyssopifolia*, typical of the pine forest, and *Micromeria lachnophylla* of the high mountain. Another example may be *Argyranthemum adauctum* and the high altitude *Argyranthemum teneriffae*. *Cistus symphytifolius*, a common shrub of the pine forest, is close to its upper limit here, but has a rare relative in Las Cañadas, while conversely, the yellow-flowered crucifer *Descurainia lemsii* occurs here but is scarce, while its relative *Descurainia bourgeauana* is a dominant species in parts of Las Cañadas.

Jara *Cistus symphytifolius*

Where the track forks, after about 2 km, you can take the right-hand route which descends a little and then after about 300 m comes out onto a ridge which has a crown of very old Canary Pines. These are lovely large spreading trees, with curtains of grey lichen (*Usnea* sp.) hanging from their branches. Between the trunks you can get a good view of the island of La Palma, and also of Mt. Teide and down into the valley of Orotava (or more probably onto a sea of cloud hiding the valley from you). You will probably see Blue Chaffinches here. If you continue along the main track rather than turning right, there is another opportunity to reach a rocky outcrop with truly magnificent views and a few remaining old pines.

Umbilicus gaditanus, *Carlina salicifolia* and *Sonchus asper* all favour open places in the pine forest

Pine forest with Escobón *Chamaecytisus proliferus* and Cañaheja blanca *Todaroa montana*

BOCA DEL VALLE, BOSQUE DE LA ESPERANZA

Fairly natural pine forest mainly on the dry southern side of the dorsal ridge of the island, at about 1500 m. Great diversity of plants on rocky outcrops and small cliffs, with many species characteristic of the higher part of the Güímar valley.

This fascinating site is reached from Las Lagunetas on the main road from La Laguna along the dorsal ridge of the island. From the road the track runs at first through a zone of Canary Pine 'repoblación', where dense planting has led to natural 'self thinning' as the trees grew, so that there are now many slender moribund trees and many dead trunks and branches on the ground. Further on there are also larger trees as we get into semi-natural but still fairly damp pine forest, part of the 'Bosque de La Esperanza'. Here there is an understorey of Brezo *Erica arborea*, with some Faya *Morella faya* and *Cistus monspeliensis*, along with the yellow pea-flowered Codeso *Adenocarpus foliolosus*, with densely hairy twigs, trifoliate leaves and sticky seed pods; among the pine needles on the forest floor are clumps of a smaller legume, *Lotus campylocladus*, also with brilliant yellow flowers.

The track soons leaves this moist pine forest and emerges at the top of the enormous valley overlooking Candelaria on the southeast coast of the island. Here there is a relatively dry pine forest, and the change is marked by the loss of Brezo and the sudden prominence of Escobón *Chamaecytisus proliferus*, the only tree that competes significantly with Canary Pine in drier parts of the island. It can grow to 6 m or more, and still forms 'escobonales' in a few places on Tenerife. The foresters tend to root it out, but here it is growing luxuriantly.

Blasting has been used in some places during construction of the forest track, and this has demonstrated the remarkable ability of pine roots to grow down cracks in layers of old lava flows, reaching many metres into massive basaltic rock. In other places the roots have found weak places and are clearly aiding the erosion of the rock faces, by penetrating cracks and gradually splitting off slabs of rock. The track runs fairly level on the northeast side of Montaña Amarilla (or Montaña de Cuchillo) above a magnificent ravine on the left with pines on very steep slopes. This area was mapped as low density natural pine forest in the middle of the 20th century. We have puzzled over the scarcity of dead or dying old trees here, and of dead wood on the slopes, and conclude that this must be partly because people have been extracting materials from this forest for centuries, and partly because the slopes are so steep that when trees die or are burnt in forest fires they fall and slip down the slopes during heavy rain.

There were large fires in the area in 1991 and 1995, which accounts for the charring on the trees and limited undergrowth. Some plants benefit from the fires, including the *Lotus* mentioned above and common pasture plants including clover *Trifolium* sp., which produce bright green 'lawns' at some seasons. The giant sow-thistle *Sonchus acaulis,* conspicuously dotting the slopes below the track, has a thick woody base and roots that doubtless also survive the passage of fires. The Codeso *Adenocarpus foliolosus* that grows under the pines in moist forest gives way here to a similar leguminous shrub, *Teline osyrioides*, endemic to Tenerife.

The slopes above the track are interspersed with small cliffs and large boulders, and the vegetation is characteristic of pine forest on the somewhat drier southeastern side of Tenerife. Three typical species are *Cistus symphytifolius,* with very large purple flowers, an aromatic herb *Bystropogon origanifolius,* and the silver-felted labiate *Sideritis oroteneriffae,* which is a Tenerife endemic. An impressive and less common species is the umbelliferous *Todaroa montana,* growing to about 2 m and with dense hemispherical heads of pale yellow flowers; it favours steep rocky places. A low-growing distant relative present in shady spots among the rocks is *Pimpinella dendrotragium,* local to the Güímar valley but here at the top of its altitudinal range; it forms a dense clump of straggly stems, with pinnate, lobed, flat, shiny green leaves and slender inflorescences with loose white umbels. Other typical plants of this area

Pines with Cerrajón de monte *Sonchus acaulis* on steep slopes with vegetation influenced by fires

Greenovia aurea and *Teline osyrioides*

Palomera Pericallis lanata *and Cerraja de Güímar* Sonchus gummifer

Sideritis oroteneriffae, *a species typical of the pine forest*

are the spreading *Echium virescens,* with multiple spikes of pale pink flowers in spring, and *Argyranthemum* 'vincentii' (not yet formally described) forming an erect bush about 1 m high, with linear lobed leaves.

    The rock faces further along the track have been colonised by a number of species typical of cliffs in the drier parts of Tenerife, several growing here at 1500–1600 m, well above their normal altitudinal limit. One of these is the crucifer *Lobularia canariensis,* covered densely with small white flowers in spring. Composites (members of the family Asteraceae) are well represented, for instance by the uncommon cliff-dwelling sow-thistle *Sonchus gummifer,* which is endemic to Tenerife; it has blue-green or purplish foliage and leaves divided right back to the midrib, leaving oval but sharp-pointed leaflets. Growing on a shady rock face beside the *Sonchus* is *Pericallis lanata,* both of them at a much higher altitude than usual. The *Pericallis* has green and white-woolly foliage and large purple flowers with a dark purple central disc. Its relative *Pericallis cruenta,* an upright plant also with purple-flowers, is scarce here but widespread in forests in the north. The thistle *Carlina salicifolia,* with striking disc-shaped flowers, occurs sparingly here, while a species of *Phagnalon*, probably *P. purpurascens,* is also present, as well as *Tragopogon porrifolius,* a possibly native thistle-relative with latex but no thorns; it has star-like mauve flowers made up only of ray florets. We also found on the rocks here a single purple orchid, *Orchis patens.*

Aeonium smithii *often grows in rock cracks*

Greenovia aizoon, *a tiny member of its genus*

Members of the houseleek family (Crassulaceae) on the rocks include the red-stalked and golden-flowered *Aeonium spathulatum,* which is sometimes abundant on lava flows, but there is also a more unusual relative, *Aeonium smithii,* better known from much higher up in Las Cañadas. Here it is growing in cracks in the rocks and through accumulations of pine needles, beside an exposed part of the track near a viewpoint at Risco Filabrés; it is easily recognised by its thick stem with dense white hairs on the section just below the leaf rosette, and the reddish slot-like glands on the succulent leaves. The genus *Greenovia* is represented mainly by the uncommon *G. aizoon,* a rock plant growing in tight clusters of small cup-shaped rosettes, with glandular-hairy leaves and yellow flowers. However, there are also a few individuals of *Greenovia aurea,* with non-hairy blue-green leaves and a tall – sometimes branched – inflorescence with brilliant yellow flowers. On sheltered vertical rock walls is *Monanthes brachycaulos,* with a tight rosette of swollen purplish and green leaves, and tiny starry flowers with purple sepals covered with glandular hairs.

If you go another 200 m you will get a fine view of the top of the Güímar valley, looking down over magnificent rock walls created by a formation of vertical dykes only a couple of metres thick, known locally as Laja de Chafa. They have been exposed by erosion following the giant landslip that created the Güímar valley. The pines nearby and just below the track are mature, but in the distance one can see large areas planted with Canary Pines at various times during the last half century.

Although the plants are the obvious things to look at along this track, those with sharp ears as well as eyes will have a chance to find several endemic birds, including the Blue Chaffinch – a heavy-billed pine-specialist found only in Tenerife and Gran Canaria – as well as the Canary Island subspecies of the ordinary Chaffinch, Goldcrest, Blue Tit, Canary Chiffchaff and Buzzard. The local Great Spotted Woodpecker now flourishes in the forest near Las Lagunetas, but is uncommon on dry southern slopes lacking dead trees.

Dyke-controlled erosion at Laja de Chafa with extensive pine 'repoblacion' behind

Monanthes brachycaulos, a tiny succulent of rock faces

Argyranthemum 'vincentii'

Volcán de Chinyero and nearby habitats

The mountain and crater formed by the most recent volcanic eruption on Tenerife, in 1909. Also access to lava flows and cinder fields produced by 15th and 18th century eruptions, which are now being colonised by Canary Pines and a number of associated plants, creating habitats that are both unusual and austerely beautiful.

The relatively gentle slopes between 1400 and 1800 m to the south of the town of Garachico are the most volcanically active areas on Tenerife, and many eruptions have occurred there within historical times. The first to be witnessed was that of Boca Cangrejo witnessed by Christopher Columbus in 1492 (see Geology chapter) but the most destructive event was in 1706 when major lava flows from Volcán de Trevejo (Volcán de Garachico) near Arenas Negras flowed northwards down the slopes, destroying the port and much of the town of Garachico. In 1909 a spectacular eruption – the most recent to occur on the island – created a new mountain, Volcán de Chinyero, with a massive summit crater. However, this event was much less destructive than the eruption of 1706, since most of the lava flowed to the west and northwest, through an uninhabited region. The flows extended over about 1.5 square kilometres and consisted mostly of viscous 'aa' lava but with some more fluid 'pahoehoe' lava.

Both Volcán de Chinyero and the lava that it produced has remained largely free of plants over the past century. When we studied the invertebrate fauna of one part of the Chinyero lava in 1984 we noticed only the lichen *Stereocaulon vesuvianum* on some rock faces and a few tiny patches of the moss *Grimmia trichophylla* in crevices. A botanical study of the whole

Canary Pines and shrubs colonising cinders and lava near Volcán de Samara below Pico Viejo

lava flow some years later recorded one Canary Pine, the fern *Cheilanthes pulchella,* 12 kinds of mosses and 18 kinds of lichens. A few higher plants were found in places with a little soil, mainly along tracks over the lava created by domestic animals. The commonest were *Echium aculeatum* and the grass *Bromus rigidus*, but other species recorded were ones that we have found on somewhat older lava flows: Escobón *Chamaecytisus proliferus*, the high altitude scabious *Pterocephalus lasiospermus* and the herbaceous perennials *Scrophularia glabrata* and *Bystropogon origanifolius*.

About 6 km up the road from Volcán Chinyero is Volcán de Samara, with an information board and marked trails. Here one can see clearly the process of colonisation of lava and cinder fields of various ages by the Canary Pine and the cushion-shaped shrubs of the high mountain zone. In between these two sites, below the point where the road running from Chío (near Guía de Isora) up to Las Cañadas passes Montaña Boca Cangrejo, there is a place where the recent lava flowed around small areas of older volcanic rock without submerging them; such areas are known as 'islotes' ('kipukas' in Hawai'i). We do not know whether all of the plants growing on them before the eruption were killed by the heat, but the pines growing there now were planted subsequently, probably in the 1960s. Other plants growing on the islotes have clearly arrived naturally, finding conditions there more suitable than on the surrounding young and unweathered lava.

The westward flowing lava from Boca Cangrejo (and one flow from Chinyero) poured down the steep slopes above Santiago del Teide, creating the dramatic patterns that provide the

Lava flows near Las Manchas from the Boca Cangrejo eruption of 1492, with colonising pines and 'islotes' of older rock surrounded by the recent lava; Montaña Bilma, surrounded by a lava flow from Volcán Chinyero of 1909, is visible at top left

name of the local village, Las Manchas. Here we are close to the boundary between the relatively recent rocks of the NW rift zone and the ancient rocks of the Teno massif just to the west. This is reflected in the presence of two shrubs associated with dry woodland rather than pine forest. One of these is 'Retama blanca' *Retama rhodorhizoides*, restricted to Teno and the Güímar valley; it is a broom with white flowers and ovoid green fruits that go black as they ripen. Also present is *Sonchus canariensis*, a western species that can grow to 2.5 m or more.

On a different road, the Arenas Negras Recreation Zone is 3 km north of Chinyero and at an altitude of about 1250 m, close to Volcán de Garachico and the adjacent Montañas Negras, where there are lava flows and large cindery areas created by the 1706 eruption. The steep clinker slopes of these 300-year-old craters are now being colonised by a few species of plants, especially Canary Pines, Brezo *Erica arborea*, the figwort *Scrophularia glabrata* and the dock *Rumex maderensis*. The last two species clearly have the ability to grow fast when water is available (the area is often within the cloud sea) but to die back to the roots in hot, dry periods.

The track to the recreation area from La Montañeta is through managed forest, with more mature planted pines and an understorey of Brezo, Faya, *Cistus symphytifolius*, *Adenocarpus viscosus* and *Lotus campylocladus*, as well as fine displays of fungi on the forest floor.

The road west from La Montañeta to Los Llanos crosses the 1706 Garachico lava (confusingly signed as Chinyero) at a slightly lower level, offering a wonderful floral display in early summer. It features

Pines on 'islote' of old lava isolated by the Chinyero flow

Retama blanca *Retama rhodorhizoides*

Cerrajón arbóreo *Sonchus canariensis* (seeds top left)

The edge of the Garachico lava flow, with colonising Aceda de Madeira *Rumex maderensis* and a few Canary Pines, while the older lava on the right has a form of fayal-brezal

dense clumps of *Aeonium spathulatum* with golden flowers and bright orange seeding heads, *Rumex maderensis* with green foliage and masses of pink flowers, and the fern *Davallia canariensis* with green and sometimes dry golden fronds. A species of *Bystropogon* also occurs here, and in shady places in the lava one can also find a delicate succulent *Aeonium* relative *Aichryson laxum*, with golden flowers and hairy leaves and stems which at this site are purple rather than green.

The western edge of the 1706 lava flow is raised above the adjacent older lavas, providing a clear contrast between the relatively fresh lava colonised by widely scattered pines, and the older lower ground with dense woodland.

Estrellitas (or Gongarillo canario) *Aichryson laxum* colonises shady places on lava

6 *Pine forest – 'pinar'*

The fern *Davallia canariensis* colonising lava from the 1706 Garachico eruption

Bejequillo canario *Aeonium spathulatum* and *Rumex maderensis* flourish on lava.

Mature pines near Madre del Agua

PINE FOREST NEAR VILAFLOR

Vilaflor, which at 1400 m is the highest village on Tenerife, is a good starting point for visits to several different types of pine forest characteristic of the south side of the island. Above the village and close to the main road there are splendid examples of old Canary Pines, showing how large these trees can grow in the right place.

A couple of kilometres above Vilaflor a fairly level track to the right gives access (on foot or bicycle, or partly by car) to areas of planted and natural forest. About six kilometres along the track a signed footpath to the left leads to the Paisaje Lunar. This passes at first through planted pines around 30–40 years old, and on our visit on the last day of November 2012, just after a year-long drought had ended, many of them were looking sick, with brown needles on many branches, while shrubs growing among the pines included many apparently dead individuals. On the track we came across groups of the iridescent blue-black scarabaeid beetle *Pachydema obscura*, an endemic species that is also common in Las Cañadas, in which the larvae feed on roots and the adults emerge after rain, appearing on the surface in large numbers and dispersing, mating and dying within a short period.

6 Pine forest – 'pinar'

Planted pine forest with undergrowth of *Cistus symphytifolius*
(on left) and Escobón (on right and inset)

Gregarious scarab beetles *Pachydema obscura* emerging for a short life above ground.

Echium wildpretii growing among planted pines south of Las Cañadas

Further up the footpath, at a height of 1800–1850 m, is an area with rough and relatively unweathered lava (including an interesting natural channel formed of pāhoehoe lava). Here planted pines are mixed with shrubs typical of Las Cañadas such as Tajinaste rojo *Echium wildpretii*, the mountain thistle *Carlina xeranthemoides* and even the broom Retama del Pico *Spartocytisus supranubius*. Felled Escobón *Chamaecytisus proliferus* bushes among the pines shows that the foresters did not want them competing with the pines.

The Paisaje Lunar itself, though smaller in scale than the tourist literature implies, is of interest since it has open natural pine forest with trees of all ages, mixed with Retama del Pico and Escobón, growing on dry rock surfaces of lava and pyroclastic deposits formed during major explosive eruptions.

The main track remains driveable as far as Madre del Agua, where a spur to the left leads to a campground (for organised parties only). This is surrounded by an interesting pine forest at a height of about 1600 m, with many old trees, some planted but with others regenerating naturally. The undergrowth here is mainly Escobón, with *Adenocarpus viscosus* and *Cistus symphytifolius*. Continuing by foot on the main track one can reach the Barranco del Río, a major ravine running from Montaña de Guajara on the edge of Las Cañadas to the coast southeast of Granadilla. The upper part of this ravine has fine old pines in a natural state, and is a good place to see Blue Chaffinches.

For those with limited time, a stop at the Área recreativa de Las Lajas may be worthwhile. The site is in open Canary Pine forest with sparse undergrowth and is at almost 2100 m, close to the transition to the treeless high mountain zone. It is an easy place to stop and wander around, and especially if you are here in early morning or late evening there is a chance of seeing a variety of local birds such as Blue Chaffinch, Turtle Dove, Hoopoe, Canary, Berthelot's Pipit and with luck also the endemic Tenerife subspecies of the Great Spotted Woodpecker. However, as at El Lagar in the north of the island, water taps are no longer left dripping intentionally to encourage the birds, so you are now unlikely to be treated to a view of a bathing Goldcrest of the local endemic subspecies, as we were in the 1980s. Also occurring here are two large beetles: the endemic *Buprestis bertheloti* is black with a yellow pattern on its elytra (wing covers) and can often be seen flying. Its larvae feed inside dead pine trunks, where they are vulnerable to the predatory *Temnoscheila coerulea* (Trogossitidae), which is brilliant black with blue highlights,

A short distance further up the road from Las Lajas you will enter a zone where pines planted at the very top of their natural altitudinal range (and perhaps some naturally established ones) were caught by the wildfire of August 2012. The land seemed devastated, but by the end of November the fire-blackened trees were sprouting clusters of new needles from improbable places on their trunks and branches; the scorched ground, however, was strikingly free of plants.

Buprestis bertheloti
Photo: Leopoldo Moro

6 Pine forest – 'pinar'

The well known Paisaje Lunar is surrounded by mature natural dry pine forest

Tabaiba majorera *Euphorbia atropurpurea* and Sabina *Juniperus turbinata* on rocky slopes. Lower down is degraded scrub with *Cistus monspeliensis*, where pine forest once grew and is starting to return.

PINE FOREST AND MOUNTAINS NEAR IFONCHE

A little known area in the Conde massif, with pine forest high above the upper end of the Barranco del Infierno, and succulent scrub on the mountains that loom over the tourist resorts of the southwest.

North of the village of Ifonche, various tracks skirt around the dark and often mist-shrouded head of the Barranco del Infierno, but there is no easy way down into it. Ecologically, this area is a zone of transition between the dry woodland below, in and around the Barranco del Infierno, and a large tract of fairly natural dry pine forest above it to the northeast. This forest – part of which has recently been burnt – reaches almost to Vilaflor, Las Lajas and Boca Tauce on the edge of Las Cañadas.

The pine forest around the rim of the Barranco del Infierno is fairly open, with an understorey of Brezo *Erica arborea*, Escobón *Chamaecytisus proliferus*, pink-flowered *Cistus symphytifolius* and white-flowered *Cistus monspeliensis*. The latter species is typical of degraded

Pine regeneration on unused land with Escobón and Jaguarzo *Cistus monspeliensis*

6 Pine forest – 'pinar'

areas but also of regenerating ones, which it often shares with young pines. *Bencomia caudata* and *Echium virescens* are also in the area, as well as various small shrubs of the pine forest, including the composites *Carlina salicifolia*, *Atalanthus sp.* and *Argyranthemum gracile*, the crucifer *Descurainia millefolia*, the endemic dock *Rumex lunaria*, and the low-growing *Micromeria hyssopifolia* and *Lotus campylocladus*.

In addition to the impact of grazing, the land around Ifonche has been changed by cultivation over several centuries, and many widespread and introduced plants can be seen there, including *Asteriscus aquaticus, Calendula arvensis, Foeniculum vulgare, Dittrichia viscosa*, the labiate *Marrubium vulgare* and campions *Silene* sp., as well as the invasives *Eschscholzia californica* and *Opuntia maxima*.

An appropriately murky view into the depths of the Barranco del Infierno, above Adeje

Matorrisco *Lavandula canariensis* is common around Ifonche

Flowers, stems and leaves of *Euphorbia atropurpurea* form a delicate tapestry with grey and orange lichens

South of Ifonche are the magnificent peaks of the Conde massif, one of the geologically oldest parts of Tenerife. In the north are Montaña de los Brezos and Roque Imoque (both just over 1100 m) while in the south there is the massive Roque del Conde (1001 m), which dominates the skyline in so much of southwest Tenerife.

Although the vegetation on these mountains has been degraded by centuries of human use, traces of the original zonation can still be discerned. The lower and relatively level areas were evidently once covered by dry pine forest, and although this has been largely replaced by agriculture and *Cistus monspeliensis* scrub, many pines are still present. On the slopes of the peaks there are now areas with regenerating pine, but also more open areas with a wide variety of shrubs and a few Sabina *Juniperus turbinata*, which is here close to its normal altitudinal limit on the island.

High up on the mountains are Canary Pines and succulent scrub including *Aeonium pseudourbicum* and Cardoncillo gris *Ceropegia fusca*, the latter with impressive red seed pods

The implication of this pattern may be that originally the pine forest extended up on to the slopes below the peaks, but was interspersed with a form of dry woodland including Sabina and Brezo *Erica arborea*. The surviving scrub community on the dry rocky slopes seems analogous to that found at a similar height around Masca in the Teno massif, with notable presence above about 1000 m of the beautiful *Euphorbia atropurpurea*. A significant difference, however, is the absence from the Conde massif of the broom *Retama rhodorhizoides*, a key species in many parts of Teno. A few of the other species now found in the scrub are *Echium virescens*, *Sonchus canariensis*, *Phagnalon* cf *saxatile*, *Ceropegia fusca*, *Aeonium pseudourbicum*, *Kleinia neriifolia*, *Micromeria hyssopifolia* and *Monanthes pallens*. Ferns in shady crevices on these mountains include *Cheilanthes pulchella* (an early colonist of fresh lava), *Polypodium macaronesicum*, *Cosentinia vellea* and *Asplenium hemionitis*; in addition, *Asplenium filare*, represented in the Canaries by an endemic subspecies occurring in the western part of the archipelago, has been found here in the past.

Asplenium hemionitis    Cheilanthes pulchella

7

High mountain shrubland – 'matorral de cumbre'

The high mountain shrubland ecosystem

The upper part of Tenerife is a unique high-altitude volcanic environment. Its importance was recognised long ago by the Spanish government, which established the Parque Nacional del Teide in 1954 (at the same time as the Caldera de Taburiente on La Palma) and extended its area in 1999. In 2007 Teide was added to UNESCO's World Heritage List of nearly one thousand sites worldwide considered to have outstanding universal value.

The high mountain region of Tenerife elicits a wide variety of reactions from visitors. It is so alien to anything we see in northern Europe that many people find it stark and unfriendly; on the other hand, many love its austere beauty and return again and again. It is a jagged mountainous region with huge areas of lava and cinders, some of which are ancient and vegetated while others are relatively recent and barren, so that it is easy to imagine the extraordinary volcanic events that produced this landscape. Although much of the rock is grey, there are black, red-brown, whitish, yellowish and even blue-green areas; the colours and shadows are especially dramatic in the light of early morning and late evening. This is a hostile environment and few plants manage to colonise it, so there is much exposed ground. Nevertheless, the vegetation and animal life of Las Cañadas and Teide provide some of the most interesting features of the natural history of Tenerife.

The Pico del Teide, which makes photographs of the island so instantly recognisable, is a young stratovolcano formed during the last 200,000 years (see Chapter 13). In winter Teide is capped with snow and rises dramatically some 1600 m from the relatively level depression known as the Las Cañadas caldera, which is itself about 2100 m above sea level. The combination reaches 3718 m and is the highest mountain in the Canaries and in Spain; indeed, measured from its base on the sea floor, the Tenerife volcano is the third highest volcanic edifice on earth, exceeded only by Mauna Kea and Mauna Loa in Hawai'i.

Teide National Park extends over 190 square kilometres, from a minimum altitude of 1650 m on the northern slopes, up to the summit of Teide and the adjacent volcano of Pico Viejo (3135 m), with its squat outline and large summit crater complementing the elegant cone of Teide. The volcanoes and the caldera together occupy a roughly circular area about 15 km from west to east, but the park boundary extends eastwards from El Portillo for several kilometres along the dorsal ridge of the island, skirting the astrophysical observatories at Izaña. In the north the slopes of the volcanoes run steeply down towards the coast, but the southern wall of the caldera is backed by a long line of peaks about 2500 m high, and only to the south of these does the land slope down towards the coast.

Caldera wall and Montaña Blanca, with stunted Retama del Teide on slopes below the Refugio

A caldera (meaning cauldron) is a roughly circular depression high on a volcano, usually caused by collapse of volcanic materials into a partially empty underground magma chamber after an explosive eruption. However, the origin of the Las Cañadas caldera is still controversial and a giant landslip may have played a part in its formation (see Chapter 13).

Las Cañadas is made up of at least three intersecting calderas, curved amphitheatres that are now almost filled with volcanic materials. Two of these bowls are very obvious: the floor of the eastern one is at about 2200 m but the lowest part of the western one is at only 2050 m. The intersection is marked by an ancient rock wall running from northwest to southeast and comprising the Roques de García, Los Azulejos, the Roques Amarillos and the mountain of Guajara (2718 m); the latter peak is exactly 1000 m below the peak of Teide and

7 High mountain shrubland – 'matorral de cumbre'

Los Roques de García and Llano de Ucanca in dawn light

Facilities in Teide National Park

The Teide National Park is well organised for visitors. There are many tourist buses from the resorts, as well as the scheduled Titsa bus running between Puerto de la Cruz and Playa de las Américas. The main attraction for tourists is the teleférico – the cable car that runs throughout the day, except in bad weather, up to near the summit of Teide. The buses also stop at the Parador de las Cañadas del Teide, the government-run hotel-restaurant, and at the cafes at the east end of the park, as well as at a selection of viewpoints. However, the distances in the park are quite large and it is really best to take a car since the pauses by the tourist buses at any one viewpoint are short, and inadequate for exploring this fascinating area. There are several parking places and information boards describing the various trails; with a good walking guide you should be able to choose a trail that enables you to see many of the special plants described in this book. First of all, however, it is worth spending some time in the extensive botanic garden at the back of the National Park Visitor Centre in the east of the park, where the growing plants are carefully labelled and there is help and advice available in the centre itself.

Greeting the dawn from the peak of El Teide is a most rewarding way to experience the island, but requires an overnight stay at the Refugio de Altavista. Booking for this is essential, and permission is needed to visit the actual summit of the mountain. The teleférico does not go as far as the peak, although its upper terminus is a good place from which to walk to the refuge; the alternative way to reach the refuge is to walk up from the base of Montaña Blanca (~4 hours).

forms the highest point on the wall of the caldera. The northern edge of the caldera has been obscured by the rise of the young volcanoes, but the southern rim is intact, with impressive inward-facing cliff walls rising in the west rise as much as 500 m above the floor of the caldera. In the east, where the caldera floor is higher, the height of the cliffs above it is less.

The Teide volcanic complex is still in an active phase; several eruptions have occurred in the recent past and more are likely in the next few centuries. The summit cone of Teide (150 m high and with its base above the level of the teleférico terminal) was formed as recently as about 850 AD, by eruptions that also produced rivers of black lava flowing down the north side of the mountain. At present, the only indications of activity in the summit crater are cracks in the ground that emit steam and other hot gases and support some plants and animals.

In 1798 Pico Viejo erupted in a spectacular manner at the Narices del Teide (Teide's nostrils): streams of black basaltic lava poured down the mountainside and spread over more than four square kilometres of the floor of the caldera. Even more recently there have been eruptions on a rift zone just northwest of the park, leaving large areas covered by black cinders deposited from the air. The last of these eruptions, Volcán de Chinyero in 1909, was the most recent to occur on Tenerife (see Chapters 6 and 13).

The 'cañadas' that give the area its name form an irregular broad band inside the southern rim of the caldera. The word cañada is roughly equivalent to drove road and the cañadas were used as gathering places for goats during the centuries when they were brought up from the lowlands to graze in the caldera each summer. The cañadas are flat dusty areas where the ground surface consists of granular and powdery volcanic particles washed and blown down from nearby slopes and now partially masking the underlying lava floor in the lowest parts of the caldera. In some places more recent lava flowing from the volcanoes has reached the cañadas, creating an abrupt transition from dusty plain to jagged basaltic rock. As a result of the intense volcanic activity of Teide in recent millennia, the ground in many parts of the caldera consists of unweathered basaltic lava or loose pumice or cinders, with soil almost absent. Since the volcanic deposits are mostly highly permeable, rainwater is quickly lost from the surface.

Pico Viejo and the lava flows of 1798 from Narices del Teide

These raw volcanic substrates are difficult habitats for plants and animals, and the challenge is increased by the extreme climatic conditions. The high mountain zone is characterised by low humidity and intense solar radiation. Figures from the meteorological station at Izaña (2367 m a.s.l) at the eastern end of the park show a mean annual temperature of 9.8°C and annual rainfall of 300–500 mm. Close to the summit of Teide, at ~3500 m, the mean annual temperature is 3.5°C and precipitation about 200 mm. These figures, however, obscure massive variations. Las Cañadas normally lies above the cloud sea that frequently envelops the lower slopes on the north face of El Teide, so there is usually intense sunlight by day and clear still nights. As a result, temperature differences between day and night may be as much as 10–20°C. Seasonal changes are also great, with mean monthly temperature of 5°C in January and 18°C in July, the latter accompanied by relative humidity of the air below 30%. In winter, frosts are common and in the past snow fell on about 13 days per year on average; in recent decades, however, the average has fallen to nine, perhaps because of climate change. Rainfall also shows a strong seasonal pattern, typically with very little rain from May to September, while the remaining months average 30–100 mm per month.

Southern wall of the caldera with pines, a cānada and pahoehoe lava

As a result of the strong seasonal changes, many of the plants remain dormant for much of the year, flowering in early summer. The best time to see the flowers is usually from late May through June and into early July, when the colour throughout this high region can be spectacular. However, variation between years is also important, with the annual rainfall as low as 50 mm in some years and as high as 800 mm in others. Between autumn 2011 and autumn 2012 there was a prolonged drought in the Teide area, with no rain for at least 10 months, and when we visited Las Cañadas in May-June 2012 there was far less flowering than usual; no rain at all had fallen since the previous October and many of the plants were showing obvious signs of stress.

The extreme environment of the high mountain zone imposes strong selective pressures on the plants that manage to colonise the area, and has led to the evolution of distinctive features. Many species have a compact, cushion-like form that minimises water loss, and several are hairy or have waxy coatings to the leaves. The adaptive changes have involved

Human use of El Teide and Las Cañadas

Goatherd in Las Cañadas in the early 20th century

The Pico del Teide has long been used as a navigational beacon by mariners, and the mountain was much revered by the aboriginal inhabitants, the Guanches. There are many archaeological remains in the caldera, especially in the cañadas north of the Parador and the Roques de García. These confirm the long history of utilization of resources of the high mountain zone, both during Guanche times and after the European conquest.

Occupation was doubtless always seasonal: the herdsmen from the settlements in dry areas to the south of the mountain brought their flocks up to Las Cañadas for the summer to make use of the grazing. Although early records are lacking, the custom of seasonal pasturing of goats (and perhaps sheep) in Las Cañadas probably started soon after the earliest human occupation and continued throughout the half millennium following European colonisation. This custom ceased, however, in the decade following the establishment of Teide National Park in 1954.

The archaeological remains in the caldera are mainly of oval or circular shelters, usually near large rocks and often associated with fragments of ceramic plates and stone utensils. Caves and overhangs in the lava functioned as both living places and burial sites, and smaller crevices were used to hide belongings such as pottery utensils, probably at the end of the summer occupation; these are referred to as 'escondrijos' (hiding places). The area was also a significant source of wood, through the collection of stems of the larger shrubs. Some wood was used by the herdsmen for their fires, while some was converted to charcoal, a custom that continued until late in the 20th century. Another traditional use has been hunting of rabbits with dogs and ferrets; rabbit hunting still occurs, but under strict regulations. Introduced mouflon (wild sheep) are also hunted in and near the park.

Another significant resource offered by the volcano was pumice, which is still in use on fields at middle levels in the south of the island. It was brought down from Las Cañadas and the southern slopes of Guajara using camels, and spread in a layer on the fields as a mulch; in the north of Las Cañadas, mules were used for transport. Studies on Lanzarote, where basaltic cinders (tephra) are used in a similar way, show that soil under the mulch remains moist throughout the year, while bare soil nearby becomes dry for long periods. The traditional use of pumice mulch in Tenerife suggests a similar beneficial effect. In the 20th century pumice was also used industrially as an insulating material and for producing abrasive powder.

Other mineral resources were also available on Teide. Obsidian or volcanic glass, which is especially common around Tabonal Negro, was used by the Guanches to make tools, and there are sites in Las Cañadas littered with worked fragments. Another resource of Teide was sulphur, which occurs in crystalline form around the fumaroles in the summit crater. In the late 19th century a business was established to collect sulphur and bring it down to La Rambleta via the Refugio de Altavista. This practice ended officially in 1918 but may have continued illegally for a while. Close to the refugio is the Cueva del Hielo (ice cave), which is at about 3300 m and usually retains snow through the summer; it was used in the past by travellers as a source of water, and for storing snow and ice that could then be carried down by mules to lower levels. There it was used for making ice cream for rich people of La Orotava and Los Realejos, and for hotel-based visitors. There was an initiative to exploit the Cueva del Hielo in a more commercial way, but the use of dynamite fractured the rock floor, leading to water leakage and less efficient ice storage, so the project was abandoned.

The profusion of flowers in the caldera in summer is important for the bee-keepers of the island. It is not known when bee-keeping in Las Cañadas started, but the custom was already established in the 16th century and gave rise to some place-names, for instance Montaña Abejera alta (mountain of the high apiary) at over 2000 m on the north slopes of El Teide. The last resident bee-keeper was Juan Évora, who lived at Boca Tauce until the late 1980's and from whom we once bought honey and very salty cheese. His house has now been refurbished for viewing by visitors.

In recent decades bee-keepers have been permitted to move up to 3000 hives into the caldera for the five or six months of the high altitude flowering season, using some 20 apiaries. This has been shown to have negative effects on native pollinators. Retama del Teide *Spartocytisus supranubius* bushes close to apiaries produce fewer seeds than those further away, because honeybees are less efficent pollinators than native insects. Tajinaste rojo *Echium wildpretii* is adapted for bird pollination, but when the beehives are brought up for the summer the birds stop visiting the nearby plants, since nectar supplies are depleted by the bees.

A variety of flowers from Las Cañadas, and also fine volcanic fragments of different colours, are traditionally collected and used during Corpus Christi celebrations (about two months after Easter) in the town of La Orotava – the municipality to which Las Cañadas del Teide belong. The flowers and coloured sands enable local people to create ephemeral artistic carpets on the ground in the plaza and along central streets. The custom was brought from Italy by a local family in the 19th century, and the necessary collections are now authorised by the national park administration, but restricted to the periphery of the park and along the roadsides.

The high mountain zone has also been used for scientific observation. The volcanic activity and clarity of the air high up on Tenerife attracted the attention of European scientists long ago. In 1856 Charles Piazzi Smyth, Astronomer Royal for Scotland, established a camp on the summit of Guajara. He also organized the building of the Altavista hut, at 3270 m on the slopes of El Teide, to facilitate his observations; in 1891 the English scientist and photographer George Graham-Toler arranged for improvements. The refugio was used by many scientists and walkers in the 20th century and was later taken over by the island authorities; it now functions as a tourist hostel.

In the early 1960s the focus for astronomical observations shifted to a new location just outside the national park. The Observatorio del Teide at Izaña, at an altitude of nearly 2400 m, is now an international centre for climatological and astrophysical studies, and is especially well suited for observation of the sun. Recently the park has been used by teams of scientists concerned with the exploration of Mars, making use of the similarity of the environment of Las Cañadas to that on the red planet.

Beehives of Juan Évora in 1984

Sand picture in Orotava in 1984

Making a carpet of flowers

Canary Pines on the southern wall of the caldera have been damaged by fire but survived, while Taginaste rojo *Echium wildpretii* and Tonática *Nepeta teydea* have benefited from the fire

Hairy leaves of *Echium auberianum*

divergence of many of the plants growing in and around Las Cañadas from their relatives lower down, and this is one of the reasons why the proportion of endemic species (and subspecies) among the plants of the park is among the highest for any region in the world. For example, in one tabulation of 139 vascular plant species found in the park, about a third were endemic to the Canaries and 15 (11%) were entirely restricted to the park.

The native high mountain vegetation is dominated by a rather small number of specialised shrubs, and now covers some 140 km². The shrub habitat is largely on land above 2000 metres in

7 High mountain shrubland – 'matorral de cumbre'

High mountain shrubland with *Echium wildpretii* in a poor flowering year

the caldera and on the mountains nearby, although some of the characteristic species can be found down to about 1800 m. No natural woodland exists on the floor of the caldera, although the pine forest laps around the rim. However, in the mid 1950's about 72 hectares of Canary Pines were planted in the caldera, as well as some Atlas Cedars *Cedrus atlantica* and ~114 ha of Monterey Pines *Pinus radiata*, another alien species. A more modern conservation policy in recent years has resulted in these trees being cut down as part of an attempt to keep the park as close to its natural state as possible.

There is much variety in the vegetation within the high mountain shrubland and ecologists have identified a long series of more or less distinct plant communities within it, but here we try simply to highlight the main differences that are easily visible within the area, and to mention characteristic plants.

When mammalian herbivores – mainly goats, sheep and rabbits – were brought to the island by humans, their selective grazing doubtless led to profound changes in the distribution of many plants, with especially palatable species being excluded from accessible areas. Even now the presence of rabbits *Oryctolagus cuniculus* and mouflon *Ovis orientalis* presumably results in more herbivore impact on the plants than before humans arrived. Previously, the caldera may have lacked vertebrate herbivores and the main plant-eating animals were doubtless the insects that are still there, including beetles, grasshoppers and bugs.

Las Cañadas with surface water after rain

Retama del Teide
Spartocytisus supranubius

Rosalito de cumbre
Pterocephalus lasiospermus

The cañadas are unique among the habitats in the park in being flat, with fine gravelly or powdery soils. For much of the year they are dry and dusty, but after storms or the snowmelt in spring their floors are channelled by running water and they sometimes become transformed into short-lived inward-draining lakes and pools. Although the water may stand for a few days it gradually seeps down into the ground and the lakes dry out. In the Llano de Ucanca below Los Roques – and near a few springs around the edges of the caldera – enough moisture persists to support a population of an endemic subspecies of sedge *Carex paniculata* ssp. *calderae*. The low-lying cañadas are surrounded by higher ground, so they form frost pockets on winter nights, holding cold, heavy air that flows down from nearby areas; in 1911 a temperature of minus 16.1°C was recorded in a cañada. The alternate freezing and thawing often leads to 'frost heave', forming small ridges and furrows and sorting the sand particles so that the ground becomes patterned.

The baking sun in summer and frequent winter frosts prevent colonisation of the cañadas by all but a few plants. The dominant species is usually Retama del Teide *Spartocytisus supranubius* (sometimes called Teide broom) a large bush with stiff, upright, straight twigs and white (occasionally pinkish) blossom, forming dense rounded clumps up to 3 m high and 10 m across. It also grows in other parts of the park and high on the slopes of Teide. Retama was resistant to the traditional seasonal grazing, but was probably extensively harvested for firewood. Now these pressures have eased, but Retama still suffers some browsing by rabbits and probably mouflon, and perhaps also from increased competition for water from more palatable plants that had become rare as a result of the grazing.

One potential competitor is Rosalito de cumbre *Pterocephalus lasiospermus,* a mauve scabious that has become much more abundant in recent decades, probably because goats are now absent and it is not eaten by rabbits. It now often dominates the vegetation in the stony and somewhat inclined areas around the cañadas, and in rocky places elsewhere, but it also gets into the edges of the cañadas themselves.

The relatively gentle slopes of weathered lava and pyroclasts (volcanic fragments deposited from the air) that occupy much of the floor of the caldera have higher plant diversity. A characteristic and highly visible plant is the crucifer Hierba pajonera *Descurainia bourgeauana,* with brilliant yellow flowering spikes. In winter these cushion-shaped plants have a halo of pale dead flowering stems and from a distance are reminiscent of resting sheep. The retained stems may well be advantageous, since they tend to condense dew, and during cold nights ice often accumulates on them and then melts in the sun during the day, draining down towards the roots. A closely related species, *Descurainia gonzalesii,* occurs on dry cinder slopes in a few places; it can be distinguished by its linear (rather than deeply dissected) upper leaves.

Luxuriant regeneration of Retama del Teide in a cañada eight years after an area was fenced against rabbits and mouflon

Descurainia gonzalezii

Hierba pajonera *Descurainia bourgeauana* growing on pumice, and in winter with ice sheaths on dead stems

Often to be found growing near *Descurainia* is the striking Alhelí del Teide *Erysimum scoparium,* a relative of the garden wallflower, with stiff gutter-like leaves and four-petalled blooms that are mostly deep blue-mauve but ususally with a few white ones interspersed; they are succeeded by long brown seed pods.

However, perhaps the most spectacular plants in Las Cañadas are two large species of *Echium* (bugloss). The best known is the tall red-flowered Taginaste rojo *Echium wildpretii*, which has done well in recent years, flourishing on steep rubbly slopes around the sides of the caldera and on well-drained rocky places within it. In contrast, the magnificent blue-flowered *Echium auberianum* is relatively hard to find, often growing singly in the cañadas or other loose gravelly situations, for instance on Montaña Blanca; in contrast to the previous species, even small individuals often flower, while large ones may produce several flowering spikes. One of the species that flourishes in rocky places in the caldera is the catmint *Nepeta teydea*, in the nettle family Labiatae, which forms clumps with many tall upright stems. It resembles the Alhelí in having blue flowers, but in *Nepeta* they

Tonática *Nepeta teydea*

Fistulera de cumbre *Scrophularia glabrata*

are in well-separated whorls on square stems, and the leaves are toothed, hairy and somewhat fleshy; it readily sprouts from the base following the passage of fire. In the same family is the attractive Tenerife endemic shrub *Sideritis eriocephala;* the genus is recognisable by the heavily white-felted stems and leaves, but the species are hard to separate. Another labiate is *Bystropogon origanifolius*, here at the top of its altitudinal range, again with opposite leaves, but with straggly growth and smelling of mint. Two other

Alhelí del Teide *Erysimum scoparium*

Taginaste picante *Echium auberianum*

Taginaste rojo *Echium wildpretii*

Codeso de cumbre *Adenocarpus viscosus*

species of aromatic herbs in Las Cañadas, *Micromeria lachnophylla* and *M. lasiophylla*, are inconspicuous and hard to distinguish.

The figwort *Scrophularia glabrata* is another square-stemmed shrub with toothed leaves that survives drought or fire by sprouting from the base; it occurs mainly in shaded places in the caldera but also in the pine forest at slightly lower levels; it has small dark purple flowers and small globular seed pods.

A little known rarity
Erigeron calderae

Margarita del Teide
Argyranthemum teneriffae

7 High mountain shrubland – 'matorral de cumbre'

The Tenerife endemic fleabane *Erigeron calderae* is a very rare daisy-relative with pinkish or whitish flowers and plantain-like untoothed leaves. More abundant, but also found mainly near the edge of the caldera as well as on the mountain slopes just outside it, is its relative *Cheirolophus teydis*, a striking yellow-flowered perennial found only on Tenerife and La Palma; it also has good resistance to fire.

Another two conspicuous plants in Las Cañadas, both of which can also be found at slightly lower levels outside the caldera, are Margarita del Teide *Argyranthemum teneriffae* and Codeso de cumbre *Adenocarpus viscosus* The first is a large daisy that occurs in open places but also grows on the cliffs and high up on Teide itself. The second is a spreading, yellow-flowered, hairy, leguminous shrub with tiny, tightly folded leaves, often occurring where lava flows intrude on more level cañadas. Another endemic composite is Cardo de plata *Stemmacantha cynaroides*, which resembles a compact globe artichoke, with jagged leaves and globular whitish flowerheads; it grows in a few places in the eastern part of the park, but is extremely rare and suffers from browsing by the mouflon (see Box).

Cabezón de la cumbre *Cheirolophus teydis*

Andryala pinnatifida ssp. *teydensis*, a high altitude subspecies

Perejil de cumbre
Pimpinella cumbrae

Two plant rescue projects in the Parque Nacional del Teide

Cardo de plata 'silver thistle' *Stemmacantha cynaroides* and Jarilla de cumbres 'peaks rockrose' *Helianthemum juliae* are two plants that are endemic to Tenerife and occur only in the Teide national park, where they are in danger of extinction. Although genetic and biological studies designed to halt their decline were started in 1986, it was not until 2006 that recovery plans for both species were approved. The plans focus on mitigating the factors that threaten them and on increasing the numbers of individuals in the wild.

Stemmacantha cynaroides is found in five places, between 2200 and 2400 m above sea level, and favours sunny sites with sedimentary deposits. Like its relative the globe artichoke of horticulture it spends the unfavourable season underground, but emerges to grow vigorously in spring and summer, sometimes reaching a metre in diameter, and produces seeds that are dispersed by the wind. The flowering heads are so appetizing to the introduced Muflón *Ovis orientalis* (= *O. aries*) that the plant nearly became extinct in the 1990s, when only about 150 individuals survived. Another threat is from rabbits *Oryctolagus cuniculus* (also introduced), which are very abundant in the park and can cause mortality in periods of drought, when they dig around the bases of the plants in search of water in the roots.

Helianthemum juliae is a perennial hermaphrodite only discovered in 1986, when the surviving population numbered less than 200 individuals. It requires greater humidity than Stemmacantha and occurs mainly in north-facing places above 1800 m. The plants flower in May and June; seeds are produced in August and disperse only short distances before falling to the ground. Seedlings appear in large numbers after autumn rains, but many die from high temperatures or drought. The species is therefore vulnerable to the effects of climate change, and it has been shown that its numbers decline in dry years and increase in wet ones.

The recovery plan for *Stemmacantha* involves raising plants in nurseries to reinforce the natural populations, and also creating new ones, so as to establish a total of at least 1500 individuals in eight localities. The case of *Helianthemum* is more complex, and it has been necessary to find new sites where the species does not now occur but where climatic conditions are suitable; plants are being propagated in nurseries, with the aim of establishing a total of 3000 individuals in nine populations, each with at least 150 plants.

By the summer of 2014 new populations of both species had been established, most of them fenced to exclude herbivores. *Stemmacantha* now has about 770 plants in eight sites, and *Helianthemum* has 2300 individuals, of which 725 are juveniles, also in eight sites. In the latter case six new populations have been established, in cool and humid places around the walls of the caldera. Propagation and planting of both species is now proceeding systematically and it is hoped that within three years they may be considered out of danger of extinction in the short term. The original studies suggested that before the rescue plan was started the probability of extinction of *Stemmacantha* within 100 years was 100%, and that of *Helianthemum* was 95%; the new situation should ensure a risk of less than 25% during the next century.

Manuel Marrero & José L. Martín

7 High mountain shrubland – 'matorral de cumbre'

Stemmacantha cynaroides, planted in straight lines to make it easy to record survival

Helianthemum juliae established in a new and relatively humid site on the caldera wall

Turgaite *Senecio palmensis*, a successful cliff plant

Grasses are not generally abundant in the caldera, but one attractive species is the tall clump forming *Arrhenatherum calderae*, with seed heads reminiscent of oats, which occurs only in Tenerife and La Palma.

The walls of the caldera are immensely important for the plants of the park. The walls take different forms in different parts, with vertical cliffs and isolated rocks interspersed with steep rubbly slopes, and often with layers of lava alternating with ash and other types of volcanic rock, providing many ledges and crevices. They thus offer a rich variety of niches for plants, and their inaccessibility renders them invaluable as refuges for some species that were banished from the floor of the caldera under the traditional grazing regime.

One shrub that may once have been widely established in the caldera is the Tenerife endemic *Rhamnus integrifolia*. At present a few struggling individuals can be found in crevices at the foot of the caldera wall, showing signs of heavy browsing by rabbits or mouflon; on the cliffs above are more bushes, some of which are able to produce fruit. However, this species also occurs in small numbers on cliffs in deep valleys in the south of the island.

A similar case is that of *Senecio palmensis*, an endemic relative of groundsel with fleshy, toothed or lobed leaves and bright yellow flowers; in the caldera it is rarely found except on cliffs, but it occurs at ground level lower down on the island in a few places.

Another of the plants that has survived in this way is the rare shrub *Bencomia exstipulata*, known only from Las Cañadas and similar habitat on La Palma, of which only a few individuals were surviving on cliffs when we were working in Las Cañadas in the mid 1980s. Since then, an intensive restoration programme has been carried out near Boca Tauce, and the species is beginning to reassert its rightful place as a significant member of the high mountain shrub community.

A commoner species found sparingly on the cliffs is Escobón *Chamaecytisus proliferus*, which is abundant near the upper edge of the pine forest but is here reaching its altitudinal limit around 2100 m. Two other legumes occur up to about this height: Tedera *Bituminaria bituminosa* is one of the commonest plants of the Canaries but occurs here in a high altitude form, while *Lotus campylocladus* is more typical of pine forest but just penetrates the caldera.

Near the top of the cliffs west of Boca Tauce there are a few specimens of Cedro *Juniperus cedrus*, providing names for the 'Roque del Cedro' and 'Montaña del Cedro', the highest points in this area; specimens can also be seen elsewhere around the edge of the caldera. The Cedro is considered to be in danger of extinction on the island, so we were pleased recently to find it growing at a much lower level in a ravine near Güímar.

At the northeastern point of the caldera the steep south face of La Fortaleza provides another stronghold of the Cedro, and it is home to a number of shrubs and smaller plants that are not widespread in Las Cañadas. To protect these, access to La Fortaleza is now prohibited. One of the rare species here is the beautiful shrub *Cistus osbeckiifolius*, endemic to the high mountain zone of Tenerife; it can be distinguished from the common *Cistus symphytifolius* of

Caldera wall at Roques de Chavao. Most of the plants are *Bencomia exstipulata*. Other species in the photograph or growing close by include *Senecio palmensis, Silene berthelotiana, Pimpinella cumbrae, Echium wildpretii, Nepeta teydea* and *Descurainia bourgeauana*. *Juniperus cedrus* and *Rhamnus integrifolia* also grow on these cliffs

the pine forest by its stiff, relatively narrow, silvery leaves and the dense but very short down covering the seed capsules; it also tends to branch near the base, forming a dense clump of upright stems. Another related and equally rare plant is the rockrose *Helianthemum juliae*, with five-petalled yellow flowers; it is endemic to the park but is now restricted to the caldera walls (see Box above).

Conspicuous on La Fortaleza and below Los Azulejos, but also occurring elsewhere on the Canaries, is the tall umbellifer *Ferula linkii*, which can grow to 3 m tall, with brilliant yellow flowers. A related species, but at the other end of the size range, is *Pimpinella cumbrae*, endemic to Tenerife and La Palma, a parsley-like herb producing a mass of tall and slender flowering stems; it occurs in only a narrow altitudinal band between about 1900 and 2100 m, but is relatively easy to find in shady places near the wall of the caldera. Although it looks

Southern wall of the caldera with *Aeonium smithii* (lower) and *Sideritis eriocephala* (in sun at top and inset). Notice the 'posadero' (roosting site) in the centre, probably of *Falco tinnunculus*.

delicate, its deep roots enable it to survive droughts and even the passage of fire. Also present here is *Silene berthelotiana,* a white-flowered endemic campion with yellow seed pods, which grows in the pine forest and on the walls of the caldera, but not higher up.

A composite associated with the walls of the caldera is *Carlina xeranthemoides*, a beautiful silvery thistle with golden flowers. In 2012 we were intrigued to find this species outside

Moralito *Rhamnus integrifolia* survives in the caldera only on cliffs

Silene berthelotiana occurs on the walls of the caldera and in pine forest

Las Cañadas among planted pines near the Paisaje Lunar, which is just south of Montaña Guajara, the highest point on the southern rim of the caldera. In the same area – at about 2000 m – we also found *Spartocytisus supranubius* and *Echium wildpretii*. This chance observation brought home to us the way in which the 'repoblación' of the Canary Pine in the middle of the 20th century sometimes extended into areas where the natural vegetation was of a different type – in this case mountain shrubland.

Other plants associated with the edge of the caldera include two species named after the English naturalist Philip Barker Webb, who published – with Sabin Berthelot – the most important 19th century work on the natural history of the Canaries. The Canary endemic Crespa *Plantago webbii* is a high-altitude plantain; it is a small shrub with an upright stance, often growing against rocks, with silvery green leaves closely pressed against the stem and slender inflorescences with flowers in whorls. *Tolpis webbii*, endemic to Tenerife and La Gomera, is a composite that forms a rosette at ground level, with green serrated or lobed leaves and tall, slender flowering heads with loose clusters of yellow flowers. Another elegant composite occurring in the caldera is *Andryala pinnatifida* ssp. *teydensis*, a Teide endemic subspecies of a Canary endemic species. Like *Tolpis*, it has

Malpica de cumbre *Carlina xeranthemoides*

Lechugilla de cumbre
Tolpis webbii

Crespa
Plantago webbii

yellow flowers on tall, slender stalks, but the flowers are distributed along the length of the stems rather than just terminally; the leaves are silvery green, and the plant has milky sap.

Several succulent species of the houseleek family Crassulaceae can be found around the fringes of the high mountain zone, but also occur at lower levels. One of these is *Aeonium smithii,* a very unusual member of the genus that thrives in a few parts of the cliff wall of the caldera; it has succulent scalloped leaves and white hairs near the top of the flowering stem. *Aeonium spathulatum* is well adapted to life on barren lava flows, and is an elegant clump-forming species with reddish ítems and brilliant yellow flowers, becoming bright orange as they die. An equally striking plant is *Greenovia aurea*, more typical of rocky places in the pine forest, with striking cup-shaped leaf rosettes at ground level. *Monanthes brachycaulos*, endemic to Gran Canaria and Tenerife, has a tiny (5 cm) rosette of succulent, pointed, purplish green leaves and starry pale pinkish or greenish flowers; it can reach 3000 m on steep rock surfaces. *Aichryson parlatorei* is a tiny succulent annual with yellow flowers that grows in spreading clumps on dry rocks.

A species that visitors are much less likely to encounter is *Ephedra major*, a strange gymnosperm shrub (distantly related to conifers) that forms dense bushes on rocky slopes in a few parts of Las Cañadas; it has thick stems topped by crowded bunches of thin, jointed twigs. It is a plant with separate sexes, so the sparse population in the caldera must be considered endangered. However, recent research on another species of *Ephedra* in Europe has revealed a remarkable adaptation: at the time of full moon the flowers exude drops of nectar that shine brilliantly in the moonlight, attracting nocturnal insect pollinators. *Ephedra major*, however, may turn out to be wind pollinated.

Ephedra major is one of the rarest shrubs on the island and can grow in extremely barren terrain

Violeta del Teide *Viola cheiranthifolia*

Five members of the campion family occur in Las Cañadas. The endemic *Cerastium sventenii* has opposite leaves and delicate white flowers with many petals, and occurs mainly in the southeast of the caldera. The native but non-endemic *Bufonia paniculata* forms a spreading clump, with linear leaves and jointed stems like a mini-bamboo; the flowers have four white petals alternating with longer, sharp, grey-green sepals; it is common near Los Roques and in some other parts of the caldera. A more prostrate, silvery-hairy dwarf shrub is *Polycarpaea tenuis,* with narrow, sharp-pointed leaves and dense inflorescences bearing small 5-parted flowers.

The campion *Silene berthelotiana,* described above, occurs round the walls of the caldera, but its rare and beautiful relative *Silene nocteolens* is one of the few plants that can gain a foothold on the high mountains that rise from the caldera. It grows on Pico Viejo and on the pumice fields of Montaña Blanca, where it is joined by another Tenerife endemic, the Violeta del Teide *Viola cheiranthifolia,* which finds niches among pumice and bare rock surfaces at up to 3500 m, belying its delicate appearance. The plant in the photograph has a flower, a ripening green seed pod and also ripe orange seeds (barely visible at upper left).

A third Tenerife endemic, the white flowered cudweed *Laphangium teydeum,* colonises sheltered cracks in the rocks and may hold the altitude record for a flowering plant on Tenerife, since in 1984 we photographed what seem to be seedlings of this species growing in moss beside a tiny gas vent inside the summit crater of El Teide, where they presumably benefited from the warming around the vent.

Canutillo del Teide *Silene nocteolens* on Montaña Blanca pumice

Silene nocteolens

The leaves of these three species – all endemic to the highest part of Tenerife – show striking similarity in their growth form, with fleshy, spear-shaped leaves covered with fine

Lava flows and pumice fields on Montaña Blanca and Teide, where some specialised plants manage to grow

down. These characteristics are shared by several of the other plants capable of colonising the sun-baked pumice on the highest slopes, such as *Echium auberianum*, *Erysimum scoparium* and *Pterocephalus lasiospermus*, and are evidently adaptations to this hostile high altitude environment, with extremes of temperature.

At slightly lower levels on the volcano other members of the mountain scrub community can be found, including *Descurainia bourgeauana* and Retama del Teide *Spartocytisus supranubius* growing in a wide-spreading stunted form less than 1 m tall, as well as several grasses in the genera *Anisantha*, *Poa* and *Vulpia*.

Laphangium teydeum (in garden) and seedlings in the crater of Teide, with black rim of cyanobacteria. The moss at top and bottom may be *Campylopus pilifer*, occurring at the highest altitudes.

ANIMALS OF THE HIGH MOUNTAIN SHRUBLAND

One of the most remarkable aspects of Las Cañadas and the area around it may be overlooked by many visitors. This is the enormous array of insects, arachnids and other invertebrates that live in the park or in similar habitat just outside it, and the number of these that occur nowhere else. An intensive survey of the area published in 2002 listed 1,337 species of invertebrates, of which about 1000 could be fully identified. Of the latter, about 50 (5%) are known only from the high mountain zone of Tenerife, while another 42% are restricted to Tenerife, the Canaries archipelago or to the wider set of Atlantic island groups known as Macaronesia. In other words, some 500 kinds of invertebrates – almost half of the total in the high mountain area – are unknown on the adjacent continents and have presumably evolved on the Canaries or other Atlantic islands nearby. Insects make up the bulk of this fauna (85%), but there are also many kinds of spiders (10%) and small numbers of centipedes, springtails and other groups.

Most of the invertebrates spend the winter in a resting stage, or are concealed and inactive, so a visitor who had walked in Las Cañadas on a cold winter day and seen not a single insect might be incredulous of the idea that so many kinds of special animals were living in the area. On a sunny day in June, however, when Retama del Teide *Spartocytisus supranubius* and other high mountain shrubs are in bloom, invertebrates are much more in evidence (about 80 kinds of arthropods have been found on Retama or in the litter underneath the bushes). The most obvious insect in the flowering season is often the introduced honeybee *Apis mellifera*, but there is also an endemic bumblebee *Bombus canariensis*, which is black with a white-tipped abdomen. Many other kinds of bees, bugs and flies visit Retama, and the bushes sometimes seem alive with a shrub-living grasshopper *Calliptamus plebeius*, a Canary endemic. A second species of grasshopper, the ground-living *Sphingonotus willemsei*, is sometimes equally evident underfoot, and it is no surprise that in this zone grasshoppers are second in importance only to lizards in the diet of the local Kestrels.

Codeso de cumbre *Adenocarpus viscosus* is also host to a diverse set of insects, including the impressive long-horned beetle *Trichoferus roridus*, which is more than 3 cm long, grey and square-fronted. Many other beetles are inconspicuous but they are the most diverse insect

Pimelia ascendens
Photo: Pedro Oromí

Cyphocleonus armitagei
Photo: Pedro Oromí

group in the park, with over 200 species including 48 endemic to Tenerife. One of the latter is the tenebrionid *Pimelia ascendens,* a large, black, slow-moving beetle; in the spring months it is hard to avoid seeing it stumbling across the pathway and in the past it was often to be found, along with Canary Lizards, trapped in empty bottles left lying around picnic sites. *Hegeter lateralis*, another flightless tenebrionid, is endemic to the park and common; it has even been found inside the crater at the summit of Teide.

A more conspicuous beetle is the large (2 cm) weevil *Cyphocleonus armitagei,* a speciality of the park, which is strongly associated with the daisy *Argyranthemum teneriffae*, although it is also found on *Cheirolophus teydis* and *Stemmacantha cynaroides*. The adult has a fairly short snout and is blackish with white blotches; it can be found on the flowers or in the foliage, while its larvae feed on dead stems. Another endemic weevil is *Cionus griseus*, which feeds only on the figwort *Scophularia glabrata* and is patterned with creamy white and black blotches and is somewhat globular in shape, with a very long, cylindrical and strongly down-curved snout. Other herbivorous invertebrates include the only mollusc that is widespread in the park, the tiny snail *Xerotricha nubivaga* in the family Hygromiidae, endemic to Tenerife; it feeds on foliage of various plants, including the rare *Cistus osbeckiifolius*, but is not thought to do significant damage.

Green-striped White
Euchloe belemia ssp. *eversi*
Photo: Pedro Oromí

Canary Blue *Cyclyrius webbianus* on *Tolpis webbii*

The commonest crucifer in the park, Hierba pajonera *Descurainia bourgeauana*, is also associated with many herbivorous insects, though few of them are restricted to it. It is the primary foodplant of the Tenerife endemic subspecies of the Green-striped White butterfly, *Euchloe belemia* ssp. *eversi*, known only since 1963; its caterpillars are green with pink lines. This butterfly reaches at least 2500 m on Teide but also occurs in the pine forest, and requires careful separation from the Bath White *Pontia daplidice*, which also occurs in the park. Other butterflies in the area include Small White, Canary Grayling, Painted Lady and the endemic Canary Blue *Cyclyrius webbianus*, which can be seen during many months of the year; its caterpillars feed mainly on Codeso de cumbre *Adenocarpus viscosus*.

A conspicuous day-flying moth is the Hummingbird Hawkmoth *Macroglossum stellatarum*, which may be seen darting from blossom to blossom and hovering in front of them as it feeds. There is also a chance of seeing *Bembecia vulcanica*, an attractive day-flying, wasp-mimicking moth with transparent wings, dark body with pale bands, and long, thick antennae; its larvae feed on legumes. The hairy caterpillars of the moth *Calliteara fortunata* are often abundant on Retama del Teide bushes; this is the species that also causes considerable destruction in the pine forests in some years. A number of other moths can be found in the caldera, but perhaps the most interesting is *Alucita canariensis*, a member of the

Calliteara fortunata

Calliteara fortunata
Photo: Leopoldo Moro

Eusimonia wunderlichi
Photo: P Oromí

family Alucitidae, the many-plumed moths, in which each of the four wings comprises six spines carrying flexible bristles and forming plumes with a striking resemblance to feathers. This moth is endemic to the park and is widespread there; the larvae feed on the scabious *Pterocephalus lasiospermus*.

Heteropteran bugs are often conspicuous on *Descurainia*, especially the black and white pentatomid *Eurydema lundbladi* (endemic to Tenerife and La Palma) which also feeds on Alhelí *Erysimum scoparium*, and the greenish yellow rhopalid *Brachycarenus tigrinus*, which is more widespread. A different bug, *Dictyla indigena* in the family Tingidae, with beautiful lace-like wings, can be found on two species of taginastes in the park, *Echium wildpretii* and *E. auberianum*. These rough and hairy plants are not rich in invertebrates, but *E. wildpretii* also supports two species of leaf-eating beetles in the huge genus *Laparocerus*, which has over 40 species on Tenerife, of which about six are found in the park.

The scabious Rosalito de cumbre *Pterocephalus lasiospermus* is visited by many insects when in flower and these attract a number of predatory species. The most striking of these – but hard to find because it spends most of its time well-concealed in the foliage – is the Tenerife endemic praying mantis *Pseudoyersinia teydeana*, with vestigial wings; it glues its distinctive tough egg capsules to the underside of rocks. If you look carefully at the flowers you may see one of the many local kinds of crab spiders of the families Philodromidae and Thomisidae, which also lie in wait for insects seeking pollen or nectar in the flowers. Several of these same predators can also be found on the composite *Cheirolophus teydis*, which also produces nectar that is attractive to many insects.

On or under the ground many more kinds of spiders can be found, including no less than six species of the genus *Dysdera*, three of which are from caves (including the aptly named *Dysdera gollumi*, known only from a cave in the park). Other arachnids include the camel spider or wind scorpion *Eusimonia wunderlichi*, endemic to the Canaries, which is a member of the order Solifugae, agile predators well suited to desert life.

Many ground-living predators and scavengers depend largely on the 'fallout' of aerially-dispersing insects and spiders, a resource that is available even in the summit crater of Teide, where over 60 non-resident species have been found. This explains the presence there of predators such as the jumping spider *Aelurillus lucasi*, a species of centipede in the genus *Lithobius*, and the harvestman *Bunochelis spinifera*, in which specimens found at this altitude are unusally large,

Southern Tenerife Lizard *Gallotia galloti* keeps cooler by lifting its feet away from hot rocks

At lower levels scavenging animals are abundant, though they usually spend the daylight hours concealed under rocks; they include a thysanuran in the genus *Ctenolepisma*, a group of elongate, flattened, primitive, fast-moving insects with scaly bodies and three 'tails'. Other scavengers include four species of millipedes, *Dolichoiulus* spp., all endemic to Tenerife. Another scavenging species worthy of mention is the earwig *Anataelia canariensis*, endemic to Tenerife and La Gomera. We found it in the 1980s in the barren lava from the Narices eruption, and during the recent survey it was found in the same place but not at any other sampling station in the park.

The vertebrate fauna of the high mountain zone is relatively limited. There are no truly native mammals except for bats, of which five species have been recorded: *Pipistrellus maderensis*, *Hypsugo savii*, *Nyctalus leisleri*, *Plecotus teneriffae* and *Tadarida teniotis*. However, introduced mammals are important. The Algerian hedgehog *Atelerix algirus*, house mouse *Mus musculus* and roof rat *Rattus rattus* occur here, but have all reached the island with the help of humans. Dogs are now uncommon in the park and probably do not breed there, but in the 1980s we came across pairs clearly associated with rocky lairs. Feral cats are present but not often seen. Introduced rabbits *Oryctolagus cuniculus* are numerous and there is well regulated seasonal hunting, providing sport and a means of keeping rabbit numbers within reasonable limits. The most recent and controversial introduction is of mouflon *Ovis orientalis*, wild sheep from the Mediterranean introduced for sporting purposes.

The Southern Tenerife Lizard *Gallotia galloti galloti* is very common in this zone. In summer the lizards are easy to watch near the visitor centre, readily coming to feed on scraps of fruit; in winter they bury themselves and hibernate. The Canary Skink *Chalcides viridanus* and Canary Gecko *Tarentola delalandii* both occur in the park, but are scarce at this altitude.

The most common small birds in the park are Berthelot's Pipit and Canary Chiffchaff; Kestrels are also abundant and the white guano of their 'posaderos' (roosting places) is noticeable in many places on the cliffs. Other birds that have been recorded breeding in the high mountain zone are Barbary Partridge, Plain Swift, Spectacled Warbler, Southern Grey Shrike and Raven, with Blackbird and Tenerife Blue Tit around the houses at El Portillo. Rock Doves nest in the walls of the caldera but forage partly outside it, while Canaries and sometimes Blue Chaffinches visit Las Cañadas to feed on seeds. Although no birds breed on the highest slopes, Berthelot's Pipit have been seen on the peak, and nests have been found on both Pico Viejo and Teide at up to 2500 m. Other native birds occasionally seen in Las

Male Ring Ouzel *Turdus torquatus* ready to disperse seeds of *Juniperus cedrus*
Photo: Rubén Barone, inset: female cones

Cañadas are Sparrowhawk, Buzzard, Turtle Dove, Long-eared Owl, Hoopoe, Great Spotted Woodpecker, Grey Wagtail, Robin and Sardinian Warbler. The populations of the Red Kite and the Egyptian Vulture that once fed on carrion in Las Cañadas are lost from the island.

Migratory birds are uncommon in the high mountain zone, but there is one intriguing exception. A recent study by local biologists has shown that small numbers of Ring Ouzels *Turdus torquatus*, probably of the British and Scandinavian race that winters mainly in southern Spain and the Atlas Mountains, spend the winter around the rim of Las Cañadas. Analysis of droppings showed that while there, over 90% of their diet consisted of female cones of Cedro Canario *Juniperus cedrus*, an endangered tree endemic to the Canaries and Madeira, although the birds also take some fruits of the rare endemic shrub *Rhamnus integrifolia,* and also invertebrates. Experimentally, germination rates of juniper seeds from droppings of Ring Ouzels were double those of seeds taken directly from ripe cones. In the past Ravens *Corvus corax* are known to have dispersed juniper seeds in the Canaries, but they are now rare in Tenerife. Lizards *Gallotia galloti* also act as dispersers, but only over short distances, so Ring Ouzels are now the most effective long-distance dispersers for this juniper population.

We were particularly intrigued by this work, since in the previous summer we had watched a family of Ring Ouzels – conceivably the birds destined to spend the winter on Tenerife – feeding on berries of *Juniperus communis* near our home in the Southern Uplands of Scotland. It is possible that Ring Ouzels are also crucial to the presence of other woody plants in Las Cañadas. In the 1980s we found Whitebeam *Sorbus aria* growing near the summit of Guajara, and we were recently startled to find Dog Rose *Rosa canina* nearby on the Piedras Amarillas, though it occurs in very few other places on Tenerife. It may be that seeds of these species originally reached Tenerife in the guts of migratory Ring Ouzels from northern Europe.

La Fortaleza trail, Teide National Park

High mountain habitat at about 2000 m, with ancient vegetated lava, a cañada and views of the isolated cliffs and slopes of La Fortaleza.

This trail starts behind the Visitor Centre. The first part is easy walking along a well marked trail through weathered fields of lava and pumice. In winter and early spring the landscape can be somewhat forbidding, with little plant or animal activity, but there are magnificent views across to Montaña Blanca and El Teide. In the distance on your right you will see the tree covered mountain El Cabezón, which is at the extreme upper limit of the pine plantations. In the spring and early summer this walk takes you through a blaze of colour which is enhanced by the sweet, almost sickly, scent of the flowers and the loud hum of insects. The ground is dotted with shrubs, with bare outcrops of dark lava and patches of pale pyroclasts (of the pumice type) in between.

Nearly all the plants you will see are found only in the Canary Islands, and all six of the commonest shrubs of Las Cañadas can be seen along this path. The most conspicuous of them may be two leguminous shrubs; Retama del Teide, appropriately named in Latin *Spartocytisus supranubius* (the broom above the clouds) with a mass of white or pale pink flowers in summer, contrasting with the bright yellow flowers of Codeso de Cumbre *Adenocarpus viscosus*. Also present are two characteristic crucifers – Hierba pajonera *Descurainia bourgeauana* – which forms domed sheep-sized shrubs and has long spikes of yellow flowers, and the beautiful Alhelí del Teide *Erysimum scoparium*, with mauvish pink flowers. You may also see Margarita del

Teide *Argyranthemum teneriffae*, growing in dense clumps and with large white daisy-like flowers. The sixth species is now abundant almost everywhere in Las Cañadas; it is the scabious Rosalito de Cumbre *Pterocephalus lasiospermus,* a clump-forming shrub with long-stemmed pink flowers. Of these six shrubs, four occur only on Tenerife and the other two also on one or two other islands.

After a substantial walk the trail descends onto the flat Cañada de los Guancheros; here you will be immediately aware of the lack of vegetation, except for a few Retama del Teide bushes at the edge and the occasional seedling that germinates in the pumice sand and then dies. This is a very inhospitable environment for most plants and animals, but some invertebrates manage to live here. Under the few rocks that are scattered over the surface you can find a wider variety of animals; there are beetles, centipedes, small cockroaches *Phyllodromica brullei* (endemic to the Canaries), bristletails and also several species of spiders in the family Gnaphosidae. This is a group that is very successful in dry places, and members of many of the species live in silken sacs when resting by day; at night they hunt for prey rather than spinning snares. Most of these cañada animals are nocturnal and avoid the heat of the day by sheltering under rocks that are large enough to remain relatively cool underneath.

Montaña Blanca and El Teide from near Izaña, with large areas of replanted pine forest

Cistus osbeckiifolius

The trail crosses the cañada with the cliffs of La Fortaleza on your right. While the southern section of the wall of the Las Cañadas caldera is more or less complete, all that remains in the north is this isolated cliff of La Fortaleza and the nearby mountain of El Cabezón. These cliffs and the slopes below them are a refuge for some very rare plants, a few of which grow here and nowhere else. Nowadays one is not allowed to climb the cliffs or the slopes below them, but in early summer you may be able to see the pink-flowered, silvery-leaved *Cistus osbeckiifolius*, a rare relative of the common *Cistus* of the pine forests. More conspicuous from a distance are the spectacular red flowers of Tajinaste rojo *Echium wildpretii*, the most famous flower of Tenerife, with stems growing several metres high. Also conspicuous on the slopes is the giant yellow-flowered *Ferula linkii*, and high on the cliff face some *Juniperus cedrus*. The latter is one of the plants that is making a recovery from the heavy grazing by goats and sheep, which was outlawed in this area half a century ago.

A large white-flowered leguminous bush likely to be visible from a distance is Escobón *Chamaecytisus proliferus*, a plant found in the understorey of parts of the pine forest and which also forms the dominant vegetation in some southern areas high on the outer slopes leading up to the rim of the caldera. Smaller plants growing here include the plantain *Plantago webbii* and the attractive *Sideritis soluta*, a sage-like plant with white woolly leaves and spikes of pale yellow flowers.

The yellow-flowered, high mountain thistle *Carlina xeranthemoides* also grows profusely here, as well as two other Canary endemic plants: the pink-flowered campion *Silene nocteolens* and the elegant composite *Tolpis webbii*, with

Ferula linkii and *Juniperus cedrus* on La Fortaleza, 1984

slender branched stems bearing small brilliant yellow flowers. There is also an attractive tall grass *Arrhenatherum calderae* which is almost as restricted in its range.

Berthelot's Pipit is common here, and there is a good chance of seeing Spectacled Warbler, Great Grey Shrike, and coveys of Barbary Partridge. The endemic Blue Chaffinch does not breed in Las Cañadas, but apparently it sometimes makes trips up from the pine forest to feed in the bushes of Retama del Teide. Rabbits are common, sometimes living in holes in the lava rather than in burrows; they are hunted during the autumn to control their numbers.

7 High mountain shrubland – 'matorral de cumbre'

Lava from Las Narices halted by the caldera wall, with colonising Canary Pines (and intruding needles from above)

LAS NARICES DEL TEIDE AND CHAVAO

One of the most recent lava flows on the island (1798) at 2000 m, easily accessible by car; many features of volcanic topography and views of plants on the caldera wall.

Las Narices del Teide – Teide's nostrils – is the name given to the twin vents high on the slopes of Pico Viejo (meaning 'old peak', although it is actually younger than the main cone of Teide itself) and the lava flow that poured out in 1798. This was the most recent eruption in Las Cañadas and one of only about half a dozen on the island in historical times. You are not allowed to walk on the recent lava, but the TF38 road northwest from Boca Tauce crosses the Narices lava flow, and many features of this area can be appreciated from the roadside; there are several small pull-offs and also the Mirador de Chío about 3 km from Boca Tauce. As you start up the road from Boca Tauce you will be able to see where the lava flowed down the flank of Pico Viejo in a broad wandering river, part of which was only halted when it reached the caldera wall to your left.

Below you on the right there is an area of older pahoehoe lava. Here there are Retama del Teide, Codeso de Cumbre, *Argyranthemum*, *Scrophularia*, *Pterocephalus* and a few patches of moss and lichen in cracks. This is a place where you might get a glimpse of a few mouflon (introduced wild sheep) grazing in the distance. Rabbits are common here and you will see droppings which are deposited in piles and are very slow to disintegrate in this climate.

Planted *Bencomia exstipulata* flourishing on a rubble slope below the wall of the caldera

The left hand border of the low area of pahoehoe is overhung by a wall of lava formed by the edge of the 1798 flow. This is unweathered and blackish aa lava, with no vegetation apart from the very occasional plant that has gained a foothold near the edge; it is extraordinarily sharp and unstable. On first glance it would seem improbable that anything lives on it at all, but we showed during our research on the island that this type of recent unvegetated lava actually supports an interesting community of invertebrates.

The paler areas visible in various parts of this flow are called kipukas – a word which comes from Hawai'i – or 'islotes': they are small islands of earlier lava which were surrounded but not covered by the most recent flow; the contrast in the vegetation is immediately obvious. The paler, older, lava has sparse shrubs and smaller plants, while the adjacent fresh dark aa lava or 'malpaís' (meaning badland) has virtually no vegetation; this is because 200 years is not enough for lava at this altitude and in this climate to become weathered and accumulate enough debris to form soil for plants.

From this area you can easily access one of the official trails which starts beside the old building at Boca Tauce, on the Vilaflor road just west of the National Park information booth, and runs along the edge of the caldera below the Roques de Chavao. Part way along you can make a diversion to the left on a path that goes over the rim of the caldera and down into the pine forest that covers the slopes below. Or you can continue on past Montaña del Cedro and rejoin the road part way across the Narices lava. You are not allowed to climb up on the slopes on your left, but you can see many of the typical plants of the caldera wall from the trail, especially in one place where the fresh Narices lava has actually piled up against the cliff and the path climbs up on to it, giving the walker first-hand experience of the extraordinary ruggedness of aa lava.

The walls of the caldera in this area have been colonised by Canary Pines that have gradually advanced over the rim from the southwest; there are now some pines growing in the clumps of Retama in the rubbly area between the Narices lava and the slopes of the caldera wall, and a few pines have even managed to get a foothold on the black lava. The pines and the other plants on the slopes suffered from a wildfire in summer 2012 that swept over the rim and down the slopes, killing most of the foliage on most of the trees, and leaving bare rocky ground between them. The path was closed for some months while the park staff cleared up after the fire. They felled a large number of the charred pines but many of them sprouted new growth from the stumps, while those that remained upright produced new shoots on the trunks.

The main site used in the restoration programme for the endangered shrub *Bencomia exstipulata* is on the lower slopes of the caldera wall here, and it had an extremely lucky escape from the fire. Young *Bencomia* seedlings have also been planted on the slopes, protected from rabbits and mouflon by steel mesh cages. Where the fire passed over the rocky slopes almost all the above-ground vegetation was consumed, but several of the native plants such as *Nepeta teydea* and *Pimpinella cumbrae* soon sprouted again from deep roots unharmed by the fire.

Birds are fairly scarce in this area, although the Canary Chiffchaff, Tenerife Blue Tit and Berthelot's Pipit are all likely to put in an appearance. Tenerife Lizards are also common and are preyed upon by Kestrels which you are likely to see hovering overhead. In summer the most conspicuous insect here is the Canary Blue butterfly, but one can also see ant-lions *Myrmeleon alternans* – members of a group that is well represented in the Canaries but much less so in northwest Europe. Ant-lion adults are rather like dragonflies – with two pairs of large wings and long thin bodies – but their flight is weaker and they have conspicuous stout antennae.

Young planted
Bencomia exstipulata
with protective cage

Pimpinella cumbrae
with roots exposed but
not killed by fire

Tonática
Nepeta teydea

Los Roques de García

Eroded remnants of an ancient volcano, with adjacent vegetated lava flows; high mountain vegetation; view down into a large cañada – the Llano de Ucanca; dramatic rock formations separating upper and lower segments of the caldera; pāhoehoe and aā lava.

Opposite the Parador towards the south of the park lie the conspicuous and much photographed Roques de García. These are the remains of a rock wall separating the two major parts of the Las Cañadas caldera, which are at different levels. They consist of a row of spectacularly eroded ancient rocks stretching from Guajara in the southern wall of Las Cañadas to the lower slopes of Teide. The rocks are part of the core of the Las Cañadas Volcano, predecessor to Teide and Pico Viejo, exposed by the subsidences and giant landslide that formed the caldera (see Chapter 13). The weird shapes of some of them are due to differential erosion, since the rocks are composed of many different kinds of material, with differing resistance to erosion.

From near the section of the parking place closest to the rocks you can look down into the Llano de Ucanca, which is part of the remains of the southwestern caldera, about 200 m lower than the eastern one. The lower caldera has been largely filled with lava and its floor is now covered with fine volcanic sand produced by thousands of years of erosion from the surrounding mountains. You can see how the tips of the tongues of dark lava, which flowed down the slope on your far right, are now submerged under accumulated sand. When the snow melts in the spring, the whole of the lowest part of the cañada can become a lake for several days. There are many clumps of Retama del Teide *Spartocytisus supranubius* in the central part, and *Pterocephalus lasiospermus,* a scabious with mauve flowers, grows around the edges, but much of the cañada is bare.

Below you on the rubble slopes of Los Roques and elsewhere grows the famous Taginaste rojo *Echium wildpretii,* forming red columns three metres high when in flower in midsummer, but remaining as white skeletons for much of the year. There are also clumps of Retama and of Alhelí *Erysimum scoparium,* the brilliant purple wallflower of Las Cañadas.

If you walk along the broad sandy track east of Los Roques you will find many of the same plants. The track passes through a cañada where there are numerous stumps of felled trees. This was one of the main places in the caldera that were planted in 1954-55 with non-native Monterey Pine *Pinus radiata*, which were subsequently removed. If you continue you will pass through a fairly well vegetated gully, where there is a chance of seeing Southern Grey Shrike and the ubiquitous Berthelot's Pipit and Canary Blue butterfly, and possibly some of the other insects described in the visit to La Fortaleza.

As you approach the northern end of Los Roques you come to an area of pāhoehoe lava which contrasts with nearby āa lava. This is a good opportunity to compare the two main types of lava: the aā is chaotic and sharp while the pāhoehoe – which was hotter and more liquid when erupted from the volcanic vent – is much smoother, either fairly flat on the surface or ropy. Some way further up the slope from Los Roques there is the mouth of a lava tube through which pāhoehoe lava once flowed. If you continue along, with the last of Los Roques on your left, you will find a fine look-out point over the Llano de Ucanca. You can see where the nearest part of the pāhoehoe flow stopped right at the edge of the cliff, while in contrast about 75 m further back the lava flowed over the lip and spread out below.

Alhelí *Erysimum scoparium* near Los Azulejos, with Taginaste rojo just coming into flower

Mouth of the Cueva de Los Roques, 1984

From any high point near Los Roques you can get a good view of much of Las Cañadas and its vegetation. In the opposite direction you can see across to Guajara, which at 2718 m is the highest mountain in the southern rim of Las Cañadas, exactly 1000 m lower than the summit of El Teide.

The dominant plant between you and Guajara is the white-flowered Retama del Teide; it even grows on the face of the mountain wherever it can get a roothold. Between you and the Parador there are also large clumps of *Pterocephalus*, a shrub that was rare in the middle of the 20th century – presumably because of grazing by goats and sheep – which has now made a spectacular recovery.

Near the road you can see *Descurainia bourgeauana* which for most of the year has straw-coloured dead flower stems that give it a characteristic cushion-like look; in early summer it has brilliant yellow flowers. Less common here is Codeso de Cumbre *Adenocarpus viscosus*, with yellow pea flowers. which also forms hummocks. Immediately around you grow local species of *Tolpis, Plantago, Argyranthemum, Scrophularia* and *Lotus*, with the rare shrub *Rhamnus integrifolia* in crevices.

The Piedras amarillas appear hostile to plants, but both *Rosa canina* (just visible in centre) and *Viola cheiranthifolia* manage to get a foothold in the crevices

Los Azulejos and Guajara, the highest point in the wall of Las Cañadas caldera

The road from Los Roques to Boca Tauce passes through Los Azulejos, blue-green crumbly formations produced by the action of superheated steam on silicates in the rocks, while below Guajara are the Piedras Amarillas, which glow in an extraordinary shade of golden red when floodlit by the evening sun.

Myrtle servicing invertebrate traps on historic lava in 1984

8

LAVA FLOWS, CINDERS AND CAVES

Most visitors to Tenerife notice and admire the extraordinary barren landscapes of lava and cinders created by successive volcanic eruptions, but few people are aware of the underground volcanic habitats. Below the surface in many parts of the island there is an extraordinary network of subterranean spaces ranging from deep volcanic pits and lava tubes big enough to walk through, down to a maze of more or less connected tiny cracks and crevices, near the surface and deep underground. Investigations by local scientists in the last few decades have shown that the subterranean cavities hidden beneath the raw volcanic terrain of Tenerife provide living space for an array of specialised invertebrate animals highly adapted to conditions in these inhospitable places.

When an eruption occurs, hot lava, cinders and gases kill all life in the immediate vicinity and produce a sterile environment. However, the existing topography inevitably affects the progress of a lava flow and there may be small areas (often higher than surrounding ground) which are not overwhelmed by lava, so that some vegetation and animals survive. These areas are known in the Canaries as 'islotes' and examples are easy to find around El Teide. Similar contrasts between raw lava and older vegetated volcanic materials can be seen along the edges of the recent lava flows.

Fresh lava from the Narices del Teide eruption of 1798 overlying an older flow that has been colonised by plants and by the invertebrates that depend on them

Plant colonisation of barren lava flows or expanses of cinders is a slow process – sometimes taking centuries – depending on the climate and on the structure of the volcanic rock, which eventually breaks down to form soil. The wind blows in spores of lichens, mosses and ferns, and seeds of other plants, and a few of these eventually succeed in growing, so that a diverse community is gradually established. However, it has now been shown that plants are not the only early colonists, and that recent lava and cinders support a small community of invertebrate animals even when they appear completely barren and inhospitable to life.

We began investigating the animals on barren lava flows while visiting La Laguna University in the 1980s. At that time very few faunal studies of these habitats had been done anywhere, but we had been intrigued by the discovery of a cricket living on new lava flows on Kilauea Volcano in Hawai'i, and by reports that this cricket disappeared from the flows as they aged and were colonised by plants. We soon found that in the

Anataelia canariensis (male), a typical 'lavicole' earwig, adapted to life on barren lava. Photo: Pedro Oromí

Cueva Sobrado, a lava tube in which many specialised endemic invertebrates have been found. Photo: Ramón Oromí

Canaries there were many invertebrates living on barren 'historic' lava (produced during the last half millennium in historically documented eruptions) and we showed that as in the case of the cricket in Hawai'i, several of the species concerned were absent from vegetated older lava. Evidently there was a process of 'faunal succession' on lava flows, with the pioneers or 'lavicoles' being replaced by a new set of animals as a plant community became established.

We also came to realise that there is a network of spaces penetrating the lava and cinders, and providing cool and moist refuges for invertebrates during the heat of the day. Furthermore, there are connections from the surface cracks to subterranean habitats, so it was not too surprising to learn that some of the invertebrates that we recorded on the recent lava, or close relatives of them, were being found in lava tubes by our entomologist colleagues in La Laguna University, who were studying the biology of underground environments in the Canaries. We later carried out joint research with them, both above and below ground,

Loboptera troglobia, a cockroach highly adapted to underground life
Photo: Pedro Oromí

Visitors on a guided tour of Cueva del Viento

A working visit to Cueva de San Marcos in the 1980s

Coaxing a cave invertebrate into a collection tube. Photo: Pedro Oromí

in other parts of the Canaries archipelago and on more distant volcanic islands.

One can now learn about volcanic caves at a modern visitor centre near La Cueva del Viento, high up above Icod de los Vinos, and go on a guided tour of part of the lava tube, which has the most diverse invertebrate fauna of any cave system in the Canary Islands. It has also yielded bones of extinct endemic vertebrates such as the Tenerife Giant Rat *Canariomys bravoi,* the Tenerife Giant Lizard *Gallotia goliath* and the recently discovered Long-legged Bunting *Emberiza alcoveri* and Slender-billed Greenfinch *Carduelis aurelioi* (see Chapter 1).

Volcanic caves are of two main types. Volcanic pits are either deep cracks in the lava, or cavities left when magma has sunk back down a volcanic vent at the end of an eruption. Lava tubes are much more common on the Canaries and can be found at all levels, from high on Teide at over 3200 m, down to sea level at Playa de San Marcos below Icod. These tubes were formed at the time of volcanic eruptions that produced lava flows of the relatively liquid 'pāhoehoe' type and are anything from a few metres to several kilometres long; they often lie within a few metres of the surface. La Cueva del Viento, with a total mapped length of more than 17 km, is probably among the five longest volcanic cave complexes in the world, in the same league as the famous Kazumura Cave in Hawai'i and Leviathan Cave in Kenya.

If one considers the barrenness of the recent lava and the dark caves and crevices below it, but knowing that invertebrates are living there, an obvious question arises: What do these animals eat? There are no green plants in the dark underground, and none on the most recent lava flows. It turns out that these pioneer animals are scavengers and predators rather than herbivores; they depend for their food on insects and other minute animals (both living and dead) and plant debris from richer habitats, flying or carried by the wind and deposited on the lava; that this airborne debris is significant can be demonstrated by leaving a tray of soapy water in the open on a barren area and collecting the contents after a while.

On Tenerife we found that the animals living on barren historic lava in or near Las Cañadas comprised mainly 'bristletails', primitive and flightless scavenging insects in the family Lepismatidae; other animals in this habitat included harvestmen and spiders (predators), earwigs (omnivores) and crickets (scavengers). On older rocks nearby, where plants had become established, the fauna was more diverse and included many invertebrates dependent on green plants for their food, and the pioneering lavicoles had almost entirely disappeared. The pioneers presumably lose out as a result of competition with newly established species, and we suspect that the arrival of ants (Formicidae) – which are potential competitors and also predators – may be especially problematic for them.

Invertebrates in the cracks and cavities within recent and barren lava have even less access to food than the pioneers on the surface. Members of some species may depend mainly on the sparse supply of debris that filters down, but others routinely move up at night through cracks to feed on the surface, and then retreat during the day; many of the same species can be found in shallow lava tubes, especially near the entrances. These animals are not obviously specialised for cave life, though they need to be able to find their way around in the dark, and they are termed 'troglophiles' (cave-lovers).

As the centuries pass, the surface cracks and crevices in lava and cinder fields become blocked by dusty material and debris blown in or washed in by rain. These deposits provide footholds for plants, but also hinder free passage of animals from below, so that nightly commuting from subterranean cavities to the surface becomes more difficult. At this stage, therefore, the animal community on the surface begins to diverge from that underground. The surface is colonised by herbivorous insects eating green plants and dead plant debris, and the slow establishment of a complex terrestrial ecosystem gets under way.

As soil develops, the blocking of the connections between the surface and the deeper cavities reduces air movements and gradually leads to conditions more typical of a deep cave, with stable temperature (close to the mean temperature above ground) and high humidity, as well as perpetual darkness and a much reduced food supply. The gradual change does not suit the pioneering subterranean animals, and sets the stage for a second – and much more gradual – replacement of the pioneers by specialised cave animals adapted to the unusual conditions. The specialised animals presumably colonise the caves from more ancient lava underlying or adjacent to the more recent flow; this is likely to be feasible since successive eruptions often occur in the same general area.

The new colonists – whose ancestors have lived permanently in caves and subterranean spaces for hundreds of thousands of years – have evolved to be very different from their surface-living ancestors. They tend to have elongated legs and highly sensitive antennae, useful in the dark, and they conserve energy and resources by reducing or losing their eyes, reducing their pigmentation (often becoming whitish) and moving very slowly. Evolutionary changes also occur in the reproductive system, since the shortage of food prevents the production of large numbers of young. As the biologists at La Laguna were showing in the 1980s, insects such as cockroaches living permanently in deep caves develop only a few large eggs at a time, thus putting all their resources into a few large young and giving them a good start in life.

The Mesocavernous Shallow Substratum

An interesting and special cave-dwelling fauna can be found in many places on Tenerife, including spiders, pseudoscorpions, woodlice, millipedes, centipedes, cockroaches, planthoppers and beetles. As elsewhere, these animals have special adaptations such as eyelessness, depigmentation of the integument, elongation of appendages and resistance to starvation, collectively known as the troglomorphic syndrome. Specialised cave-adapted species can only live underground and are unable to survive above the surface even for a short time, so the species occurring on an oceanic island are always endemics that have evolved locally from surface-living ancestors. However, it is found that widely separated caves within Tenerife frequently harbour the same species, which means that they have spread through underground habitats other than caves. One of these habitats is made up of the 'mesocaverns', a network of cracks, voids and other small spaces within the deep parent rock, with environmental conditions similar to those in 'human-size' caves ('macrocaverns').

Animals in the MSS are captured in traps comprising a perforated cylinder isolated from the surface and left in place permanently, and an inner bait container that is removable for servicing. Diagram by Heriberto López

However, a shallower underground habitat is also available to these species: the so called mesocavernous shallow substratum (MSS). A type of MSS also found in continental, non-volcanic terrain is the debris (colluvium) accumulated at the base of rocky plugs and cliffs. As soon as this debris is covered by a thin soil it constitutes a suitable habitat for cave-dwelling species (troglobionts). In Tenerife the inhabited colluvial MSS is found only in geologically old areas such as Teno and Anaga, mainly in scattered localities in the laurel forest. In such places this MSS is the only available underground habitat, since there are no lava tubes and the deep mesocaverns are usually silted up with clay due to prolonged erosion of the old lavas.

In the central part of the island, which is younger and with abundant lava flows on its surface, there is another type of MSS, exclusive to volcanic terrain and discovered in the Canaries (El Hierro and Tenerife). This is known as 'volcanic MSS' and consists of the superficial clinker of recent to medium-aged basaltic lavas that has been covered by a layer

Colluvial MSS site in an ancient part of the island

Volcanic MSS formed by layers of lava and clinker

An MSS trap in position

Volcanic MSS in dry pine forest with a thin layer of soil and pine needles over clinker

of volcanic cinders, or more often by soil formed on the surface over hundreds or a few thousand years. The interconnected spaces of the clinker are both more abundant and larger than in colluvia, and the underground habitat is rapidly established without any transformation of the rock structure. This MSS is extremely widespread over most of the island where basaltic lavas are dominant, being absent only in areas of trachytic lavas and pumitic tuffs which have fewer and unconnected spaces. It is particularly abundant on the northern slope of the island, mostly at intermediate elevations in the pine forest, allowing the dispersal of troglobionts over an extensive area.

After more than 30 years of active research, both in the lava tubes and the MSS, performed by the Grupo de Investigaciones Entomológicas de Tenerife (GIET) team at the University of La Laguna, 74 troglobiont species have been found on the island. Of these, 17 species occur both in caves and MSS, 42 are known strictly from caves and 15 have been collected only in the MSS, especially in colluvial MSS from Anaga and Teno.

It is worth pointing out that in general the species occurring in lava tubes and volcanic MSS are highly adapted for living underground (eyeless, with longer appendages), probably as a response to selective pressures that have forced them to adapt to an environment with very scarce organic matter. In contrast, the colluvial MSS is older and richer in organic matter, and the troglobiont species living there are usually less troglomorphic (i.e. still retaining tiny eyes, some pigmentation and less elongate appendages) in spite of having had longer to evolve adaptations to their underground environment.

Text and photographs: Pedro Oromí

These evolutionary changes fit the subterranean animals for their harsh environment, but naturally make them unsuited to life on the surface, where good vision, fast movement and the ability to cope with strong solar radiation, high temperatures and low humidity are all essential. These animals have thus become obligate cave dwellers, and are termed 'troglobionts'.

Lava tubes often run close to the surface, and may have several large entrances. As soil develops and gets down into the cracks, plant roots may penetrate shallow lava tubes and smaller subterranean spaces, providing a new resource. Caves with roots can support animals including sap-sucking planthoppers with varying degrees of adaptation to caves, and the presence of these homopteran bugs provides extra opportunities for predators and scavengers, as well as for fungi. Furthermore, the animal community in the shallow parts of the cave may include many unspecialised animals (the troglophiles) that can still come and go, though perhaps infrequently, so that a single species will include both surface-living and cave-living individuals; these troglophiles are often widely distributed in the archipelago and adjacent continental land.

In contrast, the deeper parts of lava tubes, and especially the cracks and crevices that still exist within the deeper parts of the lava, provide homes for the more specialised troglobionts. In the Cueva del Viento, which has segments of tube at different levels, it has been shown that the parts of the complex that lie deepest in the lava have the least diverse biological communities. Troglophiles are almost absent, and even the most efficient troglobionts are rare, evidently finding it hard to make a living in an environment where food is so extremely scarce.

Since these specialised animals cannot survive outside, and are isolated on oceanic islands, they must logically have evolved into troglobionts separately on each island. Their ancestors must have been surface-living, but with troglophilic tendencies, and at some stage a population must have become sufficiently isolated underground to start along the long evolutionary pathway that creates a specialised troglobiont. As implied above, the isolation results from soil development and the blocking of connections between the lava surface and the spaces below it, so that those members of a species that are living underground can no longer easily breed with individuals on the surface.

This situation can set in motion the long-term evolutionary divergence of a subterranean population of a species from the population on the surface, and ultimately the origin of a distinct troglobiontic species. Scarcity of food, perpetual darkness, high humidity and equable temperature all impose strong pressures of natural selection on the animals isolated underground. Genetic change in response to these pressures is no longer inhibited by continual interchange of genes with surface-living individuals subject to a different selective regime, and the small size of the cave populations facilitates rapid evolution.

This evolutionary scenario is doubtless rare, but millions of years have been available. The result that we see today is that each ancient island has a large 'pool' of endemic troglobiontic animals, which in principle may find their way into systems of cracks or deep caves anywhere on the island. In practice, some crevice systems are easier to penetrate than others, so that the diversity found within them depends not only on their suitability for troglobionts but also on the ease of access from other underground habitats, including connected systems of crevices as well as caves. The long term result, well studied in the Canary Islands, is the development of subterranean communities that include many troglophilic and accidental species near the entrances and in shallow tubes, and a smaller but still substantial number of rare troglobiontic species in the deeper parts.

In the Cueva del Viento 85 species of invertebrates have been recorded. Almost two thirds of these are not obligate cave dwellers; they include pseudoscorpions, spiders, woodlice, millipedes, centipedes, springtails, earwigs, beetles and flies. The remaining 31 species are endemic troglobionts, species found only in Tenerife caves and nowhere else in the world. They include many different kinds of insects and other invertebrates: pseudoscorpions (2), spiders (9), woodlice (3), centipedes (1), millipedes (3), cockroaches (2), homopteran planthoppers (1) and beetles (10). Some of those most highly adapted to the cave environment are the eyeless cockroach *Loboptera subterranea*, the eyeless spider *Troglohyphantes oromii* (family Linyphiidae) and the staphylinid beetles *Domene vulcanica* and *Alevonota outereloi*. The nearby Cueva de Felipe Reventón has even more known kinds of troglobiont (36) and the tally for both caves will doubtless go on increasing as speleological research continues.

Because Tenerife is one island in a volcanic archipelago, it is clearly likely that related animals on the other islands – some of them with comparable lava flows and caves – have been evolving in a similar way, by a process known as parallel evolution. It can produce specialized troglobionts on separate islands, evolved independently from surface-living ancestors but superficially very similar. Comparison of related invertebrates on different islands suggests that the usual scenario has been colonisation of the archipelago by one (or more) continental surface-living stock, which then evolves into distinct endemic surface-living species on the different islands. In some groups it is

Troglohyphantes oromii female
Photo: Pedro Oromí

Cave woodlouse *Venezillo tenerifensis*
Photo: Pedro Oromí

Domene vulcanica, a staphylinid cave beetle. Photo: Pedro Oromí

also common for several endemic species to evolve on a single island, as in the case of spiders in the genus *Dysdera*.

At the same time any surface living species with troglophilic tendencies will colonise subterranean habitats, and in some cases these will give rise to distinct troglobiontic species, as described above. In the extreme case of *Dysdera*, there are no less than 10 different troglobiontic species on Tenerife. Eight of these have a different 'sister species' (closest relative) on the surface, while the other two appear to have split into distinct species underground, and share a close relative on the surface. Disentangling such complicated evolutionary scenarios is only possible using modern molecular techniques, which can often confirm the independent evolution of the troglobiontic species.

In the case of *Dysdera* and many other troglobionts, the surface living relatives survive, but in other cases there is no obvious close relative on the surface: presumably the original colonising stock has become extinct on the surface after part of the population had colonised the underground and eventually evolved into a troglobiont. Examples are found in the staphylinid beetles of the genus *Domene* and the cockroaches in the genus *Loboptera*.

Even this short account of the rich community of invertebrates in the caves of Tenerife should give an idea of the intensity of the evolutionary ferment that characterises the fauna and flora of oceanic islands, and which is manifested below ground as well as in the open air. The cave biologists in the Canaries and their associates elsewhere have made the subterranean fauna of the islands into one of the leading illustrations of the way in which evolution operates and the origin of species occurs. Darwin would have been enthralled!

Dysdera gollumi, a specialized cave spider with long legs and pale body
Photo: Pedro Oromí

Dysdera levipes, the closest surface living relative of *Dysdera gollumi*
Photo: Pedro Oromí

Stream in the Barranco del Infierno, a good place for invertebrates;
on the right is a small-scale example of the wet rock face habitat.

9

FRESHWATER HABITATS

Natural freshwater habitats are hard to find on Tenerife, and we treat them only briefly here. Most of the streams in the 'barrancos' (ravines) on the island are now dry except during periods of heavy rainfall, but in many cases stagnant pools are left behind along the course of the stream. The flow of water in many streams was more reliable in the past. Long term data are lacking for Tenerife, but on Gran Canaria there were 285 permanently running streams in 1933, and 265 of them were drying up seasonally by 1973. On Tenerife, where the decline may have been similar, there were fewer than 10 permanent streams in 1993. Human water use is the main cause of the decline, and on Tenerife most water is now intercepted by dams and the network of subterranean 'galerías' (tunnels) and carried by concrete 'canales' (channels) and steel pipes to the agricultural areas, villages and towns. In addition, there is the effect of deforestation, which both reduces the interception of cloud moisture by foliage and speeds up run-off, thus increasing the risk that barrancos will dry out in summer.

The permanency of running water is crucial for survival of many kinds of animals. The Barranco de Tamadite (Bco. de Afur) probably still supports a population of European Eels *Anguilla anguilla*, which spend several years maturing in freshwater habitats before migrating back to the Sargasso Sea to breed; at least until recently there were other populations elsewhere on the island. One other fish, the Rock Goby *Gobius paganellus*, can spend some time in brackish or fresh water, but no fully freshwater fish colonised Tenerife naturally.

Running streams are easily the most important of the freshwater habitats for invertebrate animals. In a study of freshwater 'macroinvertebrates' two decades ago, 88 species were found in streams; 17 of these were absent from other freshwater habitats. In the laurel forest there are now only two surviving permanent streams, although there were probably many when this forest type had its fullest extent. One of the two streams is in the Barranco de Ijuana; it is small and the fauna is less diverse than in permanent streams of other habitats, but it has several species not found in those, including a recently discovered caddisfly (Trichoptera). One of the richest permanent streams outside the laurel forest is the Barranco del Infierno, where 47 kinds of invertebrates were found, including seven that occurred hardly anywhere else.

Barranco de Ijuana, with one of only two surviving permanent streams in the laurel forest

The concrete channels that follow the contours on so many mountainsides on Tenerife provide a new artificial habitat with running water, but these typically have only half as many kinds of invertebrates as natural streams, which have all of the species in the channels, plus many others. A more natural but less obvious freshwater habitat is the film of water that can be found on rock faces in some of of the barrancos. Only a few kinds of invertebrates live in these, but three Canary endemic species – the beetles *Hydroporus pilosus* and *Limnebius punctatus*, and the fly *Satchelliella binunciolata* (Psychodidae) – occur only in this habitat; since moist rock surfaces are probably becoming more scarce, these species are clearly vulnerable.

Stripeless Tree Frog *Hyla meridionalis* often occurs in moist places away from water

Trithemis arteriosa, a dragonfly typical of stream habitats

Even though water does not flow continually in most of the barrancos, stagnant pools are often left behind if continuous flow ceases at the end of a wet period. Many of these pools become dry in the height of summer, and on average they have only half of the invertebrates found in the permanent streams. Some of the animals concerned, such as the abundant blackfly *Simulium ruficorne,* have drought-resistant life stages that enable them to survive in temporary habitats. Other species may be present in these habitats mainly because they are excellent short-distance dispersers. This may apply to many water beetles, which are very diverse on the island; the family Dytiscidae has no less than 18 species (6 endemic to the Canaries) and the superfamily Hydrophiloidea has 25 (6 endemic).

Less diverse but more noticeable are the whirligig beetles in the family Gyrinidae, which form ever-moving groups on the water surface. Also on the surface one can often see some of the several kinds of pond skaters; these are predatory heteropteran bugs in seven different families that share the ability to walk on the water, having water-repellent hairs that prevent them getting wet. Other bugs often visible in pools are water boatmen (Corixidae) and backswimmers (Notonectidae) both with several species on the island. Many species of mayflies (Ephemeroptera), caddisflies (Trichoptera) and two-winged flies (Diptera) also breed in the streams and pools.

Around many freshwater habitats the most conspicuous invertebrates are often dragonflies (order Odonata). About ten species can be seen on the island; all are thought to have arrived naturally, but none are endemic to the Canaries, probably because these strong-flying insects are excellent dispersers, so that gene flow between populations on continents and various islands tends to prevent them from evolving differences in different places. The most impressive species is the Emperor Dragonfly *Anax imperator*, which has a broad distribution elsewhere; it has bright blue males and green-bodied females, and can often be seen even in built-up areas. The African species *Trithemis arteriosa* is usually found near permanent streams; the male is bright red, but the female is strongly marked with yellow, black and white, with red wings. The Scarlet Darter *Crocothemis erythraea* is another common species; the males are bright red as the name suggests, while the females are less frequently seen and have bronze-yellow bodies. There are

Wet rock face habitat with the fern *Adiantum capillus-veneris*

Pool on the Barranco de Igueste, with a species of cattail (*Typha*)

also two species of bluetail damselflies on Tenerife, *Ischnura saharensis* and *I. senegalensis*; they cannot be separated in the field, but are much smaller and more slender than any of the dragonflies; as the name suggests, the abdomen is entirely blue. These damselflies can breed in brackish water and can be seen at the salty pools on waste ground near the road into Las Galletas.

Pools are also used for breeding by the introduced frogs, but not all of the tadpoles will reach metamorphosis before their pools dry up. Both of the frogs, Perez's Frog *Pelophylax perezi* and Stripeless Tree Frog *Hyla meridionalis*, occur in barrancos and moist places in many parts of the island, but Perez's Frog is more frequent in water tanks. Tanks of all sizes are very common on Tenerife, especially in the lower areas, being used for the irrigation of farmland; there are also ponds and water gardens in parks, private gardens and golf courses. However, most tanks and small pools are not very rich in invertebrates, partly because they so often contain introduced fish, especially Guppy *Poecilia reticulata* and Western Mosquitofish *Gambusia affinis*, which have been introduced to help control mosquito larvae; the mosquitofish gives birth to live young and survives in polluted or brackish waters and at high temperatures, and has become very common on the island. Large ponds and some reservoirs contain Common Carp *Cyprinus carpio*, which have been introduced for sport fishing, and also Goldfish *Carassius auratus*. A more serious threat to native species is from the invasive Red Swamp Crayfish *Procambarus clarkii* from North America, which has now reached both Tenerife and Gran Canaria.

Streamside and poolside vegetation often includes the Canary Willow *Salix canariensis* and the introduced reed *Arundo donax*, which occurs in the lower parts of almost all barrancos. There are also several species of native and introduced sedges and rushes in the genera *Cyperus* and *Juncus*; in a few places one can also find the Southern Cattail or Cumbungi, *Typha domingensis*, which may be native. Moist barrancos are often infested with the invasive composite *Ageratina adenophora*, and introduced species of *Polygonum* can also be seen, as well as the elegant Maidenhair Fern *Adiantum capillus-veneris*.

In a few places there are relatively large permanent ponds, and although these are unnatural habitats they are especially important for birds. Just to the west of Bajamar are several ponds, where Moorhens breed and a surprising variety of birds can be seen on migration. Better known are the 'charcos' near Erjos del Tanque on the road north from Santiago del Teide; these are artificial ponds in excavations and some of them dry out in periods of drought, but they are heavily used by waterbirds such as Coot *Fulica atra*, Moorhens *Gallinula chloropus*, Grey Herons *Ardea cinerea* and many migratory species. Access is by a track starting opposite a red bus shelter close to Km16, just south of Erjos village.

Erjos ponds in mist, with Canary Willow *Salix canariensis*

10

Birds, mammals and other vertebrates

BIRDS

This chapter includes all the native birds that breed (or are known to have bred in the past) on Tenerife, together with those that fairly commonly pass through on migration or spend the winter on the island. More comprehensive lists of species recorded from the island can be found in Martín & Lorenzo (2001), Lorenzo (2007), Clarke & Collins (1996) and Rodríguez et al. (2014), which have been the source of much of the material given here. We are especially indebted to Aurelio Martín Hidalgo and Keith Emmerson for sharing with us their extensive knowledge of the birds of Tenerife.

The Orders and Families of birds are arranged alphabetically, not following a supposedly 'taxonomic' sequence, and within families the genera and species are also alphabetical. Latin names for breeding species follow the 2009 edition of the Canary Islands Government's *Lista de especies silvestres de Canarias* in almost all cases. Authors of scientific names are not given here, but are available in the official list. Scientific names for non-breeding species, along with English and Spanish names, are those used by Martín & Lorenzo (2001), which generally conform to *The EBBC Atlas of European Breeding Birds* (1997). Subspecies are mentioned if they are endemic, but we do not use separate English names for endemic subspecies.

We do not in general describe the bird species, since pictures and descriptions are easily accessible in bird books and on websites. However, for those birds that are likely to be unfamiliar to visitors from northwest Europe, we give sufficient detail to separate them from other resident birds. For breeding species, we indicate in parenthesis the islands in the Canaries on which the species breeds, using the abbreviations L (Lanzarote and the islets north of it), F (Fuerteventura), C (Gran Canaria), T (Tenerife), G (La Gomera), H (El Hierro) and P (La Palma).

As is usual on islands, the number of bird species in any one habitat is lower than would be found in a comparable mainland habitat. The Canary Islands are on the southern extremity

ENDEMICITY OF TENERIFE BREEDING BIRDS

This table lists the breeding birds of Tenerife that are special to the island, or to the Canaries, or to Macaronesia (North Atlantic islands including the Canaries, Salvages, Madeira, Azores, Cape Verdes, along with NW Morocco) and do not breed naturally anywhere else. However, it is not set in stone, as scientists often differ over the taxonomic rank of island forms; please see the systematic list for more information.

Endemic to Tenerife	Endemic to Canaries	Endemic to Macaronesia
SPECIES		
Long-legged Bunting *Emberiza alcoveri* †[1]	**Bolle's Pigeon** *Columba bollii*	**Plain Swift** *Apus unicolor*
Slender-billed Greenfinch *Carduelis aurelioi* †[1]	**White-tailed Laurel Pigeon** *Columba junoniae*	**Canary** *Serinus canaria*
	Blue Chaffinch *Fringilla teydea* [2]	**Berthelot's Pipit** *Anthus berthelotii* [4]
	Tenerife Blue Tit *Cyanistes teneriffae* [3]	
	Canary Chiffchaff *Phylloscopus canariensis*	
SUBSPECIES		
Lesser Short-toed Lark *Calandrella rufescens rufescens* †[1]	**Egyptian Vulture** *Neophron percnopterus majorensis* †[1]	**Sparrowhawk** *Accipiter nisus granti*
Blue Chaffinch *Fringilla teydea teydea* [2]	**Stone Curlew** *Burhinus oedicnemus distinctus*	**Buzzard** *Buteo buteo insularum*
Great-spotted Woodpecker *Dendrocopos major canariensis*	**Raven** *Corvus corax canariensis*	**Kestrel** *Falco tinnunculus canariensis*
	Trumpeter Finch *Bucanetes githagineus amantum* †[5]	**Yellow-legged Gull** *Larus michahellis atlantis* [6]
	Linnet *Carduelis cannabina meadewaldoi*	**Houbara Bustard** *Chlamydotis undulata fuertaventurae* †[1]
	Chaffinch *Fringilla coelebs canariensis*	**Blackcap** *Sylvia atricapilla heineken*[7]
	Southern Grey Shrike *Lanius meridionalis koenigi*	**Spectacled Warbler** *Sylvia conspicillata orbitalis*
	Berthelot's Pipit *Anthus berthelotii berthelotii* [4]	**Blackbird** *Turdus merula cabrerae*
	Grey Wagtail *Motacilla cinerea canariensis*	
	Goldcrest *Regulus regulus teneriffae*	
	Sardinian Warbler *Sylvia melanocephala leucogastra*	
	Robin *Erithacus rubecula superbus*	
	Long-eared Owl *Asio otus canariensis*	

10 Birds, mammals and other vertebrates

> **NOTES**
> 1. These birds – marked with the symbol † – are now extinct on Tenerife.
> 2. The Blue Chaffinch is a species endemic to the Canaries; the local subspecies is endemic to Tenerife.
> 3. The Tenerife Blue Tit is here treated as a species endemic to the Canaries, with the local subspecies endemic to Tenerife and La Gomera; see systematic list.
> 4. Berthelot's Pipit is a species endemic to Macaronesia; the local subspecies is endemic to the Canaries.
> 5. The Trumpeter Finch is probably now extinct on Tenerife.
> 6. This subspecies of the Yellow-legged Gull is sometimes considered endemic to Macaronesia, but some authors include Iberian and Moroccan populations within it.
> 7. This subspecies of the Blackcap may occur elsewhere; see systematic list.

of the Palaearctic zoogeographic region (which includes Europe, northern Asia and the Mediterranean part of North Africa) and the breeding birds nearly all belong to the Palaearctic avifauna, rather than the more southern Ethiopian avifauna associated with the rest of Africa.

About 52 native bird species have been breeding regularly on Tenerife recently, but a few of these (Kentish Plover, Raven, Rock Sparrow and Manx Shearwater) are reduced to dangerously low numbers. Furthermore, the Cream-coloured Courser now breeds only rarely, and the Trumpeter Finch may already be lost as a regular breeding bird; three other native species – Red Kite, Egyptian Vulture and Lesser Short-toed Lark – have certainly been lost in recent decades. (The extinct all-black Canary Islands Oystercatcher *Haematopus meadewaldoi* may never have had a breeding population on Tenerife.) On the positive side, several species of land-birds and water-birds may now have small breeding populations or be in the process of establishing them by natural colonisation (eg Common Swift, Laughing Dove, Ruddy Shelduck, Night Heron, Little Bittern and Little Egret).

The accompanying table shows that a significant proportion of the breeding birds of Tenerife are endemic, in the sense that the species or subspecies are confined to Tenerife or islands in the Northeast Atlantic, rather than being widespread continental forms. At the species level, Tenerife now has no endemic birds apart from the extinct Long-legged Bunting and Slender-billed Greenfinch. However, five species still breeding on Tenerife are endemic to the Canaries archipelago and three more are endemic to Macaronesia. At the subspecies level, where taxonomists' opinions often differ) Tenerife is currently considered to have four endemic subspecies (plus one lost recently and one at an earlier time). The island also hosts 10 subspecies (plus one certainly and another probably lost) endemic to the Canaries archipelago, and another seven (plus another now lost) endemic to Macaronesia.

In addition to the native species, there have been many introductions of birds to the island. We have listed those that may have established – or be establishing – self-sustaining populations; they fall into two groups. First, there are several game birds introduced for sporting purposes; one of these (Barbary Partridge) is now well established in various parts of the island, and two others sometimes breed in the wild. Second, there are domestic caged birds, or birds in zoos, which have escaped from captivity. These are mainly doves, parrots and finches; among them, only one (Serin) has established breeding populations in semi-natural habitats, but several others make sporadic (and inevitably poorly documented) attempts to breed, mainly in urban areas.

Order ACCIPITRIFORMES – broad-winged diurnal birds of prey

Family Accipitridae – buzzards, eagles, hawks, kites and vultures

Accipiter nisus granti, Sparrowhawk, Gavilán Común.
Subspecies endemic to Canaries (C, T, G, H, P) and Madeira. The Sparrowhawk is a fairly common breeding bird, mainly in the laurel forest and pine forest but also in agricultural areas. Generally avoids lowland areas of the south, though it is sometimes seen there outside the breeding season. It nests in a wide variety of trees, with laying in April and sometimes May or early June. Aerial interactions with Buzzards are quite common. Its food is almost entirely small to medium sized birds, and females are large enough to prey on Woodcock, Turtle Doves, Collared Doves, Rock Doves and the laurel pigeons.

Buteo buteo insularum, Buzzard, Ratonero Común, Aguililla.
Subspecies usually considered endemic to Canaries (L†, F, C, T, G, H, P), but a recent taxonomic revision assigns the Azores population to this subspecies. The Buzzard is a common breeding bird throughout most of the island, except for the extreme south. Nesting is mainly on cliffs and in barrancos, but occasionally in trees. Egg-laying starts around early March. Feeds mainly on rabbits, rats, lizards and insects and will take advantage of dead animals on the road. Slightly smaller than the Buzzard of northern Europe, *B. b. buteo*, with the underparts more streaked and less densely barred.

Circus aeruginosus, Marsh Harrier, Aguilucho lagunero; *Circus cyaneus*, Hen Harrier, Aguilucho pálido; *Circus pygargus,* Montagu's Harrier, Aguilucho cenizo.
All these occur uncommonly on passage or as winter visitors.

Milvus migrans, Black Kite, Milano Negro.
Fairly common passage migrant and very rare winter visitor on Tenerife, but has recently bred on Gran Canaria, so all records from Tenerife are of special interest. We had good views of one near the coast west of Taganana in June 2012.

[*Milvus milvus,* Red Kite, Milano Real.†
Native (C† T† G† H†) but now extinct as a breeder on the Canary Islands. The Red Kite was common on Tenerife in the past but declined rapidly during the 20th century and disappeared from its last strongholds in the pine forests of the southwest around the end of the 1960s. It is now rarely seen, although there was one reported sighting in 2007 on Teide. The probable reasons for its decline are the lack of carrion due to more modern agricultural practices, shooting, and the wide use of insecticides, especially during locust invasions. A conservation group in mainland Spain is apparently planning the reinstatement of Red Kites on Tenerife.]

[*Neophron percnopterus majorensis,* Egyptian Vulture, Alimoche Común, Guirre.†
Subspecies endemic to the Canaries, but now extinct on Tenerife (L, F, C† T† G†). The absence of old fossils and a new molecular study suggest that the Egyptian Vulture colonised the islands around the time that the indigenous Berber people arrived from Africa, around 2,500 years ago; it is thought that goats brought by the people provided for the first time an adequate food source for a scavenging bird. Since colonisation by a small founding group the population has evolved to become larger than the Iberian population, and it has also become sedentary. The species still survives on the eastern islands; on Tenerife it was common throughout the island, including some urban areas. The last record of breeding on Tenerife was in Teno and in 1984 there were still an adult and two immatures in the area.]

Family Pandionidae – ospreys

Pandion haliaetus, Osprey, Águila Pescadora, Guincho.
Resident breeder (L, F† C† T, G, H, P†). A few pairs of Ospreys nest on the cliffs of Los Gigantes. The species was more widespread in the past, but always close to the coast.

Order ANSERIFORMES – wildfowl

Family Anatidae – ducks and geese

Anas crecca, Teal, Cerceta Común.
Fairly common winter visitor.

Anas penelope, European Wigeon, Silbón Europeo.
Regular winter visitor.

Aythya ferina, Pochard, Porrón Común.
Fairly common but irregular winter visitor.

Tadorna ferruginea, Ruddy Shelduck, Tarro Canelo.
Breeds in small numbers on the eastern islands, and there have been a few breeding records on Gran Canaria and Tenerife in recent years.

Order APODIFORMES – swifts and hummingbirds

Family Apodidae – swifts

Apus apus, Swift, Vencejo Común.
Probably breeds in small numbers (L, C, T?); also regular passage migrant.

Apus melba, Alpine Swift, Vencejo Real.
Uncommon passage migrant.

Apus pallidus, Pallid Swift, Vencejo Pálido.
Breeding summer visitor (L, F, C, T, G?, H). The Pallid Swift is rare and often confused with other swifts, but is probably most frequent in the lower zone. Numbers have declined in recent years, but still breeds. Migrates to Africa for the period October to January. Differs from the Common Swift *Apus apus* in its browner, less blackish plumage, with a paler forehead and more conspicuous white throat and slower wing beats. Differs from the Plain Swift *Apus unicolor* in its larger size, more sickle-shaped wings and more gliding flight.

Apus unicolor, Plain Swift, Vencejo Unicolor.
Species endemic to the Canaries (all islands) and Madeira. The Plain Swift is a common breeder throughout Tenerife. It is a partial migrant and a large part of the population spends the winter in Africa. Nesting is in inaccessible places in rock crevices, walls of barrancos, bridges and buildings. Laying probably starts in April and continues until August. The Plain Swift is smaller and darker than the Pallid Swift and the throat is hardly paler than the breast, and it also flies faster. Its scream is a rapid trill.

Order CHARADRIIFORMES – waders and gulls

Family Burhinidae – stone curlews

Burhinus oedicnemus distinctus, Stone Curlew, Alcaraván.
Canary endemic subspecies and resident breeder (C, T, G, H, P). Another subspecies is endemic to the eastern islands; the species is widespread in Eurasia. On Tenerife the Stone Curlew is now uncommon and largely restricted to lowland and coastal areas in the south and west; it was much more abundant in the past. Stone Curlews are active mainly at dusk and dawn, and are hard to flush out in daytime since they squat close to the ground if alarmed. Nests are on open ground and laying extends from January to June, though mainly in February and March. The diet consists of insects and small vertebrates.

Family Charadriidae – plovers

Charadrius alexandrinus, Kentish Plover, Chorlitejo Patinegro.
Resident breeder (L, F, C, T). The Kentish Plover was frequent on the island in the past, but the population is now reduced to a tiny number of birds in the area of El Médano.

Charadrius dubius, Little Ringed Plover, Chorlitejo Chico.
Resident breeder (F, C, T). The first breeding records on Tenerife were in the second half of the 20th century and the population is probably less than 50 pairs. The Little Ringed Plover is found mainly in the southern coastal zone, usually beside fresh water or brackish pools. It can be distinguished from Kentish Plover *Charadrius alexandrinus* and Ringed Plover *Charadrius hiaticula* by the lack of a white wing bar in flight.

Charadrius hiaticula, Ringed Plover, Chorlitejo Grande.
Common winter visitor and passage migrant.

Pluvialis squatarola, Grey Plover, Chorlito Gris.
Common on passage and as a winter visitor.

Vanellus vanellus, Lapwing, Avefría.
Scarce on passage and as a winter visitor, but was more common in the past.

Family Glareolidae – coursers and pratincoles

Cursorius cursor, Cream-coloured Courser, Corredor.
Native but not now with a permanent population (L, F, C?, T). On Tenerife the main habitat was in the southern semi-arid plains, with records spanning two centuries. Recently there have been occasional sightings of coursers in both southern and northern coastal areas, and at least one pair bred in 2001. The food is mainly insects, but also small lizards. The courser is about the size of a Turnstone *Arenaria interpres*, but much slimmer and more active, with long legs. It is sandy coloured with a black and white eye stripe and fine curved bill. It runs fast when disturbed, and then suddenly stands very still – hence the local name Engañamuchachos, the child-teaser.

Glareola pratincola, Collared Pratincole, Canastera.
Uncommon passage migrant. Recorded mainly from freshwater sites in the south. The pratincole is a little larger than the courser, dull brown above and with long, dark, forked tail; no distinct eye-stripe (cf the courser) but adults in summer have a thin black border to the pale throat. Spends much time in the air, when it resembles a giant swallow, but with whitish rump and rufous underwing.

Family Laridae – gulls

Larus michahellis atlantis (*L. cachinnans atlantis*), Yellow-legged Gull, Gaviota Patiamarilla.
Resident breeder (all islands). The Yellow-legged Gull is usually included in the notoriously complex *Larus argentatus* group, with unstable Latin names, but new molecular data suggest a relationship with the Great Black-backed Gull *Larus marinus*. The subspecies is sometimes considered endemic to the Canaries, Madeira and Azores, but populations on western coasts of Iberia and Morocco are sometimes included; there is another subspecies in the Mediterranean. On Tenerife there are a few medium sized colonies and scattered pairs on sea cliffs and islets. Laying is in May. This species has yellow legs and a slate-grey mantle, paler than in the Lesser Black-backed Gull *Larus fuscus*.

Larus fuscus, Lesser Black-backed Gull, Gaviota Sombría.
Common winter visitor; a few pairs breed on islands north of Lanzarote. This species has a blackish-grey mantle, providing a distinction from the Yellow-legged Gull, which breeds on Tenerife. (The Herring Gull *Larus argentatus* occurs only very occasionally.)

Larus ridibundus, Black-headed Gull, Gaviota Reidora.
Regular winter visitor.

Family Recurvirostridae – stilts and avocets

Himantopus himantopus, Black-winged Stilt, Cigüeñuela.
Irregular breeder in the past (L, F, C, T), now a regular passage migrant and occasional winter visitor, and possibly irregular breeder. There is one definite breeding record of stilts in 2001, with five nests occupied at a locality in the south. Now breeds regularly on Lanzarote but only irregularly on Fuerteventura and Gran Canaria.

Family Scolopacidae – sandpipers

Actitis hypoleucos, Common Sandpiper, Andarríos Chico.
Common in winter and on passage.

Arenaria interpres, Turnstone, Vuelvepiedras.
Common in winter and on passage.

Calidris alba, Sanderling, Correlimos Tridáctilo.
Common in winter and on passage.

Calidris alpina, Dunlin, Correlimos Común.
Common in winter and on passage.

Gallinago gallinago, Snipe, Agachadiza Común.
A regular winter visitor and passage migrant.

Limosa limosa, Black-tailed Godwit, Aguja Colinegra.
Irregular passage migrant and ocasional winter visitor. (The Bar-tailed Godwit *Limosa lapponica* also occurs, but is less common; it is a regular winter visitor on the eastern islands.)

Numenius phaeopus, Whimbrel, Zarapito Trinador.
Common in winter and on passage. (The Curlew *Numenius arquatus* also occurs, but is much less common.)

Scolopax rusticola, Woodcock, Chocha Perdiz, Gallinuela.
Resident breeder (C, T, G, H, P) and also visitor. Occurs in the laurel forest, humid pine forest and chestnut plantations in the north of the island, and even in banana plantations, mainly between 600 and 1800 m. Nests are in dead leaf litter on the floor of the forests, with egg-laying extending from February to July. Feeds on worms, molluscs and other invertebrates.

Tringa glareola, Wood Sandpiper, Andarríos Bastardo.
Fairly common on passage and an occasional winter visitor.

Tringa nebularia, Greenshank, Archibebe Claro.
Fairly common in winter and on passage.

Tringa ochropus, Green Sandpiper, Andarríos Grande.
Fairly common on passage and a regular winter visitor in small numbers.

Tringa totanus, Redshank, Archibebe Común.
Fairly common in winter and on passage. (The Spotted Redshank *Tringa erythropus* occurs less frequently.)

Family Stercorariidae – skuas

Stercorarius skua, Great Skua, Págalo Grande.
The Great Skua is an uncommon winter visitor, while Arctic Skua *S. parasiticus* and Pomarine Skua *S. pomarinus* occur occasionally on passage.

Family Sternidae – terns

Sterna hirundo, Common Tern, Charrán Común, Garajao.
Native (F, C, T, G, H, P). A very few pairs of Common Terns breed on the cliffs and rocky islets of the south coast, with laying mainly in April; also occurs fairly commonly on passage. (The Arctic Tern *Sterna paradisaea* is an occasional visitor.)

Sterna sandvicensis, Sandwich Tern, Charrán Patinegro.
Common in winter and on passage.

Order CICONIIFORMES – herons, storks and ibises

Family Ardeidae – herons

Ardea cinerea, Grey Heron, Garza Real.
Native (may occasionally breed on all islands). The Grey Heron is regularly seen on the Roque de Garachico and at many inland water bodies. Occurs in variable numbers in winter, with congregations of 50–100 birds often present on the Guaza cliffs.

Ardea purpurea, Purple Heron, Garza Imperial.
Uncommon passage migrant.

Bubulcus ibis, Cattle Egret, Garcilla Bueyera.
Fairly common winter visitor and passage migrant.

Egretta garzetta, Little Egret, Garceta Común.
Native and resident, with breeding confirmed on Lanzarote and in 2005 on Tenerife. Occurs at inland freshwater sites and all along the south coast, with over 100 birds often present on Guaza cliffs.

Ixobrychus minutus, Little Bittern, Avetorillo Común.
Native (T) but rare. Breeding has been confirmed near Bajamar, but the species is also recorded on migration. Freshwater sites with thick vegetation are the normal habitat.

Nycticorax nycticorax, Night Heron, Martinete.
Native (C, T) but rare. Breeds at a few sites and also occurs on migration.

Order COLUMBIFORMES – pigeons and sandgrouse

Family Columbidae – pigeons

Columba bollii (C. trocaz bollii), Bolle's Pigeon (Dark-tailed Pigeon), Paloma Turqué.
Canary endemic species (C†, T, G, H, P). Bolle's Pigeon is sometimes considered as a Canary form of *Columba trocaz*, which also occurs on Madeira. It is distributed almost continuously along the north coast, from Anaga to Buenavista, in fragments of laurel forest and fayal brezal, but with higher density in the east. Breeding occurs almost year-round, but mainly in winter and spring. The nest is a platform of sticks in a laurel forest tree, with a preference for Brezo *Erica arborea*; only a single egg is laid. Feeds mainly on the fruit of various laurel species and also on *Ilex*, *Morella* and *Rhamnus*. Similar to Wood Pigeon *C. palumbus*, but lacks the white patches on the sides of the neck and the white bars showing on the wings in flight; the lack of dark bars on the closed wings distinguish it from Rock Dove *C. livia*. Sides of neck and upper breast pinkish; tail with a pale subterminal band, contrasting with the dark tip (cf *C. junoniae*).

Columba guinea, Speckled Pigeon, Paloma de Guinea.
Introduced (T). Widely distributed in sub-Saharan Africa. Occasionally seen in Santa Cruz and Los Cristianos, and has bred in the north of the island. This is an open-country seed-eating pigeon. Dark brown and grey, boldly speckled with white. There is a naked patch of red skin around the yellow eye.

Columba junoniae, White-tailed Laurel Pigeon (Laurel Pigeon), Paloma Rabiche.
Canary endemic species (C reintroduced, T, G, H, P). This pigeon was not recorded in Tenerife until the 1970s. It occurs in some areas of laurel forest, with greater numbers in the west (Monte del Agua, Los Silos) than the east, but present around Los Realejos, Orotava, Santa Úrsula and a few localities in Anaga; it is also often seen in surviving fragments of dry woodland. The nest is placed in rock crevices and small caves, and occasionally under rocks or tree trunks. Breeding seems to occur year-round, but mainly from April to August. Success is low because of predation by rats *Rattus rattus*. The diet is less specialised than in Bolle's Pigeon, and includes not only the fruit of laurel forest trees (especially Til *Ocotea foetens*) but also grain and cultivated plants. The plumage is rather uniformly dark brownish-grey, with the tail shading gradually to whitish near the tip, without any dark terminal bar (cf *C. bollii*).

Columba livia, Rock Dove, Paloma Bravía.
Resident breeder (all islands). Occurs throughout the island and is the most characteristic bird of deep barrancos. Nesting is in rock crevices, often in colonies. Laying seems to occur at all seasons, but mainly in spring and summer. Feeds on seeds and green shoots. Distinguished from the endemic pigeons by the strong black bars on the wings and the whitish rump.

Streptopelia decaocto, Collared Dove, Tórtola Turca.
Natural colonist; now breeds on all islands. The species has undergone a major expansion from Asia into Europe during the past few decades and appeared in the archipelago only recently. Now occurs mainly in parks and gardens. Breeding probably occurs during most of the year. Feeds mainly on seeds, but also scraps left by people, and probably fruit of *Ficus microcarpa*, an ornamental tree widely planted in urban situations. Secure identification of small doves in the genus *Streptopelia* is very difficult, since urban areas on the island also have free-flying populations of the domestic Ringneck Dove *S. 'risoria'* and also of its probable ancestor, the introduced African Collared Dove *S. roseogrisea*. Furthermore, hybridisation probably occurs between both of these and *S. decaocto*. All these doves are generally pale greyish brown above, with darker brown flight feathers. There is a conspicuous narrow black collar, but the rest of the head, neck and breast are pinkish buff, shading to white on the chin and belly.

Spilopelia (ex *Streptopelia*) *senegalensis*, Laughing Dove, Tórtola del Senegal.
Breeding populations on Lanzarote and Fuerteventura are probably derived from natural colonists; the species now also breeds on Gran Canaria, La Gomera and El Hierro, with sightings on La Palma and in several parts of Tenerife, where escapes from captivity may also have occurred; it will probably soon start to breed on Tenerife. The Laughing Dove is widespread in sub-Saharan Africa and eastwards into Asia. It is a slim, long-tailed, brown and blue-grey dove; the head and neck are pinkish, with chequered black marks on the sides and front of the neck.

Streptopelia turtur, Turtle Dove, Tórtola Común.
Native (all islands). Fairly common breeding summer visitor. Main arrival is in March-April and departure is around early August, with occasional sightings at other times. Occurs in most habitats on the island. Egg-laying is from March until July. Food is mainly seeds, but leaves of some plants are also eaten. Distinguished from the other doves on the island by the strong 'scaly' chestnut pattern on the back and wings; there are small patches of black and white barring on the sides of the neck. Normally detected by its soothing, purring song.

Order CORACIIFORMES – hoopoes, bee-eaters and kingfishers

Family Meropidae – bee-eaters

Merops apiaster, Bee-eater, Abejaruco Común.
Irregular but fairly common as a passage migrant, but seen mainly in the south.

Family Upupidae – Hoopoes

Upupa epops, Hoopoe, Abubilla, Tabobo.
Breeding resident (all islands). The Hoopoe is primarily a bird of open dry areas and those under cultivation, but it also occurs in clearings in the pine forest and sometimes at high altitude; it has become less abundant during the past century, losing ground especially in the north. Nesting is in rock crevices and walls, throughout the first half of the year. The Hoopoe feeds on the ground, largely on invertebrates, but occasionally takes small reptiles.

Order FALCONIFORMES – sharp-winged diurnal birds of prey

Family Falconidae - falcons

Falco pelegrinoides, Barbary Falcon, Halcón de Berbería.
Breeding resident (all islands). The Barbary Falcon is native to North Africa, the Middle East and southern Asia. It does not normally hybridise with the related Peregrine Falcon *F. peregrinus* (which may be present occasionally in winter). This magnificent falcon can now be seen in many parts of Tenerife, especially in the major barrancos, having been absent or extremely rare for centuries. Nesting is in inaccessible rock crevices, with laying in late February and March. Feeds primarily on Rock Doves but also attacks a wide variety of other birds. The Barbary Falcon is smaller than the Peregrine, with the upperparts paler blue-grey and the barred pale underparts with a faint buff wash.

Falco tinnunculus canariensis, Kestrel, Cernícalo Vulgar.
Subspecies endemic to the Canaries (C, T, G, H, P) and Madeira. On Fuerteventura and Lanzarote there is a separate endemic subspecies, *Falco tinnunculus dacotiae*. The Kestrel is a common sight in all parts of the island, with a population probably exceeding 1000 pairs. Nesting is mainly in rock crevices and ledges in barrancos, but also in buildings and sometimes in trees. Laying is mainly in March and April, but also in May, especially at high altitudes. On Tenerife the Kestrel feeds primarily on insects and lizards, but also mice, and occasionally frogs and birds. The local subspecies is smaller than *F. t. tinnunculus* of northern Europe. The male has more heavily streaked, darker crown, and the rest of the upper parts are deeper chestnut colour with larger spots, and the underparts are deeper cream; the female is more heavily barred, and with upper tail-coverts tending to grey-blue.

Order GALLIFORMES – game birds

Family Numididae – guineafowl

Numida meleagris, Helmeted Guineafowl, Gallina de Guinea.
This introduced species (all islands except Gomera) sometimes breeds in the wild, mainly near houses.

Family Phasianidae – pheasants, grouse and quail

Alectoris barbara, Barbary Partridge, Perdiz Moruna.
Breeding resident (all islands) introduced at some time after the Castilian conquest, probably for sport. The Barbary Partridge is native to northern Africa and Sardinia. On Tenerife it is fairly common in much of the lower zone, southern pine forest and high mountain zone. Nests are usually in the shelter of plants, and laying occurs mainly from March to June. The food consists mainly of seeds and fruits, but also buds, shoots and invertebrates. Similar to the Red-legged Partridge *Alectoris rufa* (which is also present on Tenerife in small numbers, but may not be truly wild) but slightly smaller and with chestnut and white rather than black and white markings on neck and chest.

Coturnix coturnix, Quail, Codorniz.
Resident breeder (all islands) but also occurs to some extent as a migrant. The Quail has a patchy distribution, almost entirely in the north of the island. The nest is very well hidden among plants,

including crops and vines. The breeding season is prolonged, from December into August. The diet consists of seeds and insects, along with some fruits.

***Phasianus colchicus*,** Pheasant, Faisán Vulgar.
Introduced to several islands, and sometimes breeds in the wild on Tenerife.

ORDER GRUIFORMES – BUSTARDS, RAILS AND CRANES

FAMILY OTIDIDAE – BUSTARDS

[***Chlamydotis undulata fuertaventurae*,** Houbara Bustard, Hubara.
Canary endemic subspecies (L, F, T†), now extinct on Tenerife; the only good evidence of its former presence is the finding of bones in the Cueva del Viento. This spectacular bird may once have lived in the more open parts of the pine forest in the north of the island, and perhaps also elsewhere.]

FAMILY RALLIDAE – RAILS

***Fulica atra*,** Coot, Focha Común.
Breeding resident (F, C, T, G, P?) in small numbers, and common winter visitor. The Coot became established on Tenerife in the 1980s and now breeds in several fresh water ponds such as those near Bajamar, Los Silos, Las Galletas and Erjos, where numbers are augmented by migrants. Nests in aquatic vegetation, with laying over an extended period, mainly from December to August.

***Gallinula chloropus*,** Moorhen, Polla de Agua (Gallineta Común).
Breeding resident (F, C, T, G, P). The Moorhen is usually found on ponds and reservoirs in several areas, for instance near Bajamar and Erjos. Nesting is in dense waterside vegetation, especially between February and August, but in almost any month in places where water levels are maintained. Small chicks suffer from predation by gulls and probably rats and feral cats. The food includes a wide variety of invertebrates and also leaves and buds of plants.

ORDER PASSERIFORMES – SONGBIRDS

FAMILY ALAUDIDAE – LARKS

***Alauda arvensis*,** Skylark, Alondra Común.
Fairly common in winter and to some extent on passage. Occurs in fields but also open ground near the coast in the south.

[***Calandrella rufescens*,** Lesser Short-toed Lark, Terrera Marismeña.
No longer breeds on Tenerife, but there used to be two rather distinct populations, in the humid area around Los Rodeos airport near La Laguna and in the arid southern coastal zone. The subspecies *C. rufescens rufescens* (previously thought to be confined to Tenerife) is now considered endemic to the Canaries (L, F, C, T†); the populations on the eastern islands (L, F, C) and that in the south of Tenerife were previously assigned to a separate endemic subspecies. The species occurs in North Africa and eastwards to China. Conservation measures have failed to protect the habitat of this species and both Tenerife populations have disappeared within the last two decades.]

FAMILY CORVIDAE – CROWS

Corvus corax canariensis (ex *C. c. tingitanus*), Raven, Cuervo.
Canary endemic subspecies (all islands). Close relatives occur in northern Africa. The Raven was formerly common in all zones, but has now become scarce, perhaps as a result of a reduction in carrion, and is limited to only a few areas. There are less than 10 breeding pairs left, along with several non breeders, but numbers may now be increasing. Nesting is in rocky crevices, occasionally in trees (*Pinus* or *Juniperus*). Ravens are omnivorous; they can be predatory and also feed on birds eggs and carrion, as well as fruit and

Erithacus rubecula superbus Robin
Photo: Aurelio Martín

Fringilla coelebs canariensis Chaffinch male
Photo: Aurelio Martín

Fringilla teydea Blue Chaffinch male
Photo: Rubén Barone

Lanius meridionalis koenigi Southern Grey Shrike. Photo: Aurelio Martín

Serinus canaria Canary
Photo: Aurelio Martín

Anthus berthelotii Berthelot's Pipit

Phylloscopus canariensis Canary Chiffchaff

Cyanistes teneriffae Tenerife Blue Tit

Apus unicolor Plain Swift
Photo: Aurelio Martín

Columba bollii Bolle's Pigeon
Photo: Aurelio Martín

Columba junoniae White-tailed Laurel Pigeon
Photo: Aurelio Martín

Dendrocopos major Great Spotted Woodpecker
Photo: Rubén Barone

vegetable matter; they frequent rubbish bins and congregate in the region of the Chío recreation site at the edge of the Teide National Park, exploiting food items left by visitors. The plumage is very glossy and the bill shorter than in Ravens of northwest Europe, and the call is somewhat more strident.

[*Pyrrhocorax graculus*, Alpine Chough.
Probably native in the distant past. Bones of the species have been found in the Cueva de Cosme, dated at <430,000 years ago.]

[**Pyrrhocorax pyrrhocorax**, Chough, Chova Piquirroja.
Native but now extinct on Tenerife (T† G† P). Bones of the Chough have been found in the Cueva del Viento, formed around 150,000 years ago, implying that there was once an established population on the island. There have been only a handful of recent sightings. Early attempts to re-establish the species on Tenerife were unsuccessful, although the population on La Palma is still substantial.]

Family Emberizidae – buntings

[*Emberiza alcoveri*, Long-legged Bunting, Escribano Patilargo.†
An extinct species of bunting (T†), unknown until the discovery of subfossil bones in the Cueva del Viento in the north of Tenerife. It was distinguishable by its long legs and short wings, and was probably incapable of flight.]

Emberiza calandra (ex *Miliaria calandra*), Corn Bunting, Triguero.
Partial migrant, breeding on all islands. The Canary form has sometimes been considered as a separate subspecies, *Emberiza calandra thanneri*. The Corn Bunting was once common and widespread on the island, but is now scarce, with a patchy distribution, mainly in cultivated areas in the north. Nests are well concealed on the ground, with laying from March to June. The diet consists mainly of seeds, with insects in the breeding season.

Family Estrildidae – estrildid finches

Amandava amandava, Red Avadavat (Red Munia), Bengalí Rojo.
Introduced (T). There have been records of breeding in a few parts of the island, but none since 2000. A sparrow-sized bird with a red rump, black tail and red bill. The male in the breeding season is mainly red, with brown wings and white spots on wings and breast. The non-breeding male is duller but has the red rump, while the female is greenish brown with only a few white spots.

Estrilda astrild, Common Waxbill, Pico de Coral.
Introduced (C, T). No recent breeding records. A tiny bird with finely barred plumage, bright red beak and a broad red eyestripe.

Estrilda melpoda, Orange-cheeked Waxbill, Estrilda de Carita Naranja.
Introduced (T). Escaped cage birds have bred in Puerto de la Cruz, but there are no records from recent years. A tiny bird with grey head, brown back, red base to the tail and red bill. The cheeks are orange but this may shade into red in front of the eye. (The Black-rumped Waxbill *Estrilda troglodytes* is occasionally seen on the island, but there are no recent breeding records.)

Family Fringillidae – finches

Bucanetes (ex *Rhodopechys*) *githagineus amantum*, Trumpeter Finch, Camachuelo Trompetero.
Canary endemic subspecies (L, F, C, T, G) of a species that is widespread in the Middle-east and eastwards into Asia. In the Canaries the Trumpeter Finch used to be confined to the eastern islands, but it colonised Gran Canaria in about 1850, Tenerife near the start of the 20th century and La Gomera in the 1960s; breeding may also have occurred in El Hierro near the end of the century. On Tenerife it frequented rocky arid places in the south, near both southeastern and southwestern coasts, and also Punta de Teno in the northwest. However, the population declined after the mid 1980s and it is now uncertain whether any breeding pairs survive on the island. This is a small finch, only slightly larger than the Goldfinch *Carduelis carduelis*. The male has a stout bright red bill and pinkish-brown plumage in the breeding season. The female is duller brownish with a yellowish-brown bill.

[*Carduelis aurelioi*, Slender-billed Greenfinch.†
This extinct finch is discussed in the Box in Chapter 1.]

Carduelis cannabina meadewaldoi, Linnet, Pardillo Común.
Canary endemic subspecies (C, T, G, H, P); another endemic subspecies is in the eastern islands. The Linnet used to be common in the lower zone and cultivated areas in all but the highest parts of the island, but numbers are considerably reduced in recent years, as with other species typical of such habitats. The Linnet is a popular cage bird. It nests in shrubs, usually less than 2 m above the ground, with laying from January to July, but mainly from March to May. The diet consists of seeds of wild and

cultivated plants. The subspecies is darker and more richly coloured than the birds in northwest Europe and with a larger and slightly thicker bill

Carduelis carduelis, Goldfinch, Jilguero.
Native breeding resident (L, F, C, T, G, H? P?). The Goldfinch has a patchy distribution in the lower zone up to the edge of the laurel forest. Occurs along most of the north coast (not in the extreme east or west) and in the far south, mainly in cultivated areas and places with degraded forms of dry woodland or laurel forest. It has always been a popular cage bird and was much traded in the ports, and was also eaten. However, the major decline seen during the past century has probably been due to increased control of thistles and the massive use of pesticides. Nesting is in trees, between 2–8 m above the ground, with laying from February to June. The food consists mainly of thistle seeds.

Carduelis chloris, Greenfinch, Verderón.
Breeding resident (all islands except Lanzarote). The Greenfinch has colonised the Canaries in the last half century, probably naturally, and arrived on Tenerife in about 1966. It now has a patchy distribution in the lower zone and lower forest zone along the north of the island, and occurs in a few places in the south. Nesting is in trees, mainly conifers, from 2–6 m above ground, with laying from March to June. Feeds on seeds, including those of thistles and docks.

Fringilla coelebs canariensis (ex *F. c. tintillon*), Chaffinch, Pinzón Común.
Canary Island endemic subspecies (C, T, G); separate endemic subspecies are found on El Hierro and La Palma. The Chaffinch is common in and just below the forest zone along the north side of the island. It is absent from the whole of the southern half of Tenerife, but overlaps with the Blue Chaffinch to a limited extent in areas of mixed pine forest in the north. Nests are in a wide variety of trees and between 2.5 and 9 m above ground; laying is from late March to May, and into June at high altitudes. The food includes fruits of trees in the laurel forest, pine seeds and many kinds of invertebrates. The male differs from the European subspecies in having slate-blue instead of chestnut mantle, very dark slate coloured crown and especially forehead, and paler pink breast. It differs from the Blue Chaffinch in its smaller size and less massive bill, but also in its pinkish cheeks and breast, green rump, strong white bars on the wing coverts, and yellow edges to the main wing feathers and white in the tail – all these are suffused with blue in the male Blue Chaffinch. The female is greenish-brown above and buff below; she can be confused with the female Blue Chaffinch, but is distinctly browner and smaller, and the double wing bar is pure white. The song has the same general pattern as in British populations, but individuals have a greater repertoire.

Fringilla teydea teydea, Blue Chaffinch, Pinzón Azul.
Tenerife endemic subspecies of a Canary endemic species; another subspecies occurs on Gran Canaria, but the Blue Chaffinch is – surprisingly – absent from El Hierro and La Palma. On Tenerife it is common in pine forests (including those composed of introduced Monterey Pine *Pinus radiata*), and seems to favour especially areas with an understorey of Escobón *Chamaecytisus proliferus*; it is also found in areas transitional to laurel forest (where the ordinary Chaffinch is also present) but does not occur in the lower cultivated zone. Nesting is normally in pines but occasionally in other trees, at a height between 1.4 and 20 m or more; laying is typically in early June, but sometimes in April or May, and may continue as late as August; the clutch is normally two. The diet consists primarily of pine seeds which it cracks with its sturdy bill. It is more insectivorous in the nesting season, catching invertebrates such as moths and beetles to feed its young. The Blue Chaffinch is larger than the Chaffinch *Fringilla coelebs*, and has a noticeably large bill. The male is rather uniform slate-blue, with the flight feathers dark; the blue fades to whitish on the belly and under the tail, and there is an incomplete white eye ring; it entirely lacks pink on the breast (cf Chaffinch). The female is dull grey with the wings brownish and showing only dusky pale wing bars. The song is simpler and with fewer variations than in the Chaffinch.

Serinus canaria (often spelled *Serinus canarius*), Canary, Canario.
Macaronesian endemic species (all islands); also in Madeira and Azores. The Canary occurs throughout Tenerife except in the coastal area in the extreme south. The wild Canary is slightly bigger than a Goldfinch *Carduelis carduelis* and has greyish streaked back and yellowish head and under-parts – but is never bright yellow all over like the caged variety. The rump is greenish yellow (cf Serin *S. serinus*). The song, however, is like that of the caged bird and is often delivered in flight.

Serinus serinus, Serin, Serín.
Probably introduced (C, T). First reported in the wild in 1977, and now breeds in several places between La Laguna and Puerto de la Cruz. The Serin is smaller than the Canary *Serinus canaria* but very similar, except for a bright yellow rump. The head of the male is patterned with bright yellow and dark markings.

Family Hirundinidae – swallows

Delichon urbica, House Martin, Avión Común.
Common as a passage migrant, occasionally in large numbers.

Riparia riparia, Sand Martin, Avión Zapador.
Fairly common as a passage migrant.

Hirundo rustica, Swallow, Golondrina Común.
Common as a passage migrant.

Family Laniidae – shrikes

Lanius meridionalis (ex *L. excubitor*) **koenigi**, Southern Grey Shrike, Alcaudón Meridional.
Canary endemic subspecies (L, F, C, T) of a species also occurring in southern Europe and North Africa. This shrike is sedentary and patchy in distribution in Tenerife, favouring open areas with scrub vegetation; it occurs in small numbers in arid places at low altitudes in the south, and in the high mountain zone. It is absent from the north and from Anaga, and is now very rare in Teno, where it was apparently common in the past. Nesting is in a variety of bushes, usually less than 3 m above ground. The laying period is prolonged, starting as early as December and continuing until May at high altitude. The diet is mainly of lizards and insects, but also mice and small birds. Prey are often impaled on thorny bushes such as *Lycium intricatum*, especially near the nest.

Family Motacillidae – wagtails and pipits

Anthus berthelotii berthelotii, Berthelot's Pipit, Bisbita Caminero.
Canary endemic subspecies (all islands) of a species endemic to the Canaries, Salvage Islands and Madeira. Berthelot's Pipit is very common throughout the island, but especially in uncultivated parts of the lower zone and in the high mountain zone. It is about the same size but greyer than the Meadow Pipit *Anthus pratensis*. It sometimes sings in the air. Nests are on the ground in shelter of plants or rocks, and laying occurs between December and July. The diet consists very largely of insects, though some seeds and buds are also eaten.

Anthus pratensis, Meadow Pipit, Bisbita Común.
Fairly common in winter (mainly December to January) and also occurs on passage. Most records are from the north.

Motacilla cinerea canariensis, Grey Wagtail, Lavandera Cascadeña, Alpispa.
Canary endemic subspecies, not accepted by all taxonomists (C, T, G, P). The Grey Wagtail is common in the lower zone and the forests up to 1500 m; it is usually near running water but can often be seen along roads, especially in Anaga. In contrast to northwest Europe it often also occurs in towns and gardens. Nests are in natural crevices or holes in man-made structures, usually near the ground; laying is from February to July. The Canarian subspecies is similar to populations in northwest Europe, but distinctly darker and richer yellow below.

Motacilla flava, Yellow Wagtail, Lavandera Boyera.
Fairly common on passage, especially in the south.

Motacilla alba, White Wagtail, Lavandera blanca.
Fairly common in winter and also occurs on passage.

Family Muscicapidae – Old World flycatchers

Ficedula hypoleuca, Pied Flycatcher, Papamoscas Cerrojillo.
Regular passage migrant, especially in autumn, but in variable numbers. Usually found in open places in the forest zone; we have seen it in October in replanted pine forest.

Family Paridae – titmice

Cyanistes teneriffae teneriffae (Parus teneriffae teneriffae, Parus caeruleus teneriffae), Tenerife Blue Tit, Herrerillo Común.
Canary endemic species. The relationships and taxonomic status of the blue tits of the Canary Islands are still controversial, with molecular studies continuing. One opinion considers *Cyanistes teneriffae teneriffae* as confined to Tenerife and La Gomera, with another subspecies (*C.t. hedwigae*) on Gran Canaria, and – intriguingly – a different species, *Cyanistes ultramarinus*, on La Palma, Hierro, the eastern Canaries and Northwest Africa. The Tenerife Blue Tit is common throughout the island except in the southern coastal area. It differs from the European Blue Tit in having a strikingly dark head, almost no white on the wings and a proportionally longer and thinner beak. The calls are very varied and some are deceptively like that of the Great Tit *Parus major*, which does not occur here. Other calls are similar to those of other tits. The species is found in a greater range of habitats than the European Blue Tit, including the pine forest.

Family Passeridae – sparrows

Passer hispaniolensis, Spanish Sparrow, Gorrión Moruno.
Perhaps native (all islands); the species colonised Tenerife from the eastern islands in the late 19th century. It is a bird primarily of urban areas, but colonises buildings in agricultural areas and sometimes also barrancos; numbers have decreased in the past decade. Breeding is in small colonies, in buildings, the walls of barrancos and in trees; laying is from January to July. The diet consists of seeds of wild and agricultural plants, but also insects and scraps left by humans. The male of the Spanish Sparrow differs from the House Sparrow *Passer domesticus* (introduced but rare in Gran Canaria) in its entirely chestnut brown crown and nape (House Sparrow has a broad grey cap), extensive scaly black pattern on the flanks, and white cheeks.

Petronia petronia, Rock Sparrow, Gorrión Chillón.
Resident breeder (C, T, G, H, P); the species occurs throughout Southern Europe and the Mediterranean region. The Rock Sparrow occupied the towns of Tenerife until the arrival of the Spanish Sparrow, but during the 20th century it became restricted to its more normal habitat of rocky places in the lower zone, mainly in the south of the island. It is now found only around Teno Alto in the extreme northwest, with perhaps also a colony near Santa Cruz. The Rock Sparrow is similar to a female House Sparrow *Passer domesticus*, but with a long pale eye stripe and a long pale stripe on the crown. Adults have a pale yellow spot on the upper breast, and pale markings on the tail are sometimes visible when it flies. It is very noisy and usually occurs in flocks.

Family Ploceidae – weaverbirds

Euplectes orix, Red Bishop, Tejedor Rojo.
Introduced (T); occasionally attempts to breed, in several places.

Family Pycnonotidae – bulbuls

Pycnonotus jocosus, Red-Whiskered Bulbul, Bulbul Orfeo.
Introduced (T); has bred at Los Realejos. Attempts are being made to prevent its further spread.

Family Sturnidae – starlings

Acridotheres tristis, Common Myna, Miná Común.
Introduced (C, T, P). A pair of escaped cage birds bred in La Laguna in the 1990s, leading to fears that the species might become invasive on Tenerife, as it has on many other islands and continents. However, all the wild-living birds have been captured and it is now illegal to keep or sell the species on the island.

Lamprotornis purpureus, Purple Glossy Starling, Bruñido Purpúreo.
Introduced (T). Has bred, but there are now only occasional sightings.

Sturnus vulgaris, Starling, Estornino Pinto.
A scarce winter visitor, but was commoner in the past; a few have bred in buildings in the last few decades (C, T).

Family Sylviidae – Old World warblers

Phylloscopus canariensis (ex *Phylloscopus collybita canariensis*), Canary Chiffchaff, Mosquitero Común.
Now considered to be a Canary endemic species (C, T, G, H, P), distinct from *Phylloscopus brehmii* in Iberia and *Phylloscopus collybita* elsewhere in Europe and Asia; the subspecies *Phylloscopus collybita exsul* bred in the past on Lanzarote and perhaps Fuerteventura. The Canary Chiffchaff is among the most widely distributed of birds in Tenerife, occurring everywhere except in the most arid parts of the extreme south. The spherical nest with a side entrance is placed in a wide variety of trees and shrubs and may be near the ground or up to 8 m above it; laying is from January to June. The diet is primarily small insects, but nectar is also taken from a variety of shrubs, for which it may function as a pollinator. The Canary Chiffchaff is darker olive than the Chiffchaff of northwest Europe, and more tawny buff below; the legs are black. The song is quite distinct, with considerable individual variation, and at times reminiscent of the declining scale of the Willow Warbler *Phylloscopus trochilus*.

Phylloscopus trochilus, Willow Warbler, Mosquitero Musical.
Common on passage.

Regulus regulus teneriffae, Goldcrest, Reyezuelo Sencillo.
Canary endemic subspecies (T, G) of a European and north Asian species; a separate subspecies on El Hierro and La Palma, *Regulus regulus ellenthalerae*, has recently been distinguished on molecular evidence. (In the past some workers considered the Goldcrest of the Canaries to be a form of Firecrest *Regulus ignicapillus*.) On Tenerife the Goldcrest inhabits the laurel forest and those parts of the pine forest that have a well developed understorey, especially of Brezo *Erica arborea*. Nests are placed in thin branches, at a height of 2.5 to 14 m; laying is from March to June. The diet is primarily small insects and spiders, but some seeds are also taken. The Tenerife subspecies has somewhat broader black sides of the crown than in the Goldcrest of northwest Europe, and the black continues across the forehead. The bill is slightly longer and the general colouring slightly darker. The song is shorter and harsher.

Sylvia atricapilla heineken, Blackcap, Curruca Capirotada.
Subspecies perhaps endemic to Macaronesia (L?, F?, C, T, G, H, P); the range of the subspecies outside the Canaries requires revision. On Tenerife the Blackcap is now common throughout much of the lower zone, especially in gardens and cultivated areas. Its original habitat was probably in barrancos with *Salix canariensis*, and the moist areas between the dry woodland and the laurel forest. It has clearly benefited

from changes in land use caused by humans, and it is also a common cage bird. Nests are normally within 5 m of the ground in bushes or small trees; breeding extends from February to July. The Blackcap is largely frugivorous, but also eats insects. Both sexes are darker and slightly smaller than the birds in northwest Europe and can be somewhat confusing. However, the lack of white in the outer tail feathers and the black or brown crown readily distinguish it from the other two resident *Sylvia* warblers. Its song is slightly shorter than that of the European form and it also has a quite different drawn out 'churr'.

Sylvia conspicillata orbitalis, Spectacled Warbler, Curruca Tomillera.
Macaronesian endemic subspecies occurring in all the Canary islands, Madeira and Cape Verdes; the species extends to northwest Africa and the Mediterranean. This is primarily a bird of shrublands, both near the coast and in the high mountain zone, but it also occupies open areas within the forest zone, so it is very widespread. Nests are close to the ground in a wide variety of bushes; laying starts in low areas in December and continues into May. The diet is mainly insects and the fruit of a variety of bushes, but it also takes nectar and acts as a pollinator of plants such as *Isoplexis* and *Canarina*. The Spectacled Warbler is similar to the Whitethroat *Sylvia communis* (which sometimes occurs on passage) but the male has a darker head (especially forehead) and the female is greyer; both sexes have a grey head with pale eye-ring (cf Sardinian Warbler), white throat, rufous brown wing coverts and white outer tail feathers. The song is reminiscent of the Whitethroat.

Sylvia melanocephala leucogastra, Sardinian Warbler, Curruca Cabecinegra.
Canary endemic subspecies (all islands) of a species which occurs in the Mediterranean region and northern Africa. Some writers consider that the birds in the eastern Canaries belong to *Sylvia melanocephala melanocephala*. On Tenerife the Sardinian Warbler is now found throughout most of the lower zone; it was previously absent from coastal areas in the southeast but is now beginning to colonise those parts where vegetation cover and height are increasing; it also occurs up to the lower margin of the high mountain zone, but avoiding dense forest; its original habitat was probably mainly in the dry woodland, overlapping with the Spectacled Warbler but favouring areas with taller vegetation. Nests are in bushes, normally below 5 m; breeding extends from February to July. Both sexes have a conspicuous red ring around the eye (cf Blackcap and Spectacled Warbler). The male has the head jet black (extending well below the eye), the throat white and the whole of the upper parts dark grey. The female is somewhat browner, with a dark grey head. The outer tail feathers are white in both sexes (cf Blackcap).

Family Turdidae – thrushes

Erithacus rubecula superbus, Robin, Petirrojo.
Subspecies endemic to Tenerife and Gran Canaria; the birds on La Gomera, El Hierro and La Palma conform to the continental subspecies *Erithacus rubecula rubecula*. The Robin is common throughout the laurel forest (including fayal-brezal) and humid pine forest, and also occurs in the upper parts of the lower zone. Nests are in bushes and varied other sites within 3 m of the ground. Laying extends from February to July, but is mainly in April to June. The local subspecies is darker and richer red than the birds of continental Europe, with a whiter belly. The song is richer and more varied and it often repeats short phrases in a manner reminiscent of the Song Thrush *Turdus philomelos*.

Oenanthe oenanthe, Wheatear, Collalba Gris.
Common on passage, and is seen in almost all parts of the island, including open spaces in the pine forest and in Las Cañadas.

Turdus merula cabrerae, Blackbird, Mirlo Común.
Subspecies endemic to the Canaries (C, T, G, H, P) and Madeira. The Blackbird occurs throughout the forest zones and in much of the lower zone, but avoiding open arid areas. Nests are in bushes and trees, sometimes more than 10 m above the ground; the breeding season can be very prolonged, extending from December into July. The diet consists of insects, earthworms and a wide variety of

natural and cultivated fruits. The birds on the Canaries are smaller than those in northwest Europe; the male is deeper and glossier black while the female is more blackish brown. The song is only slightly different.

Turdus torquatus, Ring Ouzel, Mirlo de Collar.
Winter visitor to Las Cañadas, feeding on fruits of the Cedro Canario *Juniperus cedrus*.

Family Timaliidae (Leiothrichidae) – babblers

Leiothrix lutea, Pekin Robin (Red-billed Leiothrix), Ruiseñor del Japón.
Introduced (T). Has bred in Tacoronte, and has been seen in other places along the north coast.

Order PELECANIFORMES – tropicbirds, boobies, etc

Family Sulidae – boobies and gannets

Morus bassanus (Sula bassana), Gannet, Alcatraz.
Fairly common winter visitor.

Order PICIFORMES – woodpeckers and toucans

Family Picidae – woodpeckers

Dendrocopos major canariensis, Great Spotted Woodpecker, Pico picapinos, Pájaro carpintero, Pájaro peto.
Tenerife endemic subspecies; another endemic subspecies is found on Gran Canaria, but the species is absent – surprisingly – from El Hierro and La Palma, both of which have plenty of the mature pine forest that is the typical habitat in the Canaries. On Tenerife restored pine forest planted in the 1950's and 60's was not occupied by breeding woodpeckers after 25 years but had become excellent habitat after 50 years. Nests are often initially near the top of a dead tree broken off by the wind, in the part that becomes softened most quickly by fungal attack; in later years new holes may be excavated lower down on the same tree. Laying is normally in April and May, and most of the young fledge between mid May and the end of June. Breeding success is rather low, and fledged young are vulnerable to predation by Sparrowhawks *Accipiter nisus*.

Order PROCELLARIIFORMES – petrels, shearwaters, etc

Family Hydrobatidae – storm petrels

Hydrobates pelagicus, Storm Petrel, Paíño Común.
Native (L, F, T, G, H). The Canaries are at the southern limit of the breeding range. Small numbers of Storm Petrels breed on the Roques de Anaga, and frequent records of the species off the southwest coast suggest the possibility of breeding on the mainland in that area. Nests in crevices and under rocks, with laying at the end of June and July. Migrates southwards for the winter. The smallest of the petrels in Canarian seas, with square tail and white rump; there is only a faint pale line on the upper wing and a diagnostic strong white stripe under the wing. It has a fluttering flight, and may be seen following ships, feeding on plankton churned up by a propeller.

Oceanodroma castro, Madeiran Storm-petrel (Band-rumped Storm-petrel), Paíño de Madeira.
Breeding winter visitor (L, F, T, H). This is a widespread warm water species; the taxonomy of it and related forms is currently subject to revision. The largest colony in the Canaries is on the Roques de Anaga, with perhaps 100 pairs; small numbers may breed on the Roque de Garachico. Nests in crevices, caves and under rocks. The species is a winter breeder, with laying poorly synchronised but mainly from October to February. It is larger and longer winged than the Storm Petrel and with a long, slightly forked tail, evenly-cut white rump band and dark underwing.

FAMILY PROCELLARIIDAE – SHEARWATERS

Bulweria bulwerii, Bulwer's Petrel, Petrel de Bulwer.
Breeding summer visitor (all islands). A widespread warm water species that breeds on all the Macaronesian island groups, migrating south for the winter. Bulwer's Petrel breeds in small colonies, mainly on offshore islets along the north coast – especially the Roques de Anaga where there are several hundred pairs – and on a few cliffs in other parts of the island, including those of Montaña de Guaza; there are also hints of breeding far inland. Egg-laying starts in May and the young fledge in September. This is a medium-sized long-winged petrel, sombre brownish above and below. Its tail is wedge-shaped but is normally held closed and thus appears pointed. It is apparently silent on the breeding grounds.

Calonectris diomedea, Cory's Shearwater, Pardela Cenicienta.
Breeding summer visitor (all islands). Cory's Shearwater is the most abundant breeding seabird on Tenerife, with a population estimated at 3000 pairs. Nesting is in burrows on cliffs in coastal regions and on the offshore islets, but also in barrancos several kilometres inland. It often feeds within sight of the shore during the day, and at dusk its haunting cries can be heard as it comes into its burrows. Like other petrels, the birds seem to become confused by lights on cloudy or stormy nights, circling above the towns and sometimes settling. Breeding is in summer, with egg-laying starting near the end of May and fledging in October. Satellite tracking has shown that during the chick-rearing period parents from western and central Canary Islands forage over the productive waters of the Canary Current within 80 km of the coast of northwest Africa; they make round trips of well over 1000 km and are away for periods of about 5-7 days (see diagrams in Chapter 1). The birds migrate south in October and November, returning again in early March. In the past shearwater chicks were collected and eaten in large numbers, but the custom became much less significant during the second half of the 20th century. Cory's Shearwater is a long-winged seabird with uniformly grey-brown upperparts and without any specially dark cap. It has pure white underparts and a heavy yellow bill. The flight is graceful and gliding, close to the water, following the contours of the waves.

Puffinus baroli (ex *Puffinus assimilis baroli*), Macaronesian Shearwater (North Atlantic Little Shearwater), Pardela macaronésica, Tahoce.
Breeding resident (L, F? C, T, G, H, P?), also occurring on Madeira, Salvage Islands and Azores; the scientific and common names are very unstable. The species is apparently sedentary, occurring in Canarian waters throughout the year, though commoner in summer. Numbers are small and breeding sites are poorly known, but include the Roques de Anaga, probably various sites along the north coast and perhaps also the southwest coast. The laying season is extended, but most eggs are probably laid in January and February. This species is smaller than other shearwaters of the area. It is blue-black above, with pure white underparts, and can be identified at sea by its bursts of quick wing-beats alternating with gliding close to the surface of the sea.

Puffinus puffinus, Manx Shearwater, Pardela pichoneta.
Breeding summer visitor (T, G? H? P). The Canaries are at the southern limit of the breeding range, which includes Britain, Madeira and the Azores. On Tenerife the Manx Shearwater declined after the end of the 19th century, and is probably now reduced to a few scattered pairs in the northern part of the Teno massif. The reduction is ascribed to predation by cats and rats, along with losses of fledglings as a result of dazzling by town lights. Nesting is on cliff ledges in wet barrancos and under rocks, and in burrows in the laurisilva. Laying is probably mainly in March and April, with fledging between mid July and mid August, when dazzled fledglings are found around Puerto de la Cruz and other towns, mainly on the north coast. The species is absent from Canarian waters from October through December.

Order PSITTACIFORMES – parrots

Family Cacatuidae – cockatoos

Nymphicus hollandicus, Cockatiel, Cacatúa Ninfa.
Introduced (T). There have been sightings in several places and breeding has occurred in Puerto de la Cruz. The Cockatiel is a small, long-tailed, grey Australian cockatoo with a mainly yellow head with orange cheek spots, and conspicuous erect crest (yellow in male, grey in female).

Family Psittacidae – parrots

Amazona ochrocephala, Yellow-crowned Amazon, Amazona Real.
Introduced (T). Native to tropical South America and Panama. Escaped birds have been seen in Puerto de la Cruz and breeding has occurred. This is a medium-sized, compact, short-tailed, green parrot with yellow forehead and crown, white eye-ring and a red patch on the wing (speculum and carpal edge) that is normally invisible when the bird is perched; there is also some red at the base of the tail.

Aratinga nenday (ex *Nandayus nenday*), Nanday Conure (Black-hooded Parakeet), Aratinga Ñanday.
Introduced (T). A South American species now also established in USA. On Tenerife small numbers are present in the south of the island and breeding has occurred in several tourist resorts. A medium-small green parakeet (or conure) with black face and beak, some black on the wings and blue tip to the tail; the thighs are red.

Melopsittacus undulatus, Budgerigar, Periquito Común.
Introduced (T). An Australian dry country species that may now be the most popular pet in the world apart from dogs and cats. Escaped cage birds have bred in Santa Cruz, but the species is now rare. A very small parrot with a very long tail. The plumage of natural wild birds is green with a yellowish head and black and white scaly markings on nape, back and wings.

Myopsittacus monachus (ex *Myiopsitta monachus*), Monk Parakeet, Cotorra Argentina.
Introduced (F, C, T, P). Native to temperate and subtropical South America but now established in many other places. On Tenerife the Monk Parakeet is now present in small numbers in towns in the east, north and south. A small, long-tailed parrot, bright green above, with a grey forehead and pale grey underparts shading to very pale green; the flight feathers are dark greenish blue and the bill is orange.

Poicephalus senegalus, Senegal Parrot, Lorito Senegalés.
Introduced (T). Native to West Africa. On Tenerife there are records of escaped birds in several places and breeding occurs in Santa Cruz. A medium-small parrot with large head and short broad tail. The head and bill are charcoal grey, the back bright green, rump yellow and the underparts with a yellow and green 'vest'.

Psittacula krameri, Ring-necked Parakeet (Rose-ringed Parakeet), Cotorra de Kramer.
Introduced (L, F, C, T). A species from tropical Africa and Asia. Escaped birds are seen in several places and there is a small small breeding colony in Santa Cruz. A small, long-tailed, bright green parrot with an orange-red bill; the male has a narrow red ring round the neck.

Order STRIGIFORMES – owls

Family Strigidae – typical owls

Asio otus canariensis, Long-eared Owl, Búho Chico.
Canary endemic subspecies (F, C, T, G, H, P). This owl occurs throughout the island, although it is less common in the laurel forest, pine forest and Teide National Park. The Canaries subspecies is

smaller than the more northern subspecies *Asio otus otus,* with more heavily mottled and streaked plumage. It also has a broader habitat range, often being found in open country. It frequently nests on the ground, for instance under clumps of Cardón, or in rock crevices in ravines and occasionally in the crowns of palm trees. Laying can occur as early as December, but is mainly from February to April. The diet consists mainly of mice and rats, though small birds are also taken, and also geckos and orthopteran insects, especially in summer. For both this species and the Barn Owl *Tyto alba,* collisions with traffic on roads are an important cause of mortality. Additional deaths of this species, and to some extent also of the Barn Owl, occur as a result of entanglement by burr bristlegrass *Setaria adhaerens,* a grass that may be native to the island although it occurs mainly in habitats created by humans.

Family Tytonidae – barn owls

Tyto alba, Barn Owl, Lechuza Común.
Breeding resident (L, F, C, T, G? H, P?). The subspecies on Tenerife is the widespread *Tyto alba alba,* but an endemic subspecies is present on the eastern islands. On Tenerife the Barn Owl has a patchy distribution, favouring steep ravines and cliffs, and occurring as much as 1000 m above sea level in the northwest and south. However, it is more frequent in the zone below the forests, often occurring in agricultural areas and near habitations. It nests in holes in the sides of barrancos, and in lava tubes. Laying is from January to June, but mainly in February and March. The diet is dominated by the House Mouse, with rats much less important; frogs are also taken, and in summer geckos form a significant part of the diet, along with insects such as earwigs and the cricket *Gryllus bimaculatus.*

MAMMALS

Apart from bats and the extinct giant rat, the terrestrial mammals on Tenerife have all been introduced by humans – accidentally or deliberately – in the last 2500 years. Prior to that time, in the absence of large mammalian herbivores, the vegetation in large parts of the island would have been very different.

Order ARTIODACTYLA – even-toed ungulates

Pig, *Sus domesticus*, Cerdo.
On Tenerife the original domestic pig breed was the Canarian Black Pig, descended from stock brought to the island by the Guanches. Pigs have been domesticated in the Old World for over 10,000 years but are ultimately derived from the Wild Boar *Sus scrofa*. The Canarian Black Pig is of interest to geneticists, resembling those from China and other parts of Asia in the abundance of bristles and wrinkles of the skin, and the large ears that cover the eyes. The Canarian breed almost became extinct in the 1980s, with numbers perhaps below 40 individuals, mainly on family farms on La Palma; however, breeding programmes were then established, including one on Tenerife. Modern breeds of pigs are now mostly kept intensively in sheds, with current numbers on Tenerife around 30,000.

Dromedary (Arabian Camel), *Camelus dromedarius*, Camello.
The one-humped camel native to southwest Asia and northern Africa was domesticated about 3000 years ago, but was brought to the Canaries from North Africa only in about 1405, with Moorish expeditions. After the Castilian conquest during the following century, camels played an important part in agricultural development, especially in the eastern islands but also in the south of Tenerife and Gran Canaria. Reinforcement of the Canarian camel population by imports from the Western Sahara continued until the 20th century; this has now ceased, and changes in agriculture have led to drastic decline in the population, though some camels are used in the tourist industry even on Tenerife. The camels of the Canaries represent the only European traditional camel population and it has recently been suggested that action should be taken to maintain genetic variability within it.

Bovine (cow), ***Bos taurus*** (***Bos primigenius taurus***), Vaca.
No cattle were present on the Canaries until the Castilian conquest, when importation occurred from Iberia. This led to development of a traditional Canarian breed, with about 5000 individuals in the archipelago in the early 1990s; La Palma had a distinct breed. On Tenerife the native cattle are referred to as 'raza basta'. They are long-lived, hardy and gentle animals, used for milk and meat but also as draught animals: both cows and bulls were used for pulling ploughs. There is now an official programme to support the traditional breed, and pulling competitions are arranged to raise its profile. Other breeds are now also present, and in total there are nearly 5000 cattle on Tenerife, though these are rarely visible.

Domestic Goat, ***Capra hircus***, Cabra.
The domestic goat is derived from the Wild Goat *Capra aegagrus* of Europe, the Middle East and central Asia. Goats were brought to the islands from Northwest Africa around 2500 years ago by the Guanches, and evidently formed a key element in their way of life. Recent molecular studies confirm that the three recognised Canarian goat breeds (Majorera, Palmera and Tinerfeña) are closely related but are very distinct from other goat breeds, implying that there has been little mixing with goats from other places. Recent study has shown that on Tenerife the goats are significantly different in the north and the south of the island.

In the past, large flocks of goats grazed most of Tenerife, including Las Cañadas in summer; goats are now excluded from Teide National Park, and numbers have declined in some other areas. However, goat farming is still an important agricultural activity; in 1990 around 150,000 goats were present in the Canaries as a whole, and even now Tenerife has about 60,000. Management is now more intensive, and supplementary food is increasingly used, but large flocks are taken out to graze on a daily basis in parts of the south of the island, and small groups can be seen in Anaga and elsewhere. 'Artesanal' cheese made from raw goat's milk is still an important commodity on the island.

Mouflon, ***Ovis orientalis*** (***Ovis aries, Ovis musimon***), Muflón.
The Mouflon, wild sheep of several Mediterranean islands, are thought to represent feral populations of ancient domestic stock, and have been widely introduced into various parts of Europe. Thirteen Mouflon were introduced to the mountains of Tenerife in 1970, amid considerable controversy, to provide additional sport for hunters; they have been subject to controlled hunting since 1977. They now number rather more than 100 and range over an area exceeding 350 km^2, both within the forests and up into the Teide National Park, where they have had a significant impact on some of the rarest plants in the world. Because of this ecological damage, it has been suggested that they should be eradicated, but the strength of the hunting lobby makes this difficult to achieve in practice.

Domestic Sheep, ***Ovis aries***, Oveja.
The origin of the modern breed of sheep in the Canaries is complex, since those brought to the islands by the Guanches were wool-less hair sheep, probably similar to West African Dwarf sheep, which are common in sub-Saharan Africa. Hair sheep are not now found in northwest Africa but may have been there around 3000 years ago, subsequently dying out as a result of increasing aridity. Columbus, on his second voyage to the New World, loaded many live domestic animals in the Canaries rather than in the Iberian peninsula, to minimise the time they would have to spend on board; this procedure became normal for later crossings of the Atlantic. The animals doubtless included local hair sheep, which are thought to have contributed to the ancestry of the Pelibüey hair sheep now present in the Caribbean region. After the Castilians conquered the Canaries they brought wool sheep to the islands, presumably from Iberia, and interbreeding resulted in the loss of the special character of the primitive hair sheep. However, we note that some sheep on Tenerife, though now with a heavy wool fleece, have retained the ancient ability to shed this seasonally, although shedding is a feature normally lost during the development of modern sheep breeds. Recently, some hair sheep have been brought back to the Canaries from Venezuela.

During recent centuries domestic sheep on the Canaries have been kept primarily for cheese production, and they are much less numerous than goats; in the early 1980s the sheep population in the archipelago was below 20,000, but it reached nearly 30,000 a decade later, with official support including that of the local government. In Tenerife there are now about 15,000 sheep. Ewe's milk is used as the primary ingredient of many Canary Island cheeses, of which the best known is 'Flor de Guía' from Gran Canaria, in which semi-skimmed or whole milk is coagulated using vegetable rennet derived from flowers of the native Cardoon thistle *Cynara cardunculus*.

ORDER CARNIVORA – CARNIVORES

Cat, ***Felis catus***, Gato.
Cats may have arrived on the Canaries along with the early human inhabitants, and feral cats are present on all islands, while domestic cats also roam widely at night. As on other islands around the world, cats have probably had a disastrous effect on native wildlife. It is clear that seabirds can now only nest in holes or on cliffs that are inaccessible to cats, and the giant lizard *Gallotia goliath* is now only found in the form of bones in lava tubes. The recent discovery of the Tenerife Speckled Lizard *Gallotia intermedia* has led to renewed efforts to control feral cats in some areas, using fences. However, effective control of cats has proved difficult all over the world, and it is only on uninhabited islands, and some with relatively small human populations such as Ascension Island in the tropical Atlantic, that feral cats have been successfully eradicated, allowing native birds to recolonize. In the eastern Canaries, the feral cats have now been eliminated from Alegranza, but there is evidence that the considerable cat population on Graciosa is responsible for the recorded decrease in the breeding population of seabirds there. It should be remembered, however, that cats also prey on introduced animals such as mice, rats and rabbits.

Dog, ***Canis familiaris***, Perro.
Dogs may have been on some of the Canary Islands before the Guanches became established, since ships doubtless visited the islands in earlier times; they have certainly been there for over two thousand years. Several breeds of Canary Island dogs are recognised in various countries, one of which is the Perro de Presa Canario (Canary catching dog). This is similar to one of the types commonly seen in the countryside in Tenerife, which is a heavy-set breed probably derived from dogs used by Guanche herdsmen; however, it is unclear whether the stock came entirely from Iberia at the time of the conquest or also involved dogs already present on the islands. The modern form was influenced by breeding for dog-fighting about two centuries ago, when local dogs were crossed with English mastiffs; they are effective guard dogs and one is quite often confronted by chained dogs of this type when walking.

The other breed most likely to be seen in the countryside is the 'Podenco Canario' or Canary hunting dog, a sturdy but slender prick-eared hound. It has often been suggested that this breed originated in Egypt and North Africa and came to the island with the earliest settlers, but recent studies suggest a common origin with other European hunting dogs, which may anyhow have an ancient Mediterranean origin. Rabbit hunting with dogs is a very popular sport, and at weekends pickup trucks full of Podencos are often encountered in remote parts of the island. When not in use, many of these hunting dogs are kept in fairly isolated sheds, and if you come near these they certainly let you know they are there and are hoping that you are coming to feed them. A third local breed is the fluffy and amiable Bichon Tenerife, probably originating from the Mediterranean region. Stray or abandoned dogs of various breeds are a fairly common sight in the countryside and they occasionally reproduce in the wild.

Ferret, ***Mustela putorius furo***, Hurón.
Ferrets, along with dogs and guns, are used for hunting rabbits, and they sometimes escape. On La Palma there is now an established feral population which is a threat to the native birds, and vigilance is needed to ensure that this does not happen in Tenerife, where there are some scattered records of escaped animals.

Order CETACEA – whales, porpoises and dolphins

The ocean around the Canary Islands is used by a remarkably large number of kinds of whales and dolphins, some of them very poorly known. Although most of the species are typical of warm-temperate waters, some cool water species reach the area under the influence of the cold Canary Current, and a few pantropical species also visit the islands. Most of the species prefer deep water and are seldom or never seen close to land, except when found stranded and dying on the beaches. Some species, however, are frequently seen close to shore and Tenerife now offers many whale-watching opportunities, with knowledgeable guides who can help with identification, which is often tricky. We include here a list of species recorded in Tenerife waters in recent years; several others are likely to occur sporadically. The English and Spanish names are mainly those used by the International Union for Conservation of Nature (IUCN); a few of the many alternative names are given in parenthesis.

Suborder MYSTICETI – baleen whales

Family Balaenidae – right whales

Eubalaena glacialis, Northern Right Whale, Ballena Franca del Norte.
This cool water species is considered to be in danger of extinction, although the population may be recovering slowly; it is seen off Tenerife only rarely. The average length is about 15 m, and it is distinguished by the lack of a dorsal fin and by the double, V-shaped spout; the tail flukes are raised in diving (in rorquals – see below – they are rarely seen).

Family Balaenopteridae – rorquals & humpback

Balaenoptera borealis, Sei Whale, Rorcual Norteño (Ballena Boba, Ballena Sei).
A migratory species occurring seasonally off Tenerife. Males average 15 m in length, females 16 m. As in the other rorquals, the dorsal fin is small and set far back on the body., but is clearly visible. The spout of the rorquals is single and tall.

Balaenoptera edeni, Tropical Whale (Eden's Whale), Rorcual Tropical (Ballena de Bryde).
Occurs off Tenerife irregularly at any time of year. It averages 12 m in males and 13 m in females. It is similar to the other rorquals but has three parallel ridges along the top of the head instead of only one; however, these are hard to see.

Balaenoptera physalus, Fin Whale, Rorcual Común (Ballena Aleta).
A migratory species occurring at some times of year. This is the largest whale normally seen off Tenerife; males average 21 m, females 22 m.

Megaptera novaeangliae, Humpback Whale, Ballena Jorobada.
Occasionally recorded. The Humpback has an average length of about 15 m. It differs from the rorquals in several ways, but especially the enormous flippers, which are about 5 m long and scalloped on the leading edge. The dorsal fin is clearly visible and the tail flukes are often raised when diving. The spout is single and bushy.

Suborder ODONTOCETI – toothed whales

Family Delphinidae – dolphins & pilot whales

Delphinus delphis, Short-beaked Common Dolphin, Delfín Común.
Occurs off Tenerife mainly in winter. A small dolphin with average length of 2.1 m, capable of swimming at over 40 km per hour. Like the *Stenella* species (see below) it has a noticeable beak sharply set off from the forehead. The body is deep grey above but lighter below and with very varied patterns on the sides, including patches of deep and light grey and yellow; this criss-cross of patterns is the best

way to distinguish it from the Striped Dolphin. Common Dolphins occur in very active schools with individuals frequently leaping, and often accompany ships.

Globicephala macrorhynchus, Short-finned Pilot Whale, Calderón Tropical (Calderón de Aleta Corta).
This species, distributed in subtropical and tropical waters around the world, occurs all year round off Tenerife and is the species most regularly seen on whale-watching trips, especially between Tenerife and La Gomera, but preferring water depths around 1500 m. It is actually a large and stocky dolphin, but it behaves more like the larger whales; average sized males are over 5 m long, females just over 4 m. The head is bulbous with no prominent beak, and the broad-based dorsal fin is set well forward, curving back with the tip almost parallel with the back. The body is black or dark grey-brown, sometimes with a grey saddle behind the dorsal fin. Pilot whales often form large family groups, and may float quietly at the surface, making observation easy. The diet includes fish and squid, and also octopus.

Grampus griseus, Risso's Dolphin (Grey Dolphin), Calderón Gris.
Occurs irregularly at any time of year. A medium-sized dolphin with average length of 3 m. The body is heavy as far as the dorsal fin, but relatively slender behind it. There is no beak, but there is a strong vertical crease or furrow running from the blowhole to the upper lip. The colour is mainly pale grey, contrasting with the dark dorsal fin.

Orcinus orca, Orca (Killer Whale), Orca.
Seen very occasionally off Tenerife. Average length 8 m in males, 7 m in females. Easily recognised by its black skin with brilliant white patches, and the enormous vertical dorsal fin of males.

Pseudorca crassidens, False Orca, Falsa Orca.
Seen very occasionallly off Tenerife. Average length 5.4 m in males, 4.6 in females. Confusable with the Pilot Whale, but fast swimming and much more active, often leaping and bow-riding with ships. The skin is black and the head relatively slender.

Stenella coeruleoalba, Striped Dolphin, Delfín Listado (Delfín Blanco y Azul).
Occurs infrequently off Tenerife. A medium-sized dolphin with average length 2.4 m. It is variable in colour from dark grey or brown to bluish grey on its back, with lighter grey flanks and a white belly. There is a dark stripe running from the eye along the side of the body and down to the anus, but often more easily seen – and a good diagnostic feature – is a dark wedge which runs down from behind the dorsal fin and forward along the pale flank, finishing at a point about half way to the eye. These dolphins are usually seen in schools, and often several will be jumping out of the water at the same time.

Stenella frontalis, Atlantic Spotted Dolphin, Delfin Moteado Atlántico (Delfín Manchado del Atlántico, Delfín Pintado).
Occurs irregularly at any time of year. This small dolphin has an average length of only 2.1 m. It has a fairly long and robust beak, with a distinctive crease between the base of the beak and the melon (forehead). Young individuals are uniform grey, shading to white on the belly, but as they mature they develop white spots all over the body, making identification straightforward at close range.

Steno bredanensis, Rough-toothed Dolphin, Delfín Dientes Rugosos (Delfín de Pico Largo).
A pantropical species that occurs irregularly at any time of year. A small dolphin averaging 2.3 m. The mouth is noticeably long and the conical beak runs smoothly into the forehead (cf *Delphinus* and *Stenella* spp.). The body is dark grey-brown, often with small pale patches.

Tursiops truncatus, Bottlenose Dolphin, Delfín Mular (Pez Mular, Tursión).
The Bottlenose Dolphin is one of the most widely distributed cetaceans, occurring worldwide in warm and temperate waters. However, it is uncommon in the open ocean, normally occurring in coastal waters and frequently accompanying ships. Around Tenerife it occurs all year, especially in

the northern part of the channel between the island and La Gomera, frequenting areas with depths less than 600 m. This is a medium-sized dolphin averaging 3 m in length. It is larger and heavier than the Striped Dolphin and Common Dolphin, with a short but well-defined beak and a tall, bulky dorsal fin. The colour is unpatterned grey, dark above but shading to pale grey or whitish below; the tip of the lower jaw is sometimes white.

FAMILY PHYSETERIDAE – SPERM WHALES

Physeter macrocephalus, Great Sperm Whale, Cachalote Común (Ballena Esperma).
Seen infrequently off Tenerife, usually in spring, when groups of females may be accompanied by calves. The average length is 15 m in males and 11 m in females. The enormous head has a vertical front, in contrast to the great baleen whales, and the dorsal fin is represented by a small hump, set well back on the body.

FAMILY ZIPHIIDAE – BEAKED WHALES

Mesoplodon densirostris, Blainville's Beaked Whale, Zifio de Blainville.
Seen very occasionally off Tenerife. A medium sized whale rarely exceeding 5 m in length, with a small dorsal fin set far back; the tail flukes are not normally seen. The distinctive feature is a large upward bulge in the lower jaw on each side, supporting a huge tooth; the bulge is often visible when blowing (cf *Ziphius cavirostris*); the spout is inconspicuous, but shoots forward at a sharp angle.

Ziphius cavirostris, Cuvier's Beaked Whale (Goosebeak Whale), Zifio Común (Ballena de Cuvier).
Occurs at some times of year off Tenerife. A medium sized whale averaging 6.4 m in length. The colour is usually pale grey or bluish, often with a dark eye ring. Only the rounded forehead shows while blowing (cf *Mesoplodon*, above); the spout is an inconspicuous fan.

ORDER CHIROPTERA – BATS

Bats arrived on the Canaries naturally: they needed no assistance from humans. They were probably rarer after the 1950s, with the widespread use of DDT, but now the populations are recovering. Mortality from wind turbines may now be important, but there is no good information. Tenerife has seven species of native bats, and the taxonomic situation has settled down to some extent after extensive revision resulting from the application of modern acoustic and molecular techniques. There is one Canary endemic species *Plecotus teneriffae,* one Canary endemic subspecies *Barbastella barbastellus guanchae* and one Macaronesian endemic species *Pipistrellus maderensis*. In addition, a non-native fruit bat escaped from captivity, but the population may now have been eradicated.

FAMILY MOLOSSIDAE – FREE-TAILED BATS

Tadarida teniotis, European Free-tailed Bat, Murciélago Rabudo.
Native (C, T, G, P, H); the species is widespread in the Mediterranean, North Africa, the Middle East and Asia. The flight of this large bat is fast, direct and high up, and on Tenerife the species occurs from sea level to the high mountain zone; it uses a wide variety of habitats, so long as they contain rock faces for roosting. The food consists mainly of large moths, and it is often attracted to street lights in villages.

FAMILY PTEROPODIDAE – FRUIT BATS

[*Rousettus aegyptiacus*, Egyptian Rousette (Egyptian Fruit Bat), Murciélago Frugívoro Egipcio.†
Introduced and potentially invasive (T). This fruit bat was reported to have set up colonies in the wild after it escaped from the Loro Parque zoo in Puerto de la Cruz, and from another zoo in the south, in about 2001. It was considered a threat to farming and the ecology of the island, so an expensive eradication campaign was undertaken by the government, which seems to have been successful.]

Family Vespertilionidae – evening bats

Barbastella barbastellus guanchae, Barbastelle, Murciélago de Bosque (Barbastela).
Canary endemic subspecies (T, G) of a species that occurs in parts of Europe (including Britain) and also in northern Africa. The Canary subspecies was discovered only recently and a mitochondrial DNA study shows it to be very distinct from continental populations, suggesting an ancient colonisation; it is considered rare and endangered. The distribution of this medium-sized bat on Tenerife is poorly known, but it seems to occur mainly in pine and laurel forest, as well as some lowland localities in the north of the island; it roosts in rock crevices and in caves, and under loose bark. The food includes moths and flies, but not large hard-bodied insects. Feeding is by both hawking and gleaning insects off leaves, and it sometimes hunts around lights in rural areas. The Canary subspecies differs from the continental populations by its uniformly blackish chestnut upper parts and by a whitish U-shaped hairy line on the lower border of the underside.

Hypsugo savii (*Pipistrellus savii*), Savi's Pipistrelle, Murciélago Montañero.
Native (C, T, G, H, P). We follow the current Canary government list in treating the *Hypsugo* bats on Tenerife as *H. savii*, since the taxonomy of the group is still somewhat fluid; however, it may be that all will prove to belong to the closely related *Hypsugo* cf *darwinii*, a taxon described from La Palma in the 19th century. *H. savii* occurs in the Mediterranean, Africa, Middle East and into Asia, and also the Cape Verdes. It is one of the commoner bats on Tenerife, occurring over a wide range of altitude, mainly in fairly open rocky habitats, around cliffs and in deep ravines. Roosting is primarily in rock crevices, but rarely underground. It has been recorded as prey of Barn Owls. The food includes a wide range of insects.

Nyctalus leisleri, Lesser Noctule (Leisler's Bat), Nóctulo Pequeño.
Native (T, P). Widespread on Tenerife, occurring mainly at middle altitudes in laurel forest and especially pine forest, and also in middle level cultivated areas and around cliffs. It is the most common bat species in Teide National Park. Roosts in tree holes and overgrown crevices, as well as in buildings, and takes readily to bat boxes. Foods include flies, moths and beetles.

Pipistrellus kuhlii, Kuhl's Pipistrelle, Murciélago de Borde Claro.
Native (L†, F, C, T). *P. kuhlii* is widely distributed in southern Europe and the Middle East, but there is evidence that more than one biological species is included under this name, and the Canary populations may really belong to the African species *P. hesperidus*. On Tenerife the range of *P. kuhlii* overlaps with that of the very similar *P. maderensis*. The two species have similar social calls and it has been shown experimentally that during aggressive encounters at feeding sites the calls of one of them elicit reactions from members of the other species as well as conspecifics. In contrast, the echolocation calls used in finding prey are distinct, suggesting divergence in diet. Roosting sites of *P. kuhlii* include crevices in rocks and under bridges.

Pipistrellus maderensis, Madeira Pipistrelle, Murciélago de Madeira.
Macaronesian endemic (T, G, H, P), probably evolved from *P. kuhlii* stock. Molecular study shows that the Madeira and Canaries populations are closely related, but with significant divergence that may justify separation at the subspecies level. With populations only on the Canaries, Madeira and the Azores, *P. maderensis* is considered endangered. On Tenerife, however, it is one of the most widespread and abundant bats, occurring in all habitats and often hunting around water tanks and near lights in populated areas. It uses a wide variety of roost sites, including rock crevices, holes in trees or cavities in buildings.

Plecotus teneriffae, Canary Long-eared Bat (Canary Big-eared Bat), Orejudo Canario.
Canary endemic species (T, H, P), previously treated as an endemic subspecies of a more widespread species; it is probably derived from *P. austriacus* of North Africa and Europe. The species is in the endangered category of the IUCN Red List because of its fragmented distribution and the recorded

decline of its largest known breeding colony (on La Palma). On Tenerife this bat lives mainly in caves in the pine forests and high mountain zone; it is probably very sedentary. It not only roosts in lava tubes but also often hunts in them, feeding on moths and flies; outside the caves, its prey includes the abundant pine forest moth *Calliteara fortunata*.

Order INSECTIVORA – insectivores

Atelerix algirus, North African Hedgehog (Algerian Hedgehog), Erizo Moruno.
Introduced and invasive (L, F, C, T). This species is native to northern Africa but also occurs in France and Spain, perhaps having been introduced by humans. It is thought that a pair from Morocco was introduced to Fuerteventura in 1892, and some were brought to Tenerife in about 1903. They are found mainly in the lower zone, but occasionally in the forests and high mountain zone. There is concern about their predation on the nests of ground-nesting birds, as in the case of the European Hedgehog in the western islands of Scotland. The North African Hedgehog is similar to the European species but usually smaller (20-25 cm long) and somewhat paler, and with a dark mask.

Crocidura russula, Greater White-toothed Shrew, Musaraña.
This shrew has been found in Santa Cruz on Tenerife, presumably having been accidentally introduced; it is a native of northern Africa and south central Europe. However, an endemic species *Crocidura canariensis* is present on the eastern Canary Islands; this was first detected in 1983 when Aurelio Martín found a skull in a Barn Owl pellet on the small island of Graciosa, north of Lanzarote.

Suncus etruscus, Pygmy White-toothed Shrew (Etruscan Shrew), Musarañita.
Presumably introduced (T). This tiny shrew, which is widely distributed in the Old World, has been found in many localities, both in the northern and southern sides of the island, from the coast up to more than 2000 m.

Order LAGOMORPHA – rabbits and hares

Oryctolagus cuniculus, European Rabbit, Conejo.
Rabbits came to all the islands around the time of the Castilian conquest, 500 years ago. They originated in the Iberian Peninsula and northwest Africa, but have been introduced to the rest of Europe and much of the rest of the world. Rabbits are now the most important wild herbivores on Tenerife, occurring in almost all parts of the island. They have been shown to have a major influence on the abundance and distribution of some of the native plants, the effects perhaps being most significant in the high mountain zone. Apart from grazing, they dig for plant roots, especially in times of drought, and their droppings can be seen even in volcanic craters and in the middle of lava flows where there is virtually no vegetation. Breeding occurs at almost all seasons, with a short break around September; hybridisation with domestic rabbits produces some large individuals of varied colours. Rabbits on Tenerife generally live in small groups, although warrens with large colonies can be found where the soil allows. Rabbits are numerous in most years, but numbers vary substantially, partly as a result of the spread of disease from farmed rabbits. Although this might be good news for the native vegetation of the island, it is also bad news since rabbit hunting with dogs, ferrets and guns is a very popular sport. In 2014 there were still over 5,600 registered hunters, although this figure is less than half what it was a few years ago, probably as a result of the crisis in the economy. Rabbit is an esteemed food in the Canaries, and 'conejo' is always on the menu in traditional restaurants.

Order PERISSODACTYLA – odd-toed ungulates

Equus africanus asinus, Donkey (Ass), Burro.
Donkeys were brought to the islands at the time of the Castilian conquest and have been used extensively in agriculture and for transport; in Fuerteventura donkeys roamed wild for many years. Mules (crosses between a male donkey and a female horse) were also used for transport of goods and for riding. Many

of the well-built trails on the island were constructed as donkey or mule tracks; they now form the basis of the modern network of hiking trails.

Equus caballus, Horse, Caballo.
Horses were brought to the island during the wars of conquest and significant numbers have probably always been kept on farms and estates since that time. There are now several stables providing riding lessons and trekking opportunities.

ORDER RODENTIA – RODENTS

[*Canariomys bravoi*, Tenerife Giant Rat, Rata gigante de Tenerife.†
This extinct rodent, endemic to Tenerife, was so different from other members of the subfamily Murinae that a new genus was created for it and a similar fossil species from Gran Canaria. (The recently discovered but also extinct *Malpaisomys* from the Eastern Canaries was very different.) *Canariomys* species are probably related to African thicket rats. Sub-fossil bones of the giant rat have been found in various parts of Tenerife, especially in lava tubes, and the species seems to have survived into the time of the early human settlers. Possible causes of its extinction include predation by introduced dogs and cats, competition with the introduced species of *Rattus*, and/or disease brought in by these. Scientists recently made a detailed comparison of a skeleton of *C. bravoi* with those of rodents in several other groups (including tree-dwellers). They concluded that *Canariomys* was considerably larger than an ordinary rat, with head & body length around 23 cm, tail about 16 cm, and a body mass around 900 g (this is about three times the weight of a Common Rat and four-fifths of that of the wild introduced rabbits). The skeletal proportions give some clues as to the probable mode of life of *C. bravoi*, suggesting a robust, unspecialised animal with some ability in both tree-climbing and digging. Careful study of the shape of the jaws and the wear on the teeth indicates a diet based on plants, fruit and invertebrates, but not grasses; we guess that the fruits, flowers and leaves of laurel-forest trees were important resources.]

Mus musculus domesticus, House Mouse, Ratón.
The House Mouse originated in the Turkestan region, but is now a worldwide species. It has been argued that if mice had colonised the Canary Islands naturally, the most likely species would have been *Mus spretus*, native to northwest Africa. *Mus musculus* is also present there now, as well as in Europe, and it may have reached the Canaries from either area, but with human help; the date of arrival on Tenerife is unknown, but probably no more than two and half thousand years ago. It is the only small rodent present on the island, and it occurs in all zones.

Rattus norvegicus, Common Rat (Norway Rat, Brown Rat), Rata Común (Rata gris).
This introduced species originated on the plains of northern China and Mongolia; it spread to Europe later than the Black Rat, perhaps in the Medieval period, and later spread worldwide; it probably reached Tenerife at the time of the Castilian conquest. It occurs mainly in towns and villages in lowland and middle areas. The Common Rat is hard to distinguish from the Black Rat, but the eyes and ears are relatively smaller, and the ears are finely furred. The tail is shorter and thicker and usually dark above and pale beneath.

Rattus rattus, Black Rat (Roof Rat, Ship Rat) Rata de Campo, Rata Negra.
The Black Rat probably originated in the Indo-Malayan region, but spread into Europe almost 2500 years ago and then travelled on ships to become a worldwide species; the date of arrival in Tenerife is unknown. This rat is found throughout the island, but rarely in large villages and towns. In the laurel forest it shows a preference for the less disturbed areas, as indicated by the fact that predation on birds nests on the ground is higher there than in areas close to roads. Apart from the impact on birds, the rats probably also have a large effect on the populations of reptiles and invertebrates. The Black Rat eats fruits of many forest trees, and experiments have shown that it prefers fruits of *Viburnum rigidum* and *Laurus novocanariensis* to those of *Persea indica*. The rats also bite off twigs (probably eating some of the bark) and stretches of forest paths are sometimes carpeted with cut leaves, often

of the laurel-relative *Apollonias barbujana*. The Black Rat is similar to the Common Rat but has relatively larger eyes and also ears, which are thinner and almost hairless; its tail is longer, thinner and uniformly coloured.

REPTILES

Order SQUAMATA – scaled reptiles

Family Gekkonidae – geckos

Hemidactylus turcicus, Turkish Gecko, Salamanquesa Rosada.
Introduced (C, T); originates in the Mediterranean and Middle East. This gecko was probably introduced inadvertently at the port of Santa Cruz. It is most easily distinguished from the Tenerife Wall Gecko (see below) by the colour of its tail, which has alternating rings of pale brown and black, and by the presence of claws on all its toes.

Family Lacertidae

The lizards of the Canary Islands belong to the endemic genus *Gallotia*; their closest relatives are the members of the genus *Psammodromus* in Europe and North Africa. The sequence of divergence of the species and subspecies has recently been worked out by studies of their mitochondrial DNA, which have revealed a complex evolutionary history.

Gallotia galloti, Tenerife Lizard, Western Canary Lizard, Lagarto Tizón.
Canary endemic species (T, P) in a Canary endemic genus. There is one subspecies on La Palma and three on Tenerife: *galloti* in the south and centre of the island including Teide, *eisentrauti* in the north and *insulanagae* confined to the Roque de Fuera off the northeast coast of Anaga. A closely related species, *G. caesaris*, is found on La Gomera and El Hierro, and has colonised the harbour area on Tenerife around Los Cristianos, perhaps after travelling on the ferry from La Gomera.

The Tenerife Lizard is widespread and common on the mainland, up to more than 3700 m (the top of Teide) and on the offshore islets. It is omnivorous, with fruit forming a large proportion of the diet, but it also eats a wide variety of invertebrates, especially ants and beetles. Seeds from fruit often pass through the gut unharmed, so lizards are probably important dispersers of seeds of *Plocama pendula*, *Rubia fruticosa* and other shrubs and climbing plants. These lizards can be a pest to vegetable growers and they are sometimes poisoned or trapped. They are active by day and though normally shy they can usually be attracted by an apple core or piece of tomato or banana.

This lizard is heavily built and shows enormous variation. Males can reach 40 cm including the tail. The adult male is dark brown or blackish, more or less spotted with blue and green; the female is rich brown above and has pale longitudinal stripes.

[*Gallotia goliath*, Tenerife Giant Lizard, Lagarto Gigante de Tenerife.†
Tenerife endemic species in a Canary endemic genus, but now extinct. This large species, which was up to 1.5 m long, has been shown by recent molecular studies to be distinct from *G. simonyi*, which survives on El Hierro. However, it does belong in the *G. simonyi* group, which also includes *G. bravoana* on La Gomera, the recently discovered *G. intermedia* on Tenerife, and probably also *G. auaritae* on La Palma (said to have been discovered in 2007). Bones from Tenerife previously described as a separate species *Gallotia maxima* probably belong to *G. goliath*. Bones and mummies of *Gallotia goliath* have been found in caves in eastern Tenerife and the species probably survived until after the arrival of the first settlers. Extinction was probably caused by predation: these large lizards would certainly have been eaten by cats and dogs, and probably also by people; predation by rats on eggs and young is also likely.]

Gallotia intermedia, Tenerife Speckled Lizard, Lagarto Canario Moteado.
Tenerife endemic species in a Canary endemic genus. This species was discovered only in 1996, and is closely related to the larger but extinct *G. goliath*. Its closest living relatives are the giant lizards of the western Canaries; it is more distantly related to *G. galloti* of Tenerife and La Palma. *Gallotia intermedia* is rare and critically endangered; its range is restricted to small parts of the cliffs and rocky slopes along the coast in the Teno massif in the northwest, and also on cliffs near the Montaña de Guaza in the far south. The total population has been estimated at around 600 individuals, but these are in about 40 separate subpopulations, some of them so small that they are likely to suffer from inbreeding. It is assumed that the species was much more widespread on the island in the past, and it seems possible that it occupied the western part of the island and was replaced by *G. goliath* in the east. The main threat to the Tenerife Speckled Lizard is predation by feral cats, along with rats; numbers of lizards have been increasing in recent years, probably as a result of fencing of some sites. These lizards are mainly herbivorous, generally eating all parts of the plants: fruit, flowers, leaves, seeds. They also scavenge around gulls' nests and predate eggs and small chicks, and doubtless also insects.

This is a heavy-bodied lizard, blackish-brown and with either a network of yellow spots or grey lines on its back; some individuals have two rows of eye-like spots, one blue and one yellow, along the sides of the body. The tail is grey-brown and the underside greyish-yellow. The eye is orange. The female is slightly smaller and has a line of seven blue eye-like spots along its back.

Family Phyllodactylidae

Tarentola delalandii, Tenerife Wall Gecko, Perenquén Común.
This gecko is currently considered to be endemic to Tenerife and La Palma, with related species on El Hierro and La Gomera, Gran Canaria and the eastern islands. Members of the genus *Tarentola* have a transatlantic distribution and are evidently excellent dispersers. They are closely related to the Helmeted Gecko *Tarentola chazaliae* of northwest Africa, which is often kept as a pet. The Tenerife Wall Gecko occurs from sea level up to about 2000 m, and can often be found by turning over rocks. It is flat-bodied, with warty skin and large head and eyes; it is generally dark ash-grey with black marks, but can change colour rapidly. The specialised adhesive pads on the toes make it an excellent climber, even on overhanging surfaces. It is often active at night, and can be seen hunting insects attracted to lights in towns. (Another species, the Moorish Gecko *Tarentola mauritanica*, has been reported on the island, but more evidence of its presence is required; it has enlarged tubercules and appears almost spiny.)

Family Polychrotidae

Anolis carolinensis, Carolina Anole.
This American species has apparently been introduced to southern Tenerife, but we have no evidence about its distribution.

Family Scincidae – skinks

Chalcides viridanus, Tenerife Skink, Lisa.
The latest studies treat this species as endemic to Tenerife, with related species on La Gomera and El Hierro, Gran Canaria, and the Eastern islands; introduced *C. viridanus* have also been found on La Palma. On Tenerife the skink is commonest at lower levels in the north, but it is widespread and occurs from sea level up to 2800 m, as well as on the Roque de Fuera and Roque de Garachico. It is secretive and is normally encountered only by turning over rocks or logs. It is a small lizard with head and body length up to 90 mm and total length roughly double this; it is rather cylindrical, with a small head and short weak legs. It is very glossy, appearing somewhat iridescent, warm brown on the back with rows of pale spots, and blackish on the sides.

Family Typhlopidae

Ramphotyphlops braminus, Brahminy Blind Snake.
Introduced (F?, C, T, G?). This tiny soil-living snake, probably native to Asia and Africa, is now the most widely distributed terrestrial snake in the world, having been distributed by the plant trade. It resembles a black, shiny and unsegmented earthworm, and is one of only two known unisexual snakes: males have never been found.

Order TESTUDINES – turtles and tortoises

Four species of sea turtles occur in the waters around the Canary Islands, although none of them now breed in the archipelago. They are hard to identify at sea, but the brief descriptions that we give should help in identification of individuals that become stranded on beaches. There is one introduced freshwater turtle and no living native land turtles (tortoises) but we include a comment on one fossil species.

Family Cheloniidae – sea turtles

Caretta caretta, Loggerhead Turtle, Tortuga Boba.
Widespread species occurring in the Atlantic, Mediterranean, Pacific and Indian Oceans. This is much the commonest of the turtles seen around the coast. It is a horny-shelled turtle with an oval and rather long shell growing to about 110 cm, although it is usually smaller than this. There are five costal plates (clearly marked areas on the upper side of the shell) along each side. The skin ranges from yellow to brown in colour, and the shell is normally reddish brown.

Eretmochelys imbricata, Hawksbill Turtle, Tortuga Carey.
A tropical turtle found in the Atlantic, Pacific and Indian Oceans, but considered critically endangered; it is rarely seen in Northern Europe. The Hawksbill is sometimes seen near the coast of Tenerife. It is horny-shelled and grows to 90 cm or more. It is easily distinguished from other species by having overlapping horny plates on its shell. There are only four costal plates along each side.

Chelonia mydas, Green Turtle, Tortuga Verde.
A species of warm seas; very rarely seen in European waters. This is a large horny-shelled turtle with the shell growing to 140 cm long. The shell is oval and brown or olive with darker markings. The plates do not overlap and there are only four costal plates along each side.

Lepidochelys kempii, Kemp's Ridley Turtle, Cotorra, (Tortuga Iora, Tortuga Marina Bastarda).
This is the rarest of all sea turtles, breeding only in Mexico and Texas, and is considered critically endangered. It has occurred as a straggler in Canarian waters.

Family Dermochelyidae

Dermochelys coriacea, Leatherback Turtle, Tortuga Laúd.
A widespread but critically endangered species occurring in the Atlantic, Mediterranean, Pacific and Indian Oceans. This is the largest sea turtle, growing to 180 cm or more. It can be distinguished from other turtles by the fact that the carapace (upper part of the shell) is formed of thick, leathery, dark brown skin instead of bony and horny scutes; its shell has 5 to 7 prominent longitudinal ridges.

Family Emydidae

Trachemys scripta elegans, Red-eared Slider Terrapin.
This North American terrapin, which is widely kept as a pet, has been found in urban and rural ponds on Tenerife; it is spreading, mainly in the north of the island, as it has recently in Iberia.

Family Testudinidae – land tortoises

[*Geochelone burchardi*, Tenerife Giant Tortoise, Tortuga Gigante de Tenerife.†
Extinct; fossil bones of this tortoise have been found in volcanic deposits of Pleistocene age in a quarry near Adeje in the south of the island, and probably also near Güímar. It seems certain that it did not survive to the time of arrival of humans, since it would undoubtedly have been eaten and remains would turn up in archaeological sites. There is no evidence of the time of its extinction, but it may well have been caused by the intensive volcanism associated with the building of the Las Cañadas edifice during the last three million years. This tortoise was probably over 80 cm long.]

AMPHIBIANS

Family Ranidae

Pelophylax perezi, Perez's Frog (Iberian Water Frog), Rana Común.
This frog was probably introduced by the Castilian colonists for food and perhaps also for mosquito control (F, C, T, G, P). It is native to the Iberian Peninsula and northwest Africa, and inhabits pools in parks and elsewhere. It grows to about 15 cm and is dark in colour: usually olive with dark markings.

Hyla meridionalis, Stripeless Tree Frog, Ranita Meridional.
This frog is found on all islands. It is officially considered introduced and invasive, but there is a possibility that it reached the islands naturally. It is native to southern France, the Iberian peninsula and northwest Africa. It is very common in the lower parts of the island in banana plantations and other agricultural land. It lives in trees, bushes and especially banana plants, often high up and quite far from water. However, it needs water for breeding, and congregates in water tanks and other permanent or semi-permanent pools in barrancos.

The tree frog grows up to 50 mm long; it has smooth skin, long legs and adhesive pads on the tips of the fingers and toes. It is variable in colour but usually green (and occasionally spotted) with a dark mark behind the eye. The voice is a deep croak, repeated at intervals of one second of less. The noise during the early part of the night, of hundreds of these frogs 'singing' to attract mates, is not easily forgotten.

FRESHWATER FISH

Family Anguillidae – eels

Anguilla anguilla, European Eel, Anguila.
Native (F, C, T, G). This is the only naturally occurring species of freshwater fish in Tenerife, although four others have been introduced. Eels used to occur on all islands, but are now considered extinct on Lanzarote, El Hierro and La Palma. Eels occur in some barrancos in north Tenerife, including Anaga, but are very rare and probably endangered. European Eels breed only in the Sargasso Sea near the West Indies. The larvae drift across the Atlantic in the Gulf Stream and find their way to fresh water where the young eels (elvers) develop until mature at 8-10 years old. They then return to the sea for their long migration back to the spawning grounds. Their colour varies according to the developmental stage. They are silvery when mature and grow to over a metre in length.

Family Cichlidae – cichlids

Tilapia tholloni.
Introduced (T). This African cichlid has been introduced for food and is considered invasive.

Family Cyprinidae

Carassius auratus, Goldfish, Carpín dorado.
Introduced (C, T, P). The goldfish is of Asian origin and was brought to the islands as a decorative fish.

Cyprinus carpio, Carp, Carpa.
Introduced (C, T, H, P). The carp is of Asian origin and was introduced as a decorative fish.

Family Poeciliidae

Poecilia reticulata, Guppy.
Introduced (C, T). This small fish (5 cm), originally from South America, has been introduced for mosquito control. It can be found in reservoirs and water tanks.

Gambusia affinis, Western Mosquitofish, Gambusia.
Introduced (C, T). This is another small fish (7.6 cm) introduced for mosquito control; it is originally from North America.

Male bush cricket *Calliphona konigi*. Photo: Pedro Oromí

11

BUTTERFLIES, DRAGONFLIES AND OTHER INVERTEBRATES

We cover first the butterflies and a few important groups of moths in some detail, and follow with a full account of the dragonflies. The rest of the chapter provides concise accounts of the diversity of other invertebrate groups, in taxonomic order. There are few references to particular species here, but we mention many conspicuous invertebrates elsewhere in the book. In these accounts, as in the rest of the book, reference to a species being 'endemic' implies that it is found only in the Canaries archipelago; many endemic species however, are restricted to the particular island on which they evolved. Names and numbers of species quoted are based on the Canary Islands government 'Lista de especies silvestres de Canarias' (2009) and will naturally be revised from time to time. Numerous endemic subspecies have also been described and are included in the government list of species living wild on the island, but we do not generally consider them here.

ORDER **LEPIDOPTERA** – BUTTERFLIES AND MOTHS

The Lepidoptera are well represented on Tenerife, with almost exactly 500 species, in 43 families. Of these, almost one third (~161) are considered to be endemic to the Canaries and many of these are restricted to Tenerife. Eight of the genera represented on Tenerife are endemic to the Canaries.

Butterflies are covered here in some detail. The upper surfaces of museum specimens of nearly all the species normally present on Tenerife are illustrated in the spread below, and there are also photos of some species in the ecosystem chapters. We discuss all species that are resident, but not those that are probably vagrants, such as *Callophrys rubi* (Green Hairstreak). We also omit *Polyommatus icarus* (Common Blue) since it is not clear that either this species or the more likely *P. celina* is established on Tenerife, though the latter is

Canararctia rufescens
Photo: Juan José Bacallado

Natural History of Tenerife

Danaus plexippus

Danaus chrysippus

Cyclirius webbianus

Zizeeria knysna

Lampides boeticus

Lycaena phlaeas

Aricia cramera

Thymelicus christi

Vanessa atalanta

Argynnis pandora

Vanessa virginiensis

Vanessa vulcania

Issoria lathonia

Vanessa cardui

11 Butterflies, dragonflies and other invertebrates

Zizeeria knysna underside
Photo: Juan José Bacallado

Cacyreus marshalli underside
Photo: Juan José Bacallado

apparently native to the easternmost islands. We also omit *Gonepteryx eversi,* the brimstone occurring on La Gomera, which is generally considered as a subspecies of *G. cleobule* but has recently been treated as a separate species; it is currently listed as occurring also on Tenerife, but we are doubtful whether this form – which is not highly differentiated at the molecular level and is not readily distinguishable from *G. cleobule* – is capable of forming a separate population on Tenerife.

Moths are not treated fully here: we mention important families and describe a few species that are especially likely to come to the attention of visitors. Among the major families endemicity is high in Geometridae, Noctuidae and Tortricidae (see below). Minor families with high endemicity are Scythrididae (all eight species on Tenerife are endemic), Nepticulidae (9 out of 12 endemic) and Blastobasidae (5 out of 8 endemic).

Family Arctiidae (sometimes included in Erebidae) – arctiids

Canararctia (ex *Rhyparia*) ***rufescens,*** Palomita peluche canaria.
Canary endemic species (T, G) and the only member of a newly established endemic genus. This is an ancient endemic species, probably mainly a moth of the laurel forest, but also occurring at El Médano and elsewhere. The larvae, like those of the distantly related *Calliteara fortunata* (Lymantriidae) have brightly coloured irritating hairs that probably deter predators; they feed on *Morella faya* but also on *Rumex, Urtica* and other plants.

Family Geometridae – geometers

This is one of the moth families in which endemicity in the archipelago is high, with 42 species on Tenerife, 17 of them endemic.

Ascotis fortunata (ex *Cleora fortunata*), Palomita de fortuna.
Macaronesian endemic (C, T, G, H, P) with an endemic subspecies (*A. fortunata fortunata*) on Gran Canaria and Tenerife. This is a common moth of laurisilva and fayal-brezal, and is well known on the island since the larvae – like other members of the family – are 'loopers' that appear to measure the ground as they move (hence Geometridae or 'ground measurer') and the adults can often be seen on the walls of houses, having been attracted to the light. The larvae feed on brambles and many other plants.

Family Hesperiidae – skippers

A group showing a number of structural differences from other butterflies. They have relatively smaller wings and larger eyes than most butterflies and a quick, darting flight; they rest with the wings partly folded over the body, or spread out.

Thymelicus christi, Lulworth Skipper, Mariposita dorada oscura.
Canary endemic species (C, T, G, H, P) with a close relative in central and southern Europe and northern Africa. It is found locally in open places in the lower and forest zones up to about 1800 m, and especially in uncultivated fields with grass and brambles. The larvae feed on grasses. Adults fly from February to September, but only in warm sunny weather. Very small (wingspan 20–24 mm). Rather moth-like and appearing very hairy, with large eyes and clubbed antennae. Coppery orange-brown with gold markings and fringes to the wings.

Family Lycaenidae – blues, coppers and hairstreaks

The members of this family are all very small and with blue, coppery or brown upperwings and spotted or barred underwings.

Aricia cramera (ex *Aricia agestis cramera*), Southern Brown Argus, Mariposa morena.
Native (C, T, G, H, P); also occurs in southern Europe and North Africa. Occurs mainly in parts of the laurel and pine forests up to about 2000 m. Larvae feed on legumes (family Fabaceae) but may also use members of the Cistaceae and Geraniaceae. Flies throughout the year. Very small (wingspan 22–26 mm). Dull brown above and pale below with fine spots, but both wings have rows of brilliant orange markings near the edge, showing above and below; the edges of the wings are fringed with white. Caterpillar green, with a purplish stripe along the back and one on each side.

Cacyreus (sic) *marshalli*, Geranium Bronze, Taladro de los geranios.
This species, which is native to southeastern Africa, is invasive in southern Europe after intentional or accidental introduction; its recent arrival in the Canaries is likely to have been with imported pelargoniums (L, F, C, T). It is now widely distributed and common in most parts of the island, including gardens. Larvae feed on Geraniaceae. A small brown butterfly with two conspicuous tails. Upperside uniform brown with brilliant white spots along the edges of the wings, almost merging on hind wing to form a white rim. Two small black spots near the tails, which are white-tipped but black at the base. Underside (especially hindwing) strongly and irregularly barred with browns and white (cf *Leptotes*). Caterpillar bristly and green with three red stripes.

Cyclyrius webbianus, Canary Blue, Manto de Canarias.
Canary endemic species (C, T, G, H, P) lacking close relatives in continental areas. Occurs all over the island except in the most arid parts, but more abundant higher up and reaching 3500 m on El Teide. The larvae feed on legumes (Fabaceae) including *Teline canariensis, Spartocytisus supranubius, Adenocarpus* spp. and *Lotus* spp. Flies in all seasons, but not in Las Cañadas between mid October and the end of April. Very small (wingspan 26–30 mm). Both the fore- and hind-wing are violet-blue above in the male and golden brown in the female, with the border somewhat darker and the extreme edge pale. In both sexes the underside of the forewing is orange-brown and the hindwing is mottled brown with a stepped white bar. Caterpillar densely covered with fine white hair.

Lampides boeticus, Long-tailed Blue, Azul rabilarga.
Native (all islands); a cosmopolitan and migratory species. Occurs mainly in the lower zone but has been found up to 2000 m. Larvae live inside the seed pods of legumes including *Adenocarpus, Chamaecytisus, Teline* and *Ulex*; they have been seen to be visited by ants (*Lasius niger*). Flies throughout the year. Small (wingspan 29–34 mm) with a distinctly hairy appearance. Upperside violet-blue in male, brown with some blue near the bases of the wings in female. Underside pale grey-brown, with white bars including a strong one across the hind wing near the margin (cf *Leptotes*). Hindwing in both sexes has a small tail with a pair of prominent black and orange spots at its base, showing above and below (cf *Zizeeria knysna*). Caterpillar dark green at first, becoming russet-brown.

Leptotes pirithous, Common Zebra Blue, Gris estriada.
Perhaps now naturally established in the island (F, L, T?, G); a migratory species occurring in southern Europe, Middle East and Africa. Larval food plants mainly legumes, including *Adenocarpus*. Very small

(21–30 mm). Coloration similar to *Lampides boeticus* and *Cacyreus marshalli*. Upperside violet blue in male, bluish brown in female. Underside pale brown, with a rather uniform pattern of fine off-white barring (cf *Lampides* and *Cacyreus*). Hindwing has a small tail with two strongly marked black and orange spots at its base. Caterpillar yellow, green or brown, with a dark dorsal stripe.

Lycaena phlaeas, Small Copper, Manto bicolor.
Native (all islands); also occurs in Europe, northern Africa and elsewhere. Common throughout the island up to 2300 m, but commonest in open pine forest. Larval food plants docks (*Rumex*). Flies throughout the year. Small (wingspan 26–40 mm). Forewing brilliant orange or copper coloured, with dark brown margin and bold black spots. Hindwing brown, with bright orange margin with brown spots. Caterpillar green, sometimes with purple line along back.

Zizeeria knysna, African Grass Blue, Niña esmaltada.
Probably native (all islands); also occurs in the extreme south of Europe and in Northwest Africa. Common at low altitudes in moist places, including gardens. Larvae feed on Fabaceae but also Amaranthaceae; they are visited by ants (*Pheidole*). Flies in all months. Very small (wingspan 22–25 mm). Upperside dark violet-blue with pale fringe in male, brown with some blue at base of wings in female. Underside silvery-grey with scattered dark spots. Caterpillars variable in colour, greenish to reddish.

Family Lymantriidae – tussock moths

A group of moths containing a number of well known pest species.

Calliteara fortunata (ex ***Dicallomera fortunata***, ***Macaronesia fortunata***). Lagarta canaria del pino.
Canary endemic species (C, T, G, H, P). This moth is a major pest in the pine forests, with cyclical outbreaks every 6–7 years; however, the effect on the trees generally amounts to pruning, with a return to health in less than a year. The caterpillars can also be found on *Chamaecytisus* and *Spartocytisus* in the high mountain zone; they provide food for many kinds of birds, including the Blue Chaffinch, while the adults are also predated by the large asilid fly *Promachus vexator*. A medium-sized hairy moth; forewing grey brown with dark wavy bars; hindwing pale. The caterpillars are very striking, with tufts of brightly coloured hairs.

Family Noctuidae – noctuids

The noctuids are heavy bodied, dull-coloured, night-flying moths with wingspan usually in the range 20–40 mm. They are the largest family in the order Lepidoptera on a worldwide basis, and on Tenerife there are 81 species, one sixth of the Lepidoptera on the island. A quarter of the species are endemic, and they include members of two endemic genera, *Mniotype* and *Paranataelia*.

Family Nymphalidae – brush-footed butterflies

The largest family of butterflies; mainly medium to large, usually patterned and with bright colours above, but with underside often cryptic; many are capable of long distance flights.

Argynnis pandora (ex ***Pandoriana pandora***), Cardinal, Pandora.
Probably native (C, T, G, H, P). Also occurs in southern Europe and northern Africa. Not common, but occurs in the north from 500–1500 m, in laurel forest and to a smaller extent in pine forests and transitional areas. The larvae feed on violets *Viola* spp. Flies from May to September. Large (wingspan 56–70 mm) and with a gliding flight. Upperside has pale greenish yellow background with many bold black spots. Underside distinctive: forewing rose-pink at base (noticeable when resting with wings closed), hindwing pale green with strong silvery bars. Caterpillar very bristly, brownish with black patches and small reddish spots.

Danaus chrysippus, Plain Tiger, Monarca Africana.
Native (F, C, T, G, P). Also occurs in Africa south of the Atlas Mountains. Occurs locally in the lower zone, especially in the east of the Anaga peninsula and in gardens. Individuals probably sometimes

arrive from Africa. The larvae feed on various Asclepiadaceae, including introduced species but also endemic *Ceropegia* spp. Flies throughout the year. Large (wingspan 62–74 mm). Upperside of forewing rich orange brown with extensive black tip marked with large white spots, and relatively inconspicuous veins (cf the larger Monarch *Danaus plexippus*). Hindwing orange with mainly black margin and several dark marks. Underside pale yellow-brown with strong black and white pattern. Caterpillar with alternating bands of black and white, pairs of oval yellow spots along the back (sometimes weakly joined across the midline) and an almost continuous yellow stripe along each side; it has a pair of black, whip-like 'horns' near each end, and another pair about a third of the way along the body (cf *Danaus plexippus*).

Danaus plexippus, Monarch (Milkweed), Monarca.
Probably native (all islands). Occurs throughout the Americas and parts of Asia. It is well known as a long-distance migrant and colonised the Canaries in the 1880s, having crossed the Atlantic perhaps by way of the Azores and Madeira. Occurs in the lower zone, mainly around towns and in gardens, but has been recorded from Las Cañadas. Foodplants are mainly Asclepiadaceae and on Tenerife they include introduced *Asclepias curassavica* but perhaps also native *Ceropegia*. Flies throughout the year. Very large (wingspan 80–120 mm). Wings mainly orange with very dark veins. There are small white spots in a dark band around the margins and also at the tip of the forewing. The caterpillars are distasteful as the result of feeding on Asclepiadaceae, which contain poisonous substances, and since the poisons are transferred to the adults as they develop, they are also distasteful. The bright colours in both stages are 'warning coloration', signalling to predators that they are not edible. The caterpillar has alternating narrow bands of yellow, black and white, and a pair of 'horns' near each end (cf *Danaus chrysippus*).

Hipparchia wyssii, Tenerife Grayling, Sátiro de Tenerife.
Tenerife endemic species (classification has been fluid; there are closely related species on each of Gran Canaria, La Gomera, El Hierro and La Palma). Widespread in the pine forest, especially in the south of the island between about 1400 and 1600 m; also occurs in the high mountain zone. Adults fly from late spring until September; they feed on grasses, and hibernate. Medium-large (wingspan 50–60 mm). Upper surface of wings uniformly blackish-brown with a narrow but conspicuous interrupted white fringe and two dark eyespots, which are more conspicuous on the lower surface. There are a few small brilliant white marks on the forewings. Caterpillar brown with darker lines.

Hypolimnas misippus, Diadem, Falsa monarca.
Probably a rare immigrant from tropical Africa (F?, T, G, P). Seen in gardens near the coast. A large butterfly in which the sexes are completely different. The male is unmistakable: velvety black with a large white spot in the middle of each forewing and hindwing, and an additional small one near the tip of the forewing. The female is a marvellous mimic of the Plain Tiger *Danaus chrysippus*, from which it can hardly be distinguished in the field.

Issoria lathonia, Queen of Spain Fritillary, Mariposa sofía.
Perhaps native (C, T, G, P). Occurs in Europe and northern Africa. The rare sightings on Tenerife may be the result of occasional immigration from Morocco; they are usually in open spaces in laurel forest, including Monte del Agua. Small (wingspan 35–42 mm). Upperside pale brownish yellow with many large dark spots. The lower surface of the hindwing has large silvery markings. The larvae feed on violets *Viola* sp., but we do not know whether they have been found on Tenerife.

Maniola jurtina, Meadow Brown, Loba.
Native (L, C, T, G, H, P). Also occurs in Europe and northern Africa. Occurs in open areas in the forest zones in the north. The larvae feed on grasses. Flies from March to September. Medium sized (44–54 mm). Male has upperside dull dark brown (slightly paler near the edges) with a single small eyespot near the tip of the forewing; female is larger, with more prominent eyespot surrounded by large patches of orange-brown, and with some orange-brown on the hindwing. Forewing underside largely

orange in both sexes, with a strong eyespot; hindwing underside grey-brown with a broad paler bar. Caterpillar hairy, pale green, with a whitish line along each side.

Pararge xiphioides (ex *P. aegeria xiphioides*), Canary Speckled Wood, Maculada canaria.
Canary endemic species (C, T, G, P), sometimes considered a subspecies of the ordinary Speckled Wood *Pararge aegeria*. Common in parts of the lower zone (including gardens) and the laurel forest, less common in the pine forest. The larvae feed on grasses and also on woodrushes (*Luzula*). Flies all year round. Small (wingspan 38–44 mm). Upperside of forewing brown with many orange-yellow patches on blackish background, and one small eyespot; hindwing brown with a row of blackish eyespots in orange patches. Underside similar but faded brown with pale patches and the eyespots visible. Caterpillar hairy, green with whitish longitudinal lines.

Vanessa atalanta, Red Admiral, Vanesa de arco.
Native (all islands); occurs in Europe and northern Africa. Widely distributed in the lower and forest zones but not in the far south; also occurs in gardens. The larvae feed on nettles (*Urtica*) and thistles. Flies throughout the year. Medium sized (wingspan 48–54 mm). Predominantly velvety black with large white spots near the tip of the forewing. A brilliant red band across the forewing is fairly straight and narrow (cf *V. vulcania*); hindwing has a broad red border with small black spots. Caterpillar solitary, bristly, with several colour forms ranging from blackish to yellowish.

Vanessa cardui, Painted Lady, Vanesa de los cardos.
Native (all islands); almost cosmopolitan. Widespread on the island but scarce in forest zones; large numbers sometimes arrive with a southeast wind from Africa. Larval food plants diverse, including thistles and nettles. Flies throughout the year. Medium sized (wingspan 46–52 mm). Upperside of forewing orange yellow with black markings near the base, brownish black with large white spots near the tip. Hindwing with inner part brown, outer part orange yellow with dark brown spots. Underside patterned with brown, pale orange and white; hindwing with five small eyespots (cf *Vanessa virginiensis*). Caterpillar solitary, usually in silk tent; spiny, black with white dots.

Vanessa virginiensis (ex *Cynthia virginiensis*), American Painted Lady, Vanesa de Virginia.
Native (C, T, G, H, P). Originated in North America but now established in the Iberian peninsula. Generally rare on Tenerife, but numbers fluctuate; most sightings in laurel forest and gardens, at any time of year. Small (wingspan 36–46 mm). Similar to *Vanessa cardui* but with noticeably rounded wings, paler and more generally yellow, the hindwing with two small eyespots; hindwing underside with two very large eye-spots (cf *Vanessa cardui*).

Vanessa vulcania (ex *Vanessa indica*), Canary Red Admiral, Vanesa de yugo.
Native (all islands); also present in Madeira. Occurs in the laurel forest, a few parts of the pine forest and lower zone, and also in parks and gardens. The main larval food plants are nettles *Urtica morifolia* and the related *Parietaria* spp., and the caterpillars and pupae can be found in rolled-up leaves. Breeds and flies throughout the year. Medium or large (wingspan 48–60 mm). Similar to *Vanessa atalanta*, but the red band on the forewing is broader and with regular invasions of black from behind, and the black spots in the red border to the hindwing are larger.

FAMILY PIERIDAE – WHITES

Medium to large butterflies with white or yellow wings, usually with some blackish or greenish markings.

Catopsilia florella, African Migrant, Migradora africana.
Probably introduced accidentally in the 1960s with its main food plant *Senna* (ex *Cassia*) *didymobotrya* (all islands). Native to Africa south of the Sahara. On Tenerife occurs in parks and gardens in the lower zone, throughout the year. Large (wingspan 54–58 mm) and distinguishable from other 'whites' at a distance by its much faster and erratic flight. Male pale yellow and whitish, female usually pale

greenish-white all over. Forewing (but not hindwing) with a very small dark dot (cf *Gonepteryx*) and with small brown marks at edges of wings. Caterpillars vary in colour, tending to match background; they are heavily parasitized by flies (family Tachinidae) and hymenopterans (Braconidae).

Colias crocea, Clouded Yellow, Azufrada.
Native (all islands); occurs in Europe and northern Africa. Widespread and abundant in the lower and forest zones. Larvae feed on legumes (Fabaceae). Flies throughout the year. Medium sized (wingspan 42–50 mm). Bright orange-yellow (yellowish-white in some females) with broad dark margin to both forewing and hindwing (with pale spots on it in female); there is a conspicuous dark spot on the forewing and a dull orange spot on the hindwing. Underside greenish orange, hindwing with a conspicuous central eyespot. Caterpillar green with a white line along each side.

Euchloe belemia, Green-striped White, Blanca verdirrayada.
Native, (F, C, T, G?); the species is widespread in southern Europe and northern Africa. The Tenerife population, representing an endemic subspecies *Euchloe belemia eversi*, was discovered only in 1963; it occurs in open areas in the pine forest and the mountain scrub zone up to at least 2500 m. Larvae feed on *Descurainia bourgeauana* and perhaps other crucifers. Flies from spring to autumn. Small (wingspan of 32–36 mm). Upperside of forewing white with black tip enclosing white spots, and one black mark almost at front edge (cf *Pieris rapae*); hindwing dusky whitish, darker near base but lacking black and white pattern (cf *Pontia daplidice*). Underside of forewing white with one dark mark near front edge (cf *Pontia daplidice*) and tip dark green enclosing white spots; hindwing dark green with white bars tending to cross the veins. Caterpillar green with a pink line along the back and on each side.

Gonepteryx cleobule, Canary Brimstone, Limonera de Tenerife.
Canary endemic (C, T, G); the related *G. cleopatra* is native to southern Europe and northern Africa. Occurs in the north between about 600–2000 m, especially in open places in laurel forest. Larvae feed on *Rhamnus* species. Flies throughout the year. Large (wingspan 56–60 mm) and bright greenish-yellow, with the upper side of the forewing orange. There is an inconspicuous orange spot near the centre of each wing (cf *Catopsilia*). Caterpillar green, with thin white and yellow lines on the sides.

Pieris cheiranthi, Canary Islands Large White, Capuchina común.
Canary endemic species (T, G, P) previously treated as a subspecies of *Pieris brassicae*, which occurs in Europe and northern Africa. Largely restricted to wet ravines in the north, between about 200–1400 m. The larvae feed on Brassicaceae and also on the introduced Nasturtium *Tropaeolum majus*, but the butterfly may have decreased recently as a result of attempts to control this invasive plant. Flies throughout the year. Large (wingspan 58–64 mm). Upperside of forewing white or yellowish, with broad dark tip; female has two large merging dark marks in the centre but these are lacking in male; hindwing white or yellowish with a black mark on the forward edge. Underside of forewing whitish with yellow tip and two merging black marks in both sexes; hindwing yellowish. Caterpillar bristly, greenish grey, with yellow stripe along back and each side.

Pieris rapae (ex *Artogeia rapae*), Small White, Blanca de la col.
Possibly native (all islands); native to Eurasia and northern Africa. Common throughout the island, especially in cultivated areas but recorded from 3000 m on El Teide. Larvae feed on Cruciferae, Resedaceae and some other plants. Medium sized (wingspan 42–50 mm). Upperside white or with a hint of yellowish, dusky near body (cf *Catopsilia*); forewing with small black tip (sometimes faint); female with two small, well separated black spots, while male has one small one (cf *Pieris cheiranthi*); hindwing white with a small dark mark on front edge. Underside of forewing white with yellowish tip and two small black spots; hindwing dusty yellow.

Pontia daplidice (ex *Pieris daplidice*), Bath White, Blanquiverdosa.
Native (all islands); occurs in southwest Europe and north Africa. Found throughout the island but commonest in dry areas. Larvae feed on Brassicaceae and Resedaceae. Flies throughout the year. Small

(wingspan 38–44 mm). Upperside of forewing white with blackish tip enclosing white spots, and one or two dark marks nearer to the body; hindwing whitish with dark and white pattern near edge. Underside of forewing white with dark green and white tip and two dark blotches near the body; hindwing blackish green with whitish blotches mainly confined between the veins (cf *Euchloe belemia*). Caterpillar grey with yellow stripes.

Family Psychidae– bagworm moths

Tenerife has only two members of this family, one of which is of special interest.

Amicta cabrerai, Polilla canaria de estuche.
Canary endemic (C, T, G, H, P). Widespread in the lower parts of the island, and also occurs on offshore rocks. The larvae of this species construct caddis-like tubes out of pieces of vegetable matter and silk, and carry them around while feeding on a variety of shrubs. The tubes can be found on the walls of houses as well as on rock surfaces or plant stems. After pupation the female remains in the tube, wingless, legless and wormlike, with reduced eyes; the male emerges as a normal small blackish moth with impressive antennae; after finding a mature female he reaches into the tube with his telescopic abdomen to fertilise her, and then soon dies; the female lays eggs in the tube.

Family Sesiidae – clear-winged moths

Bembecia vulcanica (ex *Dipsosphecia vulcania*), Bembecia de volcán.
Canary endemic (C, T, P). Recorded from the high mountain zone and now also elsewhere. A day-flying, wasp-mimicking moth. The larvae feed in stems of legumes, and chrysalises have been found in the roots of *Lotus campylocladus* near the Roques de García in Las Cañadas. Body blackish with pale bands, wings transparent but with blackish borders; antennae long, dark and rather thick.

Family Sphingidae – hawkmoths

Medium to very large moths capable of rapid flight and in some cases also hovering; caterpillars usually with a horn at the rear end.

Acherontia atropos, Death's Head Hawkmoth, Mariposa de la muerte (Esfinge de la calavera).
Probably native (L, F, C, T, G, P); occurs throughout Europe and the Middle East and Africa. Found mainly in the lower zone. Larval foodplants mainly Solanaceae. Very large (102–135 mm). Forewing mottled brown with one tiny white spot; hindwing yellow with two dark bars; thorax brown with yellow 'death's head' pattern. Abdomen with yellow and dark brown bands, and with a broad dorsal grey stripe. Caterpillar up to 12.5 cm long; greenish with seven oblique purplish sidestripes edged with yellow; tail horn rough, yellowish.

Agrius convolvuli (ex *Herse convolvuli*), Convolvulus Hawkmoth, Esfinge de la correguela.
Native (L, F, C, T, G, P); widespread in the Old World. Occurs mainly in the lower zone and laurel forest. Larvae feed typically on *Convolvulus*, but can also be found on grape vines. Very large (wingspan 89–120 mm). The forewing, hindwing, head and thorax are grey mottled with whitish. The abdomen is pink with black and white bands and a broad dark dorsal stripe. The caterpillars are apple-green or purplish, with oblique yellowish stripes on the sides and with a brownish tail horn.

Hippotion celerio, Silver-striped Hawkmoth, Esfinge de banda plateada.
Probably native (all islands); widespread in southern Europe and Asia and the rest of the Old World. Occurs mainly in the lower zone and laurel forest. Larval food includes grapevines and Rubiaceae. Large (wingspan 72–80 mm). Forewing pale brownish with a narrow silvery-white longitudinal stripe; hindwing has a pink patch close to the body and pale cells surrounded by black further out. The abdomen is tapered rather than stumpy, and is brown with no bands but with two thin white dorsal stripes. Caterpillar brown or greenish, with a narrow head end, two large eyespots a little way back and two smaller spots behind them; tail horn rough and brown.

Hyles livornica (ex *Celerio lineata*), Striped Hawkmoth, Esfinge rayada.
Probably native (L, F, C, T, H, P); worldwide distribution. Occurs throughout most of the island except the high mountain zone. Larval food plants include Rubiaceae, docks and grapevines. Very large (wingspan 78–90 mm). Forewing mainly dark brown with a pale edge, a broad pale stripe down the centre and diagnostic white veins; hindwing whitish near the body, pink further out and bordered with black. Thorax brown above with two white stripes. Abdomen pointed with black bands broken with pale spots. Caterpillar green to black with yellow dorsal and lateral stripes and a series of bold round spots; tail horn red.

Hyles tithymali, Spurge Hawkmoth, Esfinge canaria de las tabaibas.
Native (all islands). Confusing taxonomy and odd distribution, with an endemic subspecies *Hyles tithymali tithymali* on L, C, T, G, P and different, non-endemic subspecies on F and H; the species is widespread in Europe and eastwards to India. Occurs throughout Tenerife except the high mountain zone. Feeds on *Euphorbia* species. Large (wingspan 64–77 mm). Forewing greyish-brown, slightly pinkish and with bold dark brown markings and a large yellowish area with irregular edges along the centre; hindwing has a whitish patch near the body, then a broad pink band bordered with black. Thorax brown above and white at the sides. Abdomen tapering, with blackish bands and a brown dorsal stripe. Caterpillar blackish with a crimson dorsal stripe, yellowish lateral stripes and a series of bold round spots; tail horn red.

Macroglossum stellatarum, Hummingbird Hawkmoth, Esfinge colibrí.
Probably native (all islands); occurs in Europe and North Africa. Occurs throughout the island. Larval food plants Rubiaceae. Medium sized (wingspan 50–58 mm). Forewing brownish with a few dark transverse lines while the hindwing is yellow, edged with reddish-brown. The abdomen is stumpy, brownish and with pale patches on the sides. This hawkmoth is a day-flyer and darts from flower to flower hovering in front of them while it feeds. The caterpillars are green or brownish with white dots and pale longitudinal lines. The tail horn is bluish.

FAMILY TORTRICIDAE - TORTRICIDS

Small brown or grey moths with almost rectangular front wings; they rest with the wings held roof-wise over the body. In general they are fairly sedentary, and this is doubtless one reason for the high endemicity of the family in the Canaries; on Tenerife there are 39 species, and over half of these (20) are endemic.

FAMILY YPONOMEUTIDAE

Yponomeuta gigas, Arañuelo del sauce.
Canary endemic (C, T, G, P). Larvae make communal webs over the Canary Willow trees, and also on introduced poplars.

ORDER ODONATA – DRAGONFLIES AND DAMSELFLIES

At least 10 species of dragonflies are established on Tenerife. One (perhaps two) of these is a damselfly (in the family Coenagridae) while the other nine are typical dragonflies. These two groups are fairly easy to distinguish in the field. The damselflies are noticeably delicate, with very slender bodies, and at rest they hold their wings vertically over the body or partly spread. The typical dragonflies have thicker bodies and are robust, strong-flying insects, always resting with wings outspread. Both dragonflies and damselflies have aquatic larvae and need access to fresh water for breeding, but some of the larger species can be found several miles from their breeding place. They are normally seen flying between April and October. The list below includes all the dragonflies considered to be established on Tenerife. The list of the Odonata of Spain and Macaronesia at www.iberianwildlife.com mentions three additional species for the Canary Islands, *Platycnemis subdilatata* Barbary Featherleg, *Orthetrum trinacria* Long Skimmer and *Trithemis annulata* Violet Dropwing, but these are not accepted as established on Tenerife.

> **KEY TO THE DRAGONFLIES OF TENERIFE**
> (At each step, choose between the alternatives with the same number)
>
> 1. Small (body length 25–30 mm), wings much shorter than body, often held over body when at rest, abdomen conspicuously slender — *Ischnura* species (see text Coenagridae, damselflies)
> 1. Medium or large, abdomen robust
>   - 2. Large (body length 70 mm or more)
>   - 3. Blue or green and black — *Anax imperator* or *A. parthenope* (see text)
>   - 3. Yellowish brown, abdomen with dark dorsal stripe, and in male with a blue mark near base — *Anax ephippiger*
>   - 2. Medium (body length 35–60 mm)
>   - 4. Abdomen conspicuously broad, red (male) or orange (female, young male) — *Crocothemis erythraea*
>   - 4. Abdomen not conspicuously broad
>   - 5. Red or pinkish
>   - 6. Bright red, black marks on sides of abdomen — *Trithemis arteriosa* (male)
>   - 6. Pinkish or reddish, without distinct black marks on sides of abdomen (see text) — *Sympetrum fonscolombii* or *S. nigrifemur* (males)
>   - 5. Blue, green, yellow or brown
>   - 7. Oblique white stripe on sides of thorax; blue (adult male) or yellowish (female, young male) — *Orthetrum chrysostigma*
>   - 7. Without oblique white strip on side of thorax
>   - 8. Black and yellow, strongly patterned
>   - 9. With metallic blue-green highlights — *Zygonyx torridus*
>   - 9. Without metallic blue-green highlights — *Trithemis arteriosa* (female)
>   - 8. Yellow or yellowish brown, not strongly patterned (see text) — *Sympetrum fonscolombii* or *S. nigrifemur* (females)

We also include a simple key to help with identification. For the purposes of the key and list 'small' implies a body length of 30 mm or less, 'medium' 35–60 mm and 'large' 70–80 mm. There is an excellent illustrated identification guide (in Spanish) to the dragonflies of the Canary Islands by Marcos Báez.

Family Aeshnidae

Anax imperator, Blue Emperor, Emperor Dragonfly.
Native (all islands); also occurs in Madeira, Azores, south and central Europe, the Mediterranean region, Africa and part of Asia. Larvae develop in stagnant water. The adults are powerful flyers and are often seen far from water. This is one of the commonest dragonflies on Tenerife. Large (body length 75–80 mm). Mainly blue or green, the male being distinguished by the colour of the abdomen, which is bright blue throughout its length, with distinct black dorsal marks. The female is similar but only the base of the abdomen is bright blue, the rest being green or blue-green; she could be confused with *Anax parthenope*.

Anax parthenope, Lesser Emperor.
Native (L, F, C, T); occurs in Madeira, the Mediterranean region and Africa. Larvae develop in stagnant water and the adults usually stay near water. Large (body length 75–79 mm). Similar to *Anax imperator*

but the abdomen is slightly shorter and darker. The male has a small bright blue patch at the base of the abdomen, the rest being green or blackish-green. The female is generally darker.

Anax ephippiger (official list uses *Hemianax ephippiger*), Vagrant Emperor.
Native (L, C, T); also occurs in Madeira, the Iberian Peninsula, central Europe, and the desert areas of Africa and western Asia. Larvae sometimes develop in temporary waters and the adults often fly far from water. This species is very rare in Tenerife. Large (body length 70 mm) but distinctly smaller than the other two *Anax* species; generally yellowish-brown, the abdomen with a dark dorsal line and in the male also with a striking blue mark near the base.

Family Coenagridae

Ischnura saharensis, Sahara Bluetail, and *Ischnura senegalensis*, Senegal Bluetail, Ubiquitous Bluetail.
Native (L, F, C, T). These two African species of damselfly are very similar; they may also occur on Madeira. They are the only damselflies occurring on Tenerife, and the two species have not been reliably separated in the past. *Ischnura saharensis* is the species normally listed for Tenerife, but DNA analysis has shown that *Ischnura senegalensis* is present on island, while the status of *Ischnura saharensis* is unclear. The larvae probably develop in stagnant water. Small (body length 25–30 mm). Head and thorax green and black; abdomen strikingly slender and much longer than the wings, dark above and greenish-yellow below with a bright black patch near the tip in the male.

Family Libellulidae

Crocothemis erythraea, Broad Scarlet, Scarlet Darter.
Native (F, C, T, G, P); occurs in Iberia, central Europe, the Mediterranean region, Africa and into Asia. The larvae develop in either stagnant or running water, and the adults are found both close to water and far from it. Medium size (body length 40–50 mm). The whole body is conspicuously broad and the abdomen is somewhat flattened above. The adult male is bright red all over, and the young male and the female are golden yellowish. There is a dark dorsal line on the abdomen.

Orthetrum chrysostigma, Epaulet Skimmer.
Native (F, C, T, G, H, P); occurs in the south of the Iberian peninsula and the whole of Africa. A dragonfly typical of dry areas, adults often being found far from water. Medium size (body length 50 mm). The best identification character for this species is a conspicuous white stripe which runs from the base of the front wings downwards and forwards to the middle legs. The body colour is variable. Adult male mainly powder-blue, young male and female yellowish-brown; there are thin dark lines along the abdomen.

Sympetrum fonscolombii (official list uses *S. fonscolombei*), Red-veined Darter.
Native (F, L, C, T, G, P); occurs in Madeira, Azores, western Europe, the Mediterranean region, Africa and into Asia. The larvae develop in stagnant, fresh or brackish water, and the adults stay in the vicinity. Medium size (body length 35–41 mm). Male generally pinkish, with the abdomen yellowish below, female yellowish all over. In both sexes the spots at the front of the wing tips are yellow, bordered with black. The tip of the abdomen has black dorsal marks, which are absent in the very similar species *Sympetrum nigrifemur*.

Sympetrum nigrifemur (ex *Sympetrum striolatum nigrifemur*), Island Darter.
Macaronesian endemic (C, T); also occurs on Madeira. The larvae generally develop in stagnant water, and the adults do not move far away. It is very rare on Tenerife. Medium size (body length 43–48 mm), which is slightly larger than the very similar *Sympetrum fonscolombii*, and both sexes are darker. The spots at the front of the wingtips are red or reddish-yellow and the abdomen lacks black dorsal marks at the tip.

Trithemis arteriosa, Red-veined Dropwing.
Native (C, T, G, H); occurs in Africa, especially in the tropics, but not in Europe. The larvae develop in stagnant or running water and the adults are often seen on the stones nearby. Medium (body length 35–40 mm) with a slender abdomen. The entire body of the male is red, the abdomen being brightest and with distinct black patches on the sides. Female has the entire body strongly patterned black and yellow, and the yellow on the abdomen is mainly on top.

Zygonyx torridus (official list uses *Zygonyx torrida*), Ringed Cascader.
Native (F, C, T, G); also occurs in the south of the Iberian peninsula, Africa and into Asia. The larvae develop in swiftly running water and the adults usually stay near water. Medium size (body length 55–60 mm) and the largest of the libellulids on the island. The abdomen is patterned black and yellow with metallic blue-green highlights on the head and thorax.

OTHER INVERTEBRATES

PHYLUM PLATYHELMINTHES – FLATWORMS

Free-living flatworms (as opposed to parasitic forms, not considered here) are not conspicuous on Tenerife, and although a few species have been recorded, none are endemic and nearly all are thought to be introduced.

PHYLUM NEMATODA – ROUNDWORMS

Many roundworms are parasitic, but there are also many species free-living in the soil or elsewhere. They are well represented in the Canaries, with a few described as endemic, but they are not likely to be noticed by visitors and we do not consider them here.

PHYLUM ANNELIDA – SEGMENTED WORMS

Two species of leeches (Class Hirudinea) have been introduced to Tenerife, but the rest of the annelids on the island are Oligochaeta, a group that includes the earthworms and some more obscure groups. Tenerife is rich in earthworms, with 57 species recorded, but none of them are endemic. A recent study shows that many of the species are of wide distribution and that the fauna includes tropical as well as northern species. Twenty four species are considered to be definitely introduced and many others possibly so; a large proportion may have been brought in with imported plants.

PHYLUM MOLLUSCA – MOLLUSCS

All of the land-based molluscs of Tenerife are in the Gastropoda, the group including the snails and slugs. (The sole exception is the tiny freshwater clam *Pisidium casertanum*, a bivalve which may be native.) The gastropods have 20 genera on the island that are almost certainly native; another 10 genera may be native and 17 have clearly been introduced. Some of the genera have given rise to a large number of distinct forms within the Canaries, although in many cases it is not clear which of them really deserve to be considered separate species. Four of the native genera on Tenerife are endemic to the Canaries: *Monilearia* in the family Cochlicellidae, with 3 species on Tenerife; *Napaeus* in the family Enidae, with 21 species, all of them endemic to the island; *Canariella* in the family Hygromiidae, with 9 species, all endemic to the island; and *Gibbulinella* in the family Streptaxidae, with 2 species. The genus *Hemicycla*, though not endemic to the Canaries, has diversified greatly, with 15 species on Tenerife; 12 of these are endemic to the island, and many are restricted to a limited area within the island. The slug-snail genus *Insulivitrina* (ex *Plutonia*) in the family Vitrinidae has five species on Tenerife, all of them endemic; they are leaf-eaters and are fairly easy to find in the laurel forest.

PHYLUM ARTHROPODA – JOINTED-LEG INVERTEBRATES

The arthropods have jointed legs, segmented bodies and an external skeleton made of chitin that has to be moulted and replaced as the animal grows. The phylum Arthropoda contains the great majority of all animal species, primarily in the groups arachnids, crustaceans, myriapods and insects.

It is among the athropods that the extraordinary richness of animal life on Tenerife is most obvious. The most up-to-date summary (see Oromí *et al.* 2015 in Further Reading) states that 5,041 species of arthropods are currently known from Tenerife; the next most diverse of the islands (Gran Canaria) has only two-thirds as many.

Many groups of arthropods show endemicity on the islands; in other words, particular subspecies, species or genera are found only on the archipelago. The presence of an endemic genus indicates either that profound evolutionary change has occurred since colonisation, or that relatives on the mainland have become extinct, leaving the animals on the islands as the only representatives of the group; endemic species and subspecies have comparable implications. All endemic forms are therefore of particular interest, and rates of endemism are extremely high in the Canaries: 33% of all arthropod species on Tenerife (>1660 species) are found only in the Canaries and 13% of them are restricted to the island of Tenerife. At the generic level, the Canaries are currently considered to have 99 endemic genera, half of which are Coleoptera and most of the rest are Hemiptera, Lepidoptera, Orthoptera or Araneae (spiders).

Some of the endemic genera of arthropods on the archipelago have evolved a number of separate species (up to 11), often on separate islands, but others include only a single species. More spectacular diversification, creating many endemic species, has occurred within genera that are not endemic. There are 35 non-endemic genera that include more than 10 endemic species, and the total number of endemic species in these genera is 865. Maximum diversity is in the weevil genus *Laparocerus*, with nearly 150 species on the islands. Three other non-endemic genera have around 50 endemic species: *Attalus* (beetles in the superfamily Cleroidea); *Dolichoiulus* (millipedes in the family Julidae); and *Dysdera* (spiders in the family Dysderidae).

CLASS ARACHNIDA

Order SCORPIONES – scorpions

Scorpions are poor dispersers, and none have colonised the Canaries naturally. There is, however, a single introduced species on Tenerife: *Centruroides gracilis* (Family Buthidae) occurs only in the coastal area of Santa Cruz, and was presumably brought in accidentally on a ship; its sting contains a powerful poison, but it is not normally fatal to humans.

Orders ASTIGMATA, ORIBATIDA, PROSTIGMATA, IXODIDA and MESOSTIGMATA – mites

The mites are extremely diverse on the Canaries, but relatively few families have a high proportion of endemic species; we do not discuss them further here.

Order PSEUDOSCORPIONES – false scorpions

Pseudoscorpions are very small predators, resembling scorpions in having large pincers (palps) that they hold out in front as they walk, but lacking the stinging tail of scorpions. They are often found in leaf litter and small spaces such as those under tree bark. They are excellent dispersers, often attaching themselves to flying insects, and there is a rich fauna in the Canaries, many of the species being endemic. Tenerife has the greatest diversity, with 32 species, 18 of them endemic. These include the genus *Canarichelifer,* endemic to the Canaries (and Salvages), with the species *Canarichelifer teneriffae.*

ORDER OPILIONES – HARVESTMEN

There are five species of harvestmen on the Canaries, including the remarkable endemic species *Maiorerus randoi*; this is the sole member of its genus, in the recently established family Pyramidopidae (otherwise restricted to tropical Africa); *M. randoi* is known only from a small part of a single cave in Fuerteventura. Four other harvestmen occur in the Canaries (two of them endemic) and all are present on Tenerife. One is the only member of another endemic genus, *Parascleropilio*, which is of special interest in that its closest relatives are now found in the Balkans; presumably its ancestors also occurred in northwest Africa. *Bunochelis spinifera* (family Phalangiidae) in a genus endemic to the Canaries and Salvages, is widespread on Tenerife and can be found at the summit of El Teide.

ORDER ARANEAE – SPIDERS

Spiders are by far the most important group of arachnids on the Canaries, with 260 species recorded from Tenerife, in 35 families. These include a number of species with wide distribution outside the Canaries, but also 142 species endemic to the Canaries, an extraordinary proportion of 55%. This high incidence of endemicity probably reflects the ability of spiders to reach islands by air, and perhaps also on natural rafts. (Some have also arrived recently on ships and aircraft.) Once a colonising stock has reached the archipelago, spiders in some families show a strong propensity to form new species, both on separate islands and in different habitats – including caves – on single islands.

The most diverse spider family on Tenerife is the Linyphiidae, with 36 species, of which 28 (78%) are endemic. However, even higher proportions of endemics are found in the Pholcidae (21 species, 20 of them endemic) and the Dysderidae (24 species, 23 of them endemic to the Canaries and 19 of these occurring only on Tenerife). All the Dysderidae are in the genus *Dysdera*, ground-dwelling spiders that are typically red-brown in colour. They have undergone a remarkable evolutionary radiation in the Canaries, and on Tenerife members of the genus occur in a wide variety of habitats, including many lava tubes, but can most easily be found under rocks or logs, often with their large silky white egg sacs. Other spider families showing high diversity are Gnaphosidae (24 species, 10 of them endemic and the rest all probably native); Salticidae (21 species, 6 of them endemic); Theridiidae (21 species, 4 of them endemic); and Araneidae (17 species, 4 of them endemic). The spiders of Tenerife also include two genera endemic to the Canaries, *Canarionesticus* (Nesticidae) recorded only from Tenerife, and *Cladycnis* (Pisauridae) which is found on several of the Canary Islands but nowhere else.

One of the most conspicuous spiders on Tenerife is *Cyrtophora citricola*, an orb-web spider (family Araneidae) that is possibly introduced, in which females are large and black with white markings, but males are much smaller. Bushes in many parts of the lower zone, including gardens, are often festooned with the communal webs of this species; the webs sometimes also contain the much smaller and silvery *Argyrodes argyrodes* (family Theridiidae) which feeds on unguarded silk-wrapped prey of *Cyrtophora*.

ORDERS PALPIGRADI, SCHIZOMIDA AND SOLIFUGAE

These three relatively obscure orders of arachnids each has one species on Tenerife. The palpigrade is unlikely to be noticed. The schizomid is worth mentioning only because it is *Stenochrus portoricensis*, found in volcanic caves in Tenerife and also on La Palma; only females are present, these being capable of asexual reproduction. This species was previously unrecorded on the eastern side of the Atlantic, except in an orchid house in Cambridge, England.

The solifugid (camel spider or wind scorpion) *Eusimonia wunderlichi*, in the family Karschiidae, is endemic to the Canaries. The solifugids are a relatively little known Old World group related to the pseudoscorpions but superficially more like spiders. They are fast-running, usually desert-living and nocturnal; on Tenerife *Eusimonia* can be found in and around Teide National Park.

CLASSES BRANCHIOPODA, OSTRACODA AND COPEPODA

A number of species of crustaceans that live as plankton in freshwater have been found in the Canaries. Most of them are North African species and have probably reached the island naturally.

CLASS MALACOSTRACA

ORDER DECAPODA – CRAYFISH ETC

The red swamp crayfish *Procambarus clarkii* (family Cambaridae) is the only non-marine decapod found living wild in Tenerife. It is an invasive species native to southeast United States but illegally introduced to Spain and found on Tenerife in the Barranco del Cercado de Andrés in 1997; it quickly colonised the whole length of the catchment and will probably spread eventually to most permanent streams on the island.

ORDER ISOPODA – WOODLICE ETC

The woodlice have diversified extensively within the Canaries archipelago, although there are some species that also occur in the Mediterranean region. Tenerife has 30 species, and although more than half are thought to be introduced, 10 of the remainder are endemic, and with the other native species they form an important element in the fauna of the island.

ORDER AMPHIPODA – AMPHIPODS

Most amphipods are found in salt water, but some can live in damp places on land. Eight species (in three different families) have been recorded from Tenerife, of which 6 are endemic; some of these can be found in leaf litter in forests along the north of the island.

CLASS SYMPHYLA – SYMPHYLANS

Symphyla are minute arthropods restricted to damp places. Five species have been found on Tenerife but all are probably introduced.

CLASS PAUROPODA – PAUROPODS

Pauropods are another group of minute soil living arthropods. Ten species are found on Tenerife and are considered to be probably native.

CLASS DIPLOPODA – MILLIPEDES

The millipedes are diverse in the Canaries, in the orders Polyxenida, Glomerida, Polyzoniida and Julida. On Tenerife there are 44 species, of which 25 are endemic. Most of these are in the genus *Dolichoiulus* (family Julidae), a group in which the great majority of endemic species in the archipelago occur only on a single island, implying a strong tendency for isolated populations to evolve differences and eventually to become separate species.

CLASS CHILOPODA – CENTIPEDES

There are 21 species of centipedes on Tenerife, in the four orders Scolpendromorpha, Geophilomorpha, Lithobiomorpha and Scutigeromorpha. Only four species are endemic; one of the latter is *Lithobius speleovulcanus*, an obligate cave-dweller found in several caves on Tenerife. Most of the non-endemic species are widely distributed outside the archipelago. They include *Lithobius pilicornis*, which is able to live even in barren lava flows and which we were surprised to find within a few metres of the summit of El Teide. A more noticeable species is *Scolopendra valida* (previously recorded as *S. morsitans*) a large and brightly coloured centipede that also occurs in North Africa and eastwards into Asia. *Scolopendra valida* is common in the lower zone and can also be found in the pine forest, usually under rocks or logs; it can inflict a painful bite if handled, but will run fast for cover if given a chance. In and around houses one can sometimes find *Scutigera coleoptrata*, a long-legged centipede capable of a remarkable turn of speed.

CLASS COLLEMBOLA – SPRINGTAILS

The springtails are diverse on Tenerife, with 80 species recorded in the four orders Symphypleona, Entomobryomorpha, Neelipleona and Poduromorpha. Nearly a quarter of the species (19) are endemic.

CLASS DIPLURA – DIPLURANS OR TWO-PRONGED BRISTLETAILS

Four species of these tiny arthropods occur in Tenerife but only one is probably native.

CLASS PROTURA – PROTURANS

Proturans are small and inconspicuous soil-living animals. There are 12 species on Tenerife; one of them is endemic and most of the rest may be native.

CLASS INSECTA – INSECTS

Order MICROCORYPHIA – bristletails

(This group and the next were previously grouped together in the order Thysanura).
    Three species of Microcoryphia in the family Machilidae are found on Tenerife, and two of these are endemic. The machilids are superficially similar to silverfish but the thorax is humped and they can jump as well as running fast; they are found mainly on rocky shores.

Order ZYGENTOMA – silverfish

Tenerife has eight species of silverfish, all but one in the family Lepismatidae; one is endemic and the rest are probably native. These insects are of special interest since their adaptation to warm and dry conditions enables some of them to live on the most barren recent lava flows; one species has been found very near the summit of El Teide.

Order EPHEMEROPTERA – mayflies

In spite of the scarcity of freshwater on Tenerife, six species of mayflies are probably native and two of these – in the genus *Baetis* – are endemic.

Order MANTODEA – mantids

Tenerife has seven (perhaps eight) species of mantids, four of them endemic and the others probably also native. Both the well known Praying Mantis *Mantis religiosa* and *Blepharopsis mendica* occur in many parts of the island, but like all mantids they are hard to see. In Las Cañadas and the pine forest there is a chance to see an interesting but scarce endemic species *Pseudoyersinia teydeana*, which has only vestigial wings even when fully adult; it lives on low vegetation but deposits its egg masses under rocks.

Order BLATTARIA – cockroaches

The 20 species of cockroach occurring on Tenerife are almost evenly divided between those that are endemic and those that have been introduced by humans (many of the latter, such as *Periplaneta australasiae* and *P. americana,* are invasive in various parts of the world). The nine endemic species include a series of members of the genus *Loboptera* (family Blattellidae) that occur only in Tenerife and show varying degrees of adaptation to subterranean life, including blindness in some species. In contrast, the Canary endemic *Phyllodromica brullei* (family Ectobidae) is a delicate cockroach that is a characteristic member of the fauna of the pine and laurel forests, but also occurs under stones at up to 2500 m on El Teide; wings are absent in the female and only forewings are present in the male, so that both sexes are actually flightless.

Left: Ground beetle *Carabus faustus* female; middle: Cockroach *Phyllodromica brullei* male; right: Bush cricket *Canariola nubigena* or *C. willemsei*, female. Photos: Pedro Oromí

ORDER ISOPTERA – TERMITES

Three species of termites have been found on Tenerife and one of these – *Bifiditermes rogierae* (family Kalotermitidae) – is endemic. It was discovered fairly recently and is an isolated outpost of a group that is otherwise confined to tropical Africa, southern Asia and Australia; it occurs in the rotting trunks of *Euphorbia* species.

ORDER ORTHOPTERA – GRASSHOPPERS AND CRICKETS

Orthopterans are diverse in the Canaries and Tenerife has 54 species; 13 of these are endemic, this relatively low proportion reflecting the wandering tendencies of many orthopterans. Acridid (short-horned) grasshoppers make up almost half of the total with 23 species. These include two well known species of locust. The Desert Locust *Schistocerca gregaria* reaches the island only sporadically, mainly in autumn with warm winds from the Sahara; at long intervals there are major invasions, with serious damage to crops. The well known Migratory Locust *Locusta migratoria* is a widespread resident in the lower parts of the island; it is often found in agricultural areas, but does not cause serious damage. The Acrididae also include several other interesting insects including two species of the endemic genus *Arminda* (with species on all the islands), the Canary endemics *Calliptamus plebeius* and *Oedipoda canariensis* (the latter characterised by pale bluish hind wings); and the Tenerife endemic *Sphingonotus willemsei*, which is sometimes abundant in the volcanic deserts of Las Cañadas.

The bush crickets or long-horned grasshoppers (family Tettigoniidae) are less diverse, with 11 species on Tenerife, but these include *Ariagona margaritae*, the sole member of an endemic genus occurring only on Tenerife, La Gomera and El Hierro; on Tenerife it is restricted to grassy forest edges in the north. The genus *Calliphona* is represented by three species on Tenerife, including *Calliphona konigi* (endemic to Tenerife) a large predatory but flightless species of the laurel forest. Another laurel forest species is *Canariola nubigena*, small and wingless and with extremely long antennae; it is an endemic member of a genus that was thought until recently to be endemic to the Canaries, but is now known to have some relatives in southern Spain.

There are 16 species of crickets (Gryllidae) on Tenerife, but the only endemic is *Gryllomorpha canariensis*, a small grey brown and wingless cricket that occurs in many parts of the island including barren lava fields in Las Cañadas. More conspicuous is *Gryllus bimaculatus*, a widespread species capable of long distance flight. A far less conspicuous species is the mole cricket *Gryllotalpa africana*, in a separate family Gryllotalpidae, which burrows in moist sandy places including El Médano.

Among the members of smaller groups of orthopterans is the rare Tenerife endemic *Acrostira tenerifae* in the family Pamphagidae; these are slow-moving, cryptic, wingless grasshoppers in which the male is much smaller than the female.

Order EMBIOPTERA – webspinners

Webspinners are tiny, fast-moving insects with a unique ability to spin silk with their front legs, forming protective canopies under which they live. Two species occur on Tenerife, but both are thought to be introduced.

Order PHASMATODEA – stick insects

A single species of stick insect occurs on Tenerife but is certainly introduced.

Order DERMAPTERA – earwigs

The earwigs of the Canaries are fascinating to animal geographers, and on Tenerife there are 14 species, of which no less than eight are endemic. The genus *Guanchia*, comprising wingless earwigs in the family Forficulidae, has undergone a spectacular adaptive radiation on the Canaries archipelago, giving rise to 11 endemic species, six of which occur on Tenerife, mainly in wooded areas in the north. The genus *Anataelia* (family Pygidicranidae) is also of great interest, being endemic to the Canaries and with its closest relatives in China and Korea. *Anataelia canariensis* is a wingless earwig typical of the coastal spray zone along the north of the island, but some specimens of the same species have been found in deep crevices in recent lava flows in Las Cañadas.

Order THYSANOPTERA – thrips

Thrips are minute insects, many of them living in flowers. We found that they form a major part of the 'fallout' of aerially dispersing insects on the snow of El Teide, implying that they are both abundant and excellent dispersers. The group has been well studied in the Canaries and 64 species are known from Tenerife, of which nearly a quarter (15) are endemic.

Order PSOCOPTERA – booklice

The booklice are tiny insects that are not easy to distinguish. Forty species are known from Tenerife and 15 of them are endemic.

Order HEMIPTERA – true bugs, hoppers, aphids etc

The hemipteran bugs are extraordinarily diverse on the Canaries, with 51 families represented on Tenerife. Of these families, 28 are in the suborder Heteroptera (the true bugs) 9 in the Auchenorrhyncha (the planthoppers, leafhoppers, froghoppers and cicadas) and the remaining 14 in the Sternorrhyncha (the jumping plantlice, whiteflies, aphids, scale insects and mealy bugs.

The total number of hemipteran species on Tenerife is well over 600, of which nearly a quarter (145) are endemic. However, the endemic species are very unequally spread among the taxonomic groups. The aphids, for instance, have 101 species but only one of them – a species typical of the high mountain zone – is endemic. In contrast, the six families of planthoppers in the superfamily Fulgoroidea have a total of 33 species on Tenerife, and over half of these (19 species) are endemic.

Contrasts of this kind reflect the different dispersal ability of different groups, since insects in which population mixing is frequent cannot easily diverge genetically, forming separate species (aphids are adapted for dispersal in air currents to find new ephemeral food sources, but they are also often accidentally transported by humans). The somewhat ironic result is that groups of animals (or plants) that are diverse on continents and are good dispersers, tend also to be diverse on oceanic islands (because many different continental species may arrive) but they may not readily form new endemic species on the islands. Conversely, members of continental groups with limited dispersal ability rarely reach oceanic islands, but if they do so they tend to form many new endemic species, because divergence is not inhibited by continual arrival of additional immigrants from the continents.

The 28 families of Heteroptera found on Tenerife are very diverse in their body form and habits and include many colourful and conspicuous insects. In some families a high proportion of the species are endemic, and among the Miridae there are two endemic genera (*Canaricoris* and *Lindbergopsallus*) that have undergone adaptive radiation on the archipelago, with different species utilising different food plants. The nine families of Auchenorrhyncha comprise generally less conspicuous insects, many of them feeding on native plants. They include some subterranean species that live by sucking the sap of roots that penetrate caves from above. The Sternorrhyncha are generally smaller insects but have great economic importance, since they include a number of important pests on cultivated plants. A large proportion of the species seem to have been introduced by humans along with plants.

Order PLANIPENNIA (or Neuroptera) – lacewings etc

There are 34 species of neuropterans on Tenerife, in four families, and almost half of these (14 species) are endemic to the archipelago. Many of the species are easily recognised as lacewings, but the ant-lions (family Myrmeleontidae) may be unfamiliar to visitors from northern Europe. Seven species of ant-lion have been found on the island, both in sandy places in the lower zone and in Las Cañadas. The pits in which the larvae live are inverted cones dug in loose sand, into which prey animals fall, only to be grabbed by the larva lying concealed at the bottom; the pits can be found at any time of year, and in summer there is a chance of seeing the adults, which are like small dragonflies with lazy flight.

Order COLEOPTERA – beetles

Beetles are among the most diverse group of insects in the Canaries, and Tenerife alone has about 1470 species; these represent 66 different families and include over 570 species endemic to the Canary Islands. The extraordinary diversity of the beetle fauna of the archipelago has fascinated entomologists for a very long time. Thomas Vernon Wollaston, whose work on the beetles of Madeira was a strong influence on Charles Darwin, made visits to the Canaries in the late 1850s, personally describing more than 550 new species of beetles from the archipelago. An interesting comparison is with the beetles of St Helena, an oceanic island older than Tenerife, but much more isolated and not a member of an ancient archipelago. Darwin visited St Helena briefly and Wollaston spent six months there, studying its beetle fauna. However, this fauna comprises only about 260 species – in contrast with Tenerife which has more than five times as many.

On Tenerife the ground beetles (Carabidae) have been the focus of most attention. They include a large number of endemic species and genera, and are particularly associated with the laurel forest, where they are important predators of a wide range of invertebrates; some species can be found in rotting wood or under the flaky bark of Tejo *Erica platycodon*. Some genera (for instance *Calathus*) have now been studied at the molecular level, providing greater insight into the pattern of colonisation and speciation in the Canaries archipelago.

However, both the rove beetles (Staphylinidae) and weevils (Curculionidae) are more diverse than the ground beetles, each with about 200 species on Tenerife and over half of them endemic. Within the latter family the genus *Laparocerus*, which is endemic to the Northeast Atlantic (Macaronesian) islands, has evolved extraordinary diversity in the Canaries, with Tenerife now home to 41 endemic species and some subspecies.

The darkling beetles (Tenebrionidae) have 60 species on Tenerife and just half of these are endemic. Studies of the tenebrionid genus *Pimelia* (large beetles often to be seen in Teide National Park) and of *Nesotes* and *Hegeter* have shown that as might be expected, colonisation has generally been from the African continent to old islands (eg Fuerteventura) and from these to younger islands further west. However, over-water colonisation is a rare and largely random process, so it is no surprise to find that the patterns are often complex, involving multiple arrivals from the continent and occasional colonisations of old islands from younger ones.

Another group of beetles of special interest is the less well known superfamily Cleroidea; the taxonomy is fluid but it includes Melyridae, Malachiidae, Dasytidae, Gietellidae, Trogossitidae and Cleridae. These are mainly small and slender beetles with soft wing cases; most of the species are predaceous. On the Canaries the Cleroidea include four endemic genera, and on Tenerife alone there are 41 species, with a remarkable 83% of them endemic.

As well as showing adaptive radiation into a variety of habitats above ground, beetles prove to be one of the groups of animals most able to adapt to life in volcanic caves, one of the special habitats of the Canaries. In a single cave on Tenerife, the Cueva del Viento, no fewer than eight species of troglobiontic (obligately cave dwelling) beetles have been recorded, a number rivalled only by the spiders.

ORDER STREPSIPTERA – TWISTED-WING PARASITES

Three species of these uncommon parasitic insects have been found on Tenerife, and are considered as native.

ORDER TRICHOPTERA – CADDISFLIES

Fourteen species of caddis occur on Tenerife; all are considered native and eight are endemic.

ORDER DIPTERA – FLIES

At the family level the order Diptera on the Canaries has the greatest diversity of all insect groups, with 77 families represented (71 of them occurring on Tenerife) and with extremely diverse habits. Endemicity is generally lower than in the Coleoptera, but is strikingly high among the predatory and scavenging flies in the superfamilies Asiloidea and Empidoidea (suborder Brachycera). The eight families within these groups have 93 species on Tenerife, of which 59 (63%) are endemic. The other groups of Diptera have much lower levels of endemicity, with hardly any exceptions. In general few of the Diptera (or other insects) are troublesome to humans on Tenerife, although mosquitoes can be annoying at night in places near standing water.

ORDER SIPHONAPTERA – FLEAS

There are ten species of fleas on Tenerife, but all of them are likely to have been brought to the island by humans.

ORDER HYMENOPTERA – WASPS, BEES AND ANTS

The Hymenoptera are diverse on the Canaries, with about 50 families represented on Tenerife. These insects are in general very good dispersers, which has probably inhibited the evolution of endemic species in the archipelago; many hymenopteran species are also found in North Africa. However, endemism is frequent in some groups of bees (now sometimes considered as subfamilies of Apidae but previously as separate families Anthophoridae, Colletidae, Megachilidae and Halictidae); out of 59 bees in these groups occurring on Tenerife, almost half (29) are endemic species and a good many more are endemic at the subspecies level. Endemism is also frequent in some wasps (Crabronidae and Eumenidae). Among the parasitic wasps, many families have high diversity on Tenerife, but endemics are few except in Pompilidae (the spider wasps), Chrysididae (external parasites of bee and wasp larvae) and the related family Dryinidae, and to a lesser extent in the large family Ichneumonidae. The honeybee *Apis mellifera* is possibly native to Tenerife but has been managed by humans for a very long time.

12

Geology

The Canary Islands and Macaronesia

In the eastern part of the North Atlantic Ocean, to the west of the Iberian Peninsula and Northwest Africa, lie five groups of volcanic islands: Azores, Madeira, Salvage Islands (Selvagens), Canaries and Cape Verdes (Cabo Verde). Collectively these archipelagos are known to biogeographers as Macaronesia (Greek for happy or blessed islands) a grouping that implies a relationship among the archipelagos as well as with the continents to the east. Although some of the islands are close to continental land, they are oceanic in origin and have never been part of a continent, and although some of the archipelagos are distant from one another, their flora and fauna have much in common.

The origins and development of Macaronesia have recently become clearer, as investigations of the sea bed have shown that many other islands existed in the region in the past, and others are forming even now (see Box in Chapter 1). The oldest extant island (Selvagem Grande) is around 29 million years old, but there have been large, high islands continuously present in the region for at least twice as long. Furthermore, some volcanoes that are now reduced to seamounts, which are much closer to the Iberian Peninsula and the northwest tip of Africa than any extant islands, were above sea level and available – between 30 and 70 million years ago – for colonisation by animals and plants from the continents. The complex geographical history, coupled with the long period available for establishment and subsequent evolution, means that the plants and animals of the region have immensely complex relationships. However, advances in molecular biology have made it possible to unravel many of these relationships, and to gain insight into the sequence of movements involved in colonisation of the islands.

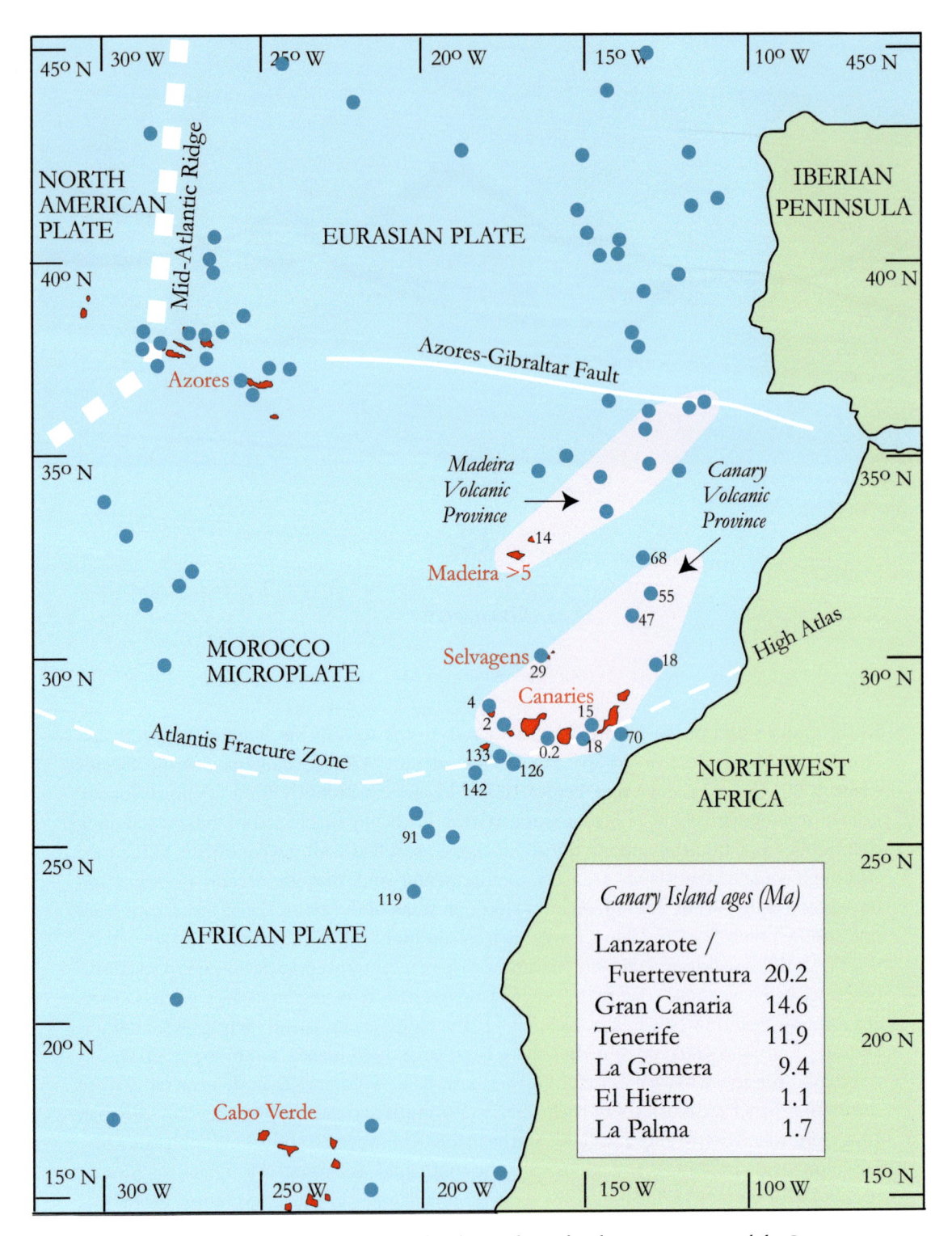

Islands and seamounts in the eastern North Atlantic; figures beside seamounts around the Canaries show their estimated ages in millions of years. Redrawn from several sources.

Island formation and plate tectonics

The Canary Islands lie between 100 and 500 km from the coast of southern Morocco, and along with some associated seamounts, they comprise a loose chain of volcanic islands west of the continental margin of Northwest Africa. The islands and seamounts are all volcanic pimples on ancient ocean crust more than 155 million years old (>155 Ma) which was formed soon after the African tectonic plate began to diverge from the American plates during the opening of the Atlantic Ocean, late in the Jurassic period.

New oceanic crust is still being added to the plates at the Mid-Atlantic Ridge, so the Atlantic is slowly widening, and the African plate (including the crust on which the islands lie) is moving northeast at a rate of just over 2 cm per year. However, the Canary Islands now lie towards the middle of the African plate, since it includes both the young and continually extending oceanic crust to the west and the ancient and much thicker continental crust of Africa to the east. The distance of each island from the coast of Northwest Africa was fixed at the time that it began to form, as a volcano erupting through the crust of the ocean floor, but its distance from the Mid-Atlantic Ridge is very slowly increasing as the Atlantic widens.

It has been accepted for several decades that the Canaries and the associated seamounts were formed primarily by a stationary 'hotspot' or 'mantle plume' rising from a depth of several hundred kilometres in the earth's mantle, with the African tectonic plate moving northeastwards over it. The idea is that the plume forms vertical channels through the oceanic crust at irregular intervals, through which magma erupts; this creates a chain of volcanoes (seamounts and islands) riding on the plate and showing a simple progression in age, the oldest ones having moved furthest from the hotspot. A model of this type provides a convincing explanation for the origin of the Hawaiian Islands, where there is a long line of islands and seamounts in the north Pacific, decreasing in age from west to east.

In the Northeast Atlantic, however, there is no clear line of volcanoes. The impression given by the map is of an irregular band of islands and seamounts some 700 km wide to the west of the continental coasts, from the north of the Iberian Peninsula down along Northwest Africa to the Cape Verdes. To the west of this is a broad area largely free of seamounts, and then clusters of them associated with the Mid-Atlantic ridge around the Azores and further south. However, geologists usually group together the Canaries, Salvage Islands and associated seamounts as the Canary Volcanic Province, and Madeira, Porto Santo and nearby seamounts as the Madeira Volcanic Province, implying that these are separate volcanic systems. Within each of these provinces there are relatively young islands in the southwest and older seamounts to the northeast, which supports the idea of a stationary hotspot or plume in each province, punching a succession of holes through the African plate as it moves to the northeast.

Although this model for the origin of the archipelagos has been favoured by most geologists, it now seems to require some modification. Within the Canaries archipelago the fixed mantle plume model correctly predicts that the youngest islands (La Palma and El Hierro, <2 Ma) are in the southwest and the oldest (the Lanzarote-Fuerteventura unit, ~20 Ma) are in the northeast. Furthermore, Fuerteventura is about 400 km from the current activity on the seafloor near El Hierro, which is a good match for the distance predicted if the hotspot is fixed and the African plate (carrying the islands) is moving over it at just over 2 cm per year.

VOLCANIC MATERIALS

Gases

Volcanic eruptions release vast quantities of mixed gases in variable proportions. Gas emission often precedes eruptive activity, providing a warning signal, but explosive eruptions suddenly release vast quantities of gas. Emission of gases quickly diminishes after the start of an eruption, but some gas may continue to escape for several centuries, as in the summit crater of El Teide. Water vapour normally makes up around two thirds of the mix and carbon dioxide (CO_2) is the other major component. High temperature eruptions (which are often explosive) contain sulphur dioxide (SO_2) while low temperature eruptions have hydrogen sulphide (H_2S). Many other gases – including additional toxic ones such as hydrogen chloride (HCl), hydrogen fluoride (HF) and bromine monoxide (BrO) – occur in small quantities.

Pyroclastic ('fire-broken') fragments

Pyroclasts are formed when gas escaping with great force breaks erupting molten magma into fragments and throws them violently into the air. The fragments have a somewhat frothy texture and may be deposited near the vent or travel considerable distances downwind before falling to earth. The various kinds of particle are named according to size: ash (<2 mm), lapilli (2–64 mm) and bombs (>64 mm). Relatively dense rubbly fragments that accumulate near the vent during fountaining of magma are termed scoria (or cinders); they are basically grey, but become deep red-brown – in the same way as lava – if oxidised by sustained high temperature during the eruption. Particles of any size may consist of pumice, which is extremely frothy (and therefore light in weight) and pale-coloured; it has a high content of silica (60-75% SiO_2), so that it is almost a glass. Deposits of unconsolidated pyroclastic particles of any size are known as tephra, while mainly consolidated deposits are referred to as pyroclastic rock. If this rock is formed of ash or lapilli it is called tuff, while if the dominant fragments are the size of bombs it is termed volcanic breccia. Particles falling close to a vent are size-sorted by gravity and tend to cover the ground in a deposit with even thickness. Pyroclastic flow deposits, which originate from a fast-moving current of gas and particles, are poorly sorted and tend to be thicker in valleys and depressions. A common type of flow deposit on Tenerife is termed ignimbrite; it is rich in pumice and glass shards and may be unconsolidated or with particles welded together as they are deposited.

Lava

The third type of volcanic material, lava, is formed when molten magma flows out of a crater or fissure. Chemically, lavas have widely different composition, but the primary distinction is between mafic and felsic lava. The term mafic is derived from the words 'magnesium' and 'ferric', referring to the high concentrations of magnesium (>5% by weight) and iron, while silica is lower than in other types of magma (45–55% by weight). The term felsic refers to the high levels of feldspars (aluminium silicates with sodium, potassium or calcium) and of silica (65–75% by weight), while amounts of iron and magnesium are low.

Morphologically, lava is of three main types. 'Pāhoehoe' lava (the Hawaiian word can be translated as 'lava on which one can walk barefoot') is usually formed when the magma is very hot and liquid, and it can move at several kilometres per hour, either as a sheet or by the advancement of 'toes'. The surface of this lava is sometimes billowy, or wrinkled and ropy, giving rise to the Spanish term 'lava cordada', meaning corded lava. Aā lava (meaning rough, stony lava) is jagged and chaotic; it forms from cooler, slower-moving magma and is the commonest type of lava on Tenerife, forming terrain referred to in Spanish as 'malpaís'. It often has very deep cracks and can be incredibly rough and sharp, so one is unwise to walk on it without stout gloves. Aā lava flows are denser and more viscous than pāhoehoe, and move more slowly and in a peculiar way: the front of the flow steepens due to pressure from behind until it breaks off and tips forwards, after which the general mass behind it follows on. These flows can measure 2–20 m thick, and because they are thick, the top and bottom cool into clinker-like rubble, insulating the interior, which cools more slowly into a dense, fine-grained rock. The rubbly layers make it easy to distinguish successive lava flows. Pāhoehoe lava can change into aā lava as it loses gas, cools and becomes more viscous, but aā lava never turns into pāhoehoe. The third type, block lava, is composed of large chunks of lava that continued to move after they had cooled, because of the pressure of molten lava behind.

Sometimes a fragment of semi-solid aā lava starts rolling like a ball down the surface of an inclined active lava flow. More lava may then stick to its surface, so that it grows like a large snowball rolling down a hill covered by a layer of soft, sticky snow. The result is a roughly spherical mass known as an accretionary lava ball, which may roll off the parent lava flow and come to rest on a completely different surface, as in the case of the 'Huevos del Teide' (Teide's eggs) on the pumice slopes of Montaña Blanca.

Ropy 'pāhoehoe' lava

Lava balls – 'Huevos del Teide'

Dykes and plugs

These are not strictly volcanic materials, but are internal constituents of volcanoes. Dykes are more or less vertical sheets of magma connecting sources of magma deep in a volcano to eruptive fissures on the surface through which lava or pyroclasts are emitted. The dykes feed molten magma to the vents, but when activity ceases the remaining magma solidifies into rock, and in a subsequent eruption the magma finds new routes along which it can force its way to the surface, so that a new set of dykes is created. The solidified magma is usually resistant to erosion, so that dykes are often exposed when weaker materials are eroded away. A volcanic plug, also called a volcanic neck or lava neck, is a mass of rock formed when magma hardens in a conduit to a large vent on a volcano. In the case of a stratovolcano, it will often be composed of phonolite. If a plug has formed in a vent and the volcano then becomes active again, the plug can cause an extreme build-up of pressure when rising gas-charged magma is trapped beneath it, and this can sometimes lead to an explosive eruption. Like dykes, plugs are often harder than the enclosing rocks, so that erosion may eventually leave the plug standing up in bold relief, as in the case of many of the peaks in the Anaga peninsula, where they provide telltale evidence of the impressive volcanoes that formed it.

However, a difficulty arises from new and controversial estimates of the ages of a number of seamounts in the Northeast Atlantic, obtained from dredged rock samples. These indicate young and old islands and seamounts intermingled in a complex pattern, with old seamounts in the northeast but apparently even older ones southwest of the Canaries (see map). The youngest island (El Hierro) seems to have an ancient seamount adjacent to it (Hierro Ridge, 133 Ma) while the oldest seamount in the region is even further to the southwest (La Bisabuela or Las Hijas, 142 Ma). This pattern is hard to reconcile with the idea of a fixed plume controlling volcanism in the region. However, it is possible that formation of the very old seamounts, at a time when the relevant part of the seafloor was still close to the Mid-Atlantic Ridge, was related to activity associated with rifting at the ridge rather than to a deep-seated hotspot.

Another difficulty with the fixed mantle plume model is the modern volcanism on Lanzarote, since renewed activity would not be expected after such a long time and so far from a hotspot, although the slow movement of the African plate could partly account for it. Another explanation is that the Canaries hotspot is actually a remnant of a 'fossil' plume, producing more dispersed hot regions, rather than a strong and well focused convective flow.

A more fundamental reappraisal of the fixed plume model has recently been proposed, based on the concept of 'shallow mantle upwelling' or 'edge-driven upper mantle convection'. The idea is that convection in the uppermost and fluid part of the mantle produces recurrent upwelling of hot magma from relatively shallow levels, within a broad zone offshore from the continental margin. This convection is related to the higher temperature of the upper layer of the mantle below the thin oceanic crust than below the thicker and cooler continental crust. This implies that the active volcanic zone is associated with the trailing edge of the African continental mass and moves to the northeast with the African plate. This contrasts with the older model of a hot plume fixed deep in the mantle, over which the plate moves steadily, leaving behind the plume and the active volcanic zone where it creates islands progressively further from the receding continent that rides on the plate.

A prediction of the recent model is that upwelling of magma and volcanism capable of forming seamounts will persist indefinitely in the oceanic zone to the west of the continent, and eruptions will occur at irregular intervals and in a complex spatial pattern. This kind of activity could account for the renewed activity of Lanzarote, since the crust below the island might have been weakened by the original island-building upwelling of magma, resulting in the new activity being focused there. It might also account for some of the anomalous seamount ages. However, the earlier model of fixed deep-seated plumes seems also to be required to account for the steady age progression within the Canary archipelago and the region around Madeira. It seems best, therefore, to accept that the origin of the islands probably requires an interaction of two distinct sources of upwelling, one fixed in the mantle and the other associated with convection currents and turbulence in the wake of the African continental mass, as it ploughs its way slowly across the fluid mantle.

The life cycle of oceanic islands

Oceanic volcanic islands such as the Canaries are normally 'shield volcanoes' formed by upwelling of mainly 'mafic' magma beneath the ocean floor (see Box: Volcanic materials).

12 Geology

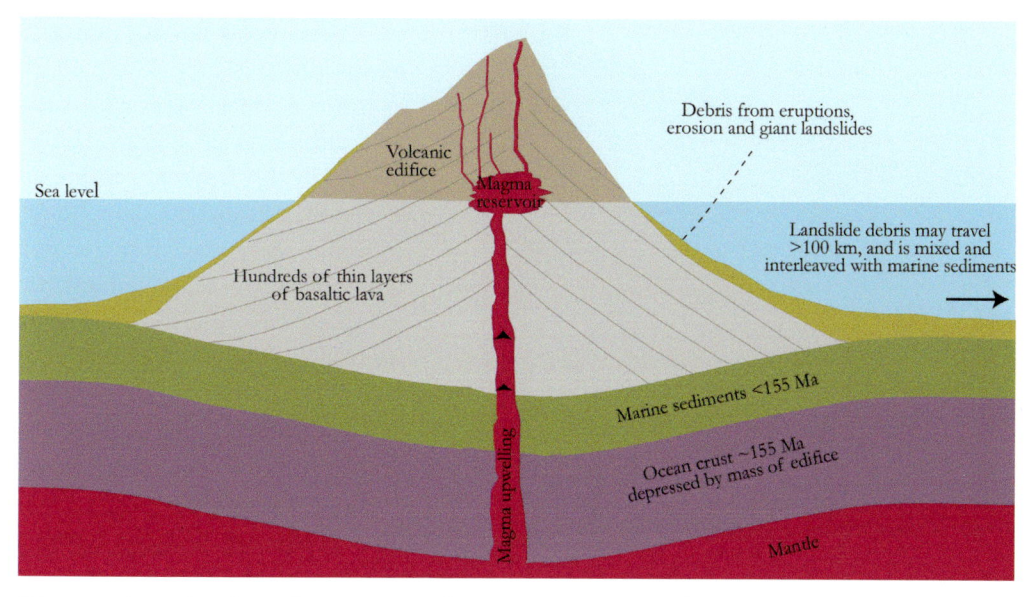

Diagramatic section through an oceanic volcano, resting on ocean floor depressed by mass of edifice

The magma is derived from material melted from the mantle which is forced up through a weakened spot in the oceanic crust and emitted through rifts in hundreds of eruptions at relatively short intervals. The viscosity of the lava is initially low, so it spreads in layers only about three metres thick, but it cools quickly in seawater and solidifies as basaltic lava flows or rounded blocks (pillow lava) or shatters into masses of glassy fragments (hyaloclastites).

The successive eruptions create a conical pile on the ocean floor, seen in scans as a seamount. A small one named El Hijo (the son) between Tenerife and Gran Canaria is probably only about 200,000 years old and is a potential island that has not yet emerged above the sea. The young island of El Hierro is a shield volcano that may have broken through the sea surface little more than a million years ago. Although no onshore eruptions on El Hierro have occurred in historic times, there have been submarine eruptions and seismic shocks around the island in the last few years. In 2011/12 activity became so intense that the authorities evacuated people temporarily from some parts of Hierro, for fear of a major earthquake or eruption. Although the area is now quieter, the eventual emergence of a new Canary Island (or a new peninsula on El Hierro) is clearly on the cards.

Seamounts in the Northeast Atlantic have slopes averaging 13°, so that if they build up from seafloor about 4000 m below the surface, they will be around 35 km across at the base if they reach the surface. The great width of the volcanic pile means that an enormous mass of lava is needed to create an island. The volume required is increased by the fact that the mass of the pile pushes down the oceanic crust beneath it, so that eventually up to half the total volume of the volcano is below the level of the surrounding ocean floor. Nonetheless, shield volcanoes typically build up quickly (hundreds of thousands or a few million years) through a series of intense episodes of volcanic activity. The eruptions of the volcano may be complemented by intrusive activity, when a mass of magma is pushed up into the base of the volcanic pile but solidifies there rather than being erupted, forming a 'plutonic' core in the volcano.

Western wall of the Güímar valley on Tenerife, a collapse embayment about 1 km deep

After breaking the surface, newly created islands may continue to grow as shield volcanoes, with periodic bursts of intensive activity involving frequent eruptions of mainly fluid basaltic lava. These eruptions are typically from fissures along linear 'rift zones' rather than from large volcanic cones. The rift zones often form a triple-arm 'Mercedes star', usually with one ray much weaker than the other two; they probably arise at the start of development of the volcano when upward pressure of rising magma pushes up and eventually fractures the brittle oceanic crust, but they may then persist throughout the active life of the volcano.

Eruptions along rift zones are fed by swarms of 'dykes', vertical sheets of molten rock that force their way up to eruptive fissures from magma reservoirs far below. Lava flows emitted from vents above the sea surface are exposed to air rather than seawater and thus cool more slowly than during the submarine phase, forming jagged flows of 'aā' lava or sheets of ropy or wrinkled 'pāhoehoe' lava, the latter often containing volcanic caves in the form of lava tubes.

Throughout their growth, island volcanoes are also subject to the destructive forces of erosion by waves, wind, running water and ice, which continually take their toll, slowly in places with massive rock formations, fast in the case of looser deposits.

However, it is now known that volcanoes suffer more dramatic setbacks during their upward growth, both before and after they emerge above the ocean surface as new islands. As the ridges formed along the rift zones increase in height as a result of magma pushing up from below, unstable structures are created, and this eventually leads to giant landslides or 'flank collapses' in which large sections of the sides of the volcano slip away into the ocean depths. Some slides are rooted deeply in the side of the volcano and are relatively slow-moving slumps, while others are more superficial debris avalanches that move rapidly and spread further. There is a hint in the geological record that flank collapses on islands are especially frequent in periods when sea level is falling rapidly, removing some of the support provided by the water.

Catastrophic events of this kind create impressive topographic features in the form of huge troughs or 'collapse embayments' on the sides of the island, which may be more than a kilometre deep. Sometimes a landslide is quickly followed by eruptions of lava sufficient to fill a large part of the embayment, and occasionally a new volcanic cone may be formed at the site of the original failure, as seems to have occurred with El Teide (see below).

Collapses of island flanks have major effects below the sea surface. High resolution scans of the ocean floor can often indicate where collapses have originated and where the materials have ended up. The debris from a collapse flows down over the submarine slopes of the volcano,

Sea level, sediments, seamounts and islands

The floor of the Northeast Atlantic is dotted with large numbers of seamounts. Some are volcanoes that have never built up to sea level, but others are 'guyots'. These are well-submerged, flat-topped seamounts representing the remains of ancient volcanic islands in the last stage of their life cycle. Guyots have suffered millions of years of erosion, some of it occurring during glacial periods when sea level was low and the island remnants were exposed and thus subject to the full force of the waves; subsequent rise in sea level leaves the summits of some guyots a hundred metres or more below the surface.

There is also another reason why ancient island remnants may have summits well below the surface. When new oceanic crust is created at the Mid-Atlantic ridge, sediments immediately start to accumulate on it. The crust on which the Canaries lie was formed when the Atlantic was narrow and the ridge was close by, and sediments have been accumulating for about 155 million years. The sediments are derived from many sources: biological activity in the ocean, transport of particles in currents, drifting of dust from the nearby Sahara, and the submarine spread of materials produced by volcanoes. They are now several kilometres thick in many places, and their mass slowly depresses the ocean floor, so that the sediments and the islands and seamounts resting on them sink slowly to greater depths.

The younger volcanoes have erupted through the thick layers of sediments and built their volcanic piles on top of them, but old volcanoes – whether they are now islands, guyots, or seamounts that never reached the sea surface – are mantled and partly submerged by the sediments; they thus represent much larger volcanic piles than is apparent from simple bathymetric maps. For instance the El Hierro ridge, about 133 Ma, is a relatively unimpressive feature of the ocean floor south of El Hierro, but if the island and the adjacent ridge (measured down to the base of the volcanic pile, where it rests on only the thin layer of sediments that had formed when its volcanic pile began to develop) are considered as a single volcanic structure, it has a total height of over 7000 m, approaching that of Teide.

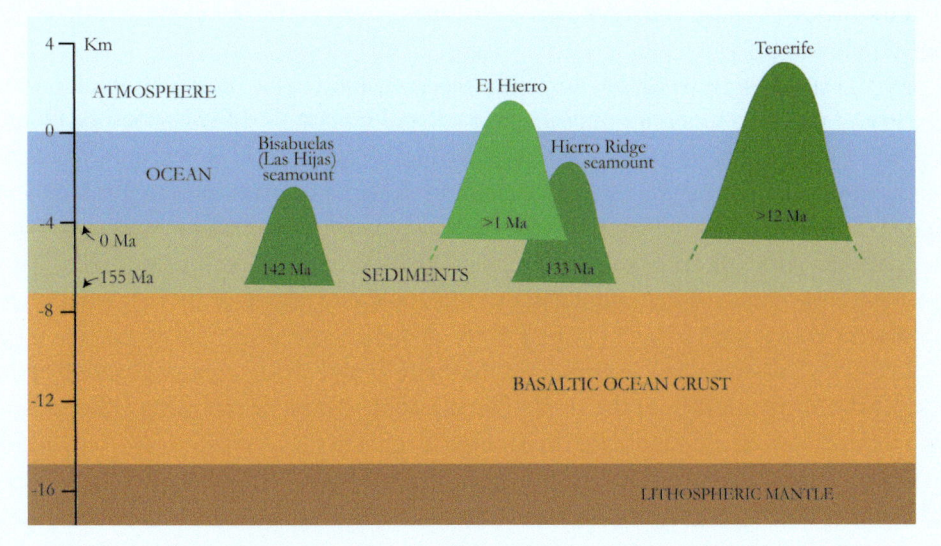

The implications of accumulation of sediment on the seafloor around some islands and seamounts near Tenerife; tentative minimum ages are indicated in millions of years.

Eruption of Mayon volcano in the Philippines in 1984, with pyroclastic flows rushing down the slopes of the mountain. Photo courtesy of the United States Geological Survey.

sometimes excavating a deep trough in older landslide debris, but spreading more widely as it travels further over more gentle slopes, sometimes reaching a distance of more than 100 km from the island. It may form deposits more than 700 m thick, including volcanic blocks a kilometre or more across, and may cover an area larger than the island itself. Since many collapses may occur within the lifespan of an island, their offshore debris is often partially superimposed, forming complex debris aprons.

The volcanoes on some islands are fed by relatively fluid mafic magma throughout their active lives; they are composed primarily of basalt and remain shield volcanoes; La Gomera is a good example in the Canaries. Other islands, including Tenerife, go through a transition that superimposes the steep and rugged summit cone of a 'stratovolcano' on the gently sloping shield volcano. Tenerife made this transition around two million years ago, but La Palma has only recently been going through it. The volcano enters this 'post-shield' stage when upwelling of magma beneath the island is reduced and eruptions become less frequent and different in character.

Volcanoes have a magma reservoir within them, above which there is a conduit system through which magma rises at the time of an eruption. In the case of stratovolcanoes, the long intervals between eruptions give time for the magma in the reservoir to cool and become 'differentiated'. This is a process in which magma changes from its original mafic composition (derived from melting of mantle material) by partial crystallisation followed by separation of layers of crystals with differing densities. The remaining magma tends to be enriched in silica, sodium and potassium. Volcanic materials derived from differentiated magma are termed 'felsic' (see Box on Volcanic materials). The relative density (specific gravity) of felsic magma is low but the viscosity is often extremely high, setting the stage for potential explosive eruptions. When erupted as lava, the flows are generally thick and slow-moving; they solidify quickly as trachytic or phonolitic rock, often failing to reach the base of the mountain. This alters the profile of the volcano, which typically forms a tall symmetrical cone with steeply sloping sides (20–40° on Teide).

Volcanoes with reservoirs containing differentiated magma occasionally show dramatic periods of activity when magma is expelled under enormous pressure as a result of renewed activity at greater depths. Recent study of past explosive events on Tenerife has led to the suggestion that they may be triggered by intrusion of hot mafic magma into a chamber containing cooler, partially crystallised and stratified felsic material, leading to overturn and mixing of the contents of the chamber. The effect has been likened to 'putting a red-hot rock into a cup of ice'. The result is an explosive eruption, in which the material ejected includes partly solidified crystalline nodules from all parts of the magma chamber. A 'Plinian' eruption column is then formed above the volcano, comprising enormous clouds of small volcanic fragments that are eventually deposited to form layers of pumice or other kinds of 'pyroclastic' rocks.

The more violent explosions also have the capacity to create spectacular 'pyroclastic density currents' or PDCs, which are the most dangerous of all volcanic phenomena. A flow of this kind is generated when an eruption column above a volcano becomes so heavily loaded with pyroclasts that it collapses back down towards the vent. It then forms a fluidised current of incandescent airborne rock fragments and gases, at temperatures up to 700°C and powered by the kinetic energy generated by the vertical collapse; it rushes down the flanks of the volcano like an avalanche, hugging the ground and reaching speeds up to several hundred km/hour. Such a flow may travel tens of kilometres from the vent and tends to be channelled down any valley it encounters; as it slows and loses its gas it drops its load of particles as a thick and fairly homogeneous carpet of tephra.

If a major explosive eruption of a volcano partially or completely empties the magma chamber, the overlying strata may then collapse back into the chamber, leading to its disintegration and a dramatic climax in which any remaining contents are ejected through a ring of newly created vents. This may be followed by massive subsidence at the surface, forming a wide summit crater known as a 'caldera', with a more or less horizontal floor.

Over millions of years, as volcanic activity of the island is further reduced, the rate of destruction by erosion overtakes the rate of growth. The upper surface is gradually reduced in height, and the coasts develop impressive cliffs as a result of the incessant beating of the waves. These erosive processes are complemented by the effect of gravity in moving rock masses downhill (a process termed 'mass wasting', which includes giant landslides). Under these influences, volcanic islands undergo a long period of decline. Some islands have a 'rejuvenated' period of volcanic activity after a hiatus lasting up to several million years, but sooner or later the destructive forces become dominant.

In the Canaries archipelago, Gran Canaria and La Gomera provide examples of islands in decline. Gran Canaria had stratovolcanoes some four million years ago, with differentiated volcanic complexes such as Roque Nublo, but it later suffered from landslides and long periods of erosion; although there have been relatively recent eruptions on the island (the latest only about 2000 years ago) they were of mafic magma and it is unlikely that Gran Canaria will ever have another stratovolcano. La Gomera – though close to Tenerife – is a separate and ancient volcano, on which activity ceased almost entirely some four million years ago. Destructive forces have long been dominant and the island now has an eroded central plateau from which deep 'barrancos' (ravines formed by running water) radiate, while the coast consists largely of cliffs, some of them as much as 800 m high.

As destructive processes continue over many millions of years, the sea-cliffs move inexorably inwards until the whole of the land has gone, and the island slowly disappears beneath the ocean. The Canaries archipelago does not have an example of this stage, since although the eastern islands of Lanzarote and Fuerteventura comprise a single and ancient geological unit, this has shown rejuvenated activity over the past few million years, including massive eruptions on Lanzarote in the 18th and 19th centuries. A perfect example, however, is provided by the Salvage Islands northeast of Tenerife. They are around 29 million years old and were once high islands, but have suffered from the multiple forces of erosion for so long that they now remain only as platforms a few metres above the waves.

In tropical regions the upward growth of corals around the edges of a slowly submerging volcano may create an atoll that can persist indefinitely, as originally proposed by Charles Darwin, but this does not happen in the cooler subtropical waters around the Canaries, where the most ancient volcanoes survive only as passive seamounts.

Early development of Tenerife

Tenerife, unlike the islands just discussed, is still in the prime of life. The origins of the modern island lie in the eroded remains of three ancient shield volcanoes. These were built up by multiple layers of basaltic lava emitted along fissures, initially under water but continuing after the volcanic piles had emerged above the sea surface, becoming islands.

The first of the volcanoes to emerge, about 12 million years ago, was the Central Shield, an enormous volcano that occupied an area encompassing the central part of the modern island (including the area around Teide) and extending northeastwards at least to where the city of La Laguna now stands. The rocks of the Central Shield are now almost entirely covered by later volcanism, except for the Roque del Conde massif near Adeje.

The Conde massif is named after Roque del Conde, a massive 1000 m mountain – strikingly flat-topped when viewed from the south

Roque del Conde viewed from the north, with *Euphorbia atropurpurea* in foreground

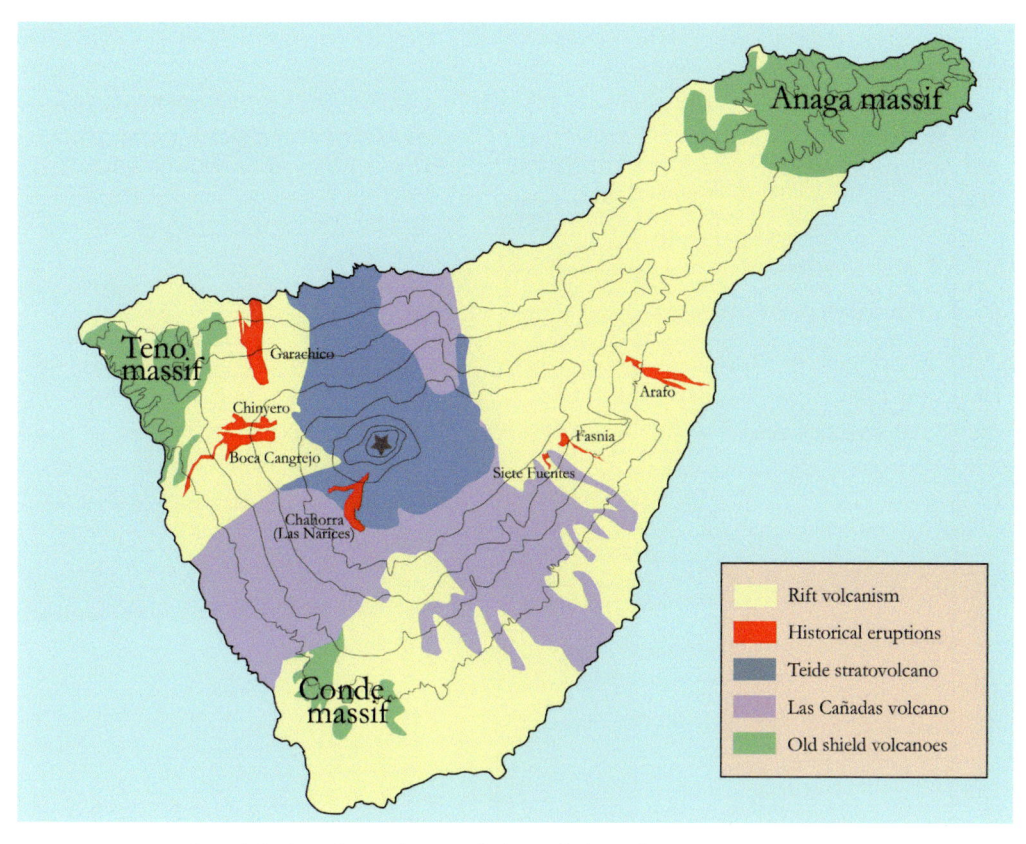

Simplified geological map of Tenerife, based on various sources

– which is a dominant feature in the landscape of the southern part of Tenerife. Roque del Conde and several other impressive peaks to the north of it (including two that exceed 1100 m in height) emerge from a high plateau above the town of Adeje. The spectacular Barranco del Infierno, with its mouth at the top of the town, has cut down into the edge of the plateau through hundreds of metres of rock, exposing a long series of thin (<4 m) mafic lava flows; these are occasionally interrupted by bands of 'almagre', soil formed during a quiet period between eruptions and burned red by a lava flow when activity recommenced. The lowest accessible flow in the barranco is dated to about 11.9 million years old (11.9 Ma) the oldest known rock on Tenerife. However, since this is 590 m above sea level, it is likely that the volcanic pile had broken the sea surface to create an island considerably earlier.

Activity of the Central Shield probably continued until at least 8.9 Ma, the age of rock strata at the summit of Roque del Conde itself. Although so little of the Central Shield is now visible, gravimetric studies suggest that the remains of a massive volcano – known as the Boca Tauce Volcano – lie concealed beneath the southwestern rim of the modern Las Cañadas Caldera; this rock mass seems to include a large element of dense intrusive (plutonic) material, and it may well represent the eroded core of the Central Shield Volcano.

The next major event in the development of Tenerife was the building of the Teno massif, to the northwest of the Central Shield. Activity of the Teno Volcano started more than six million

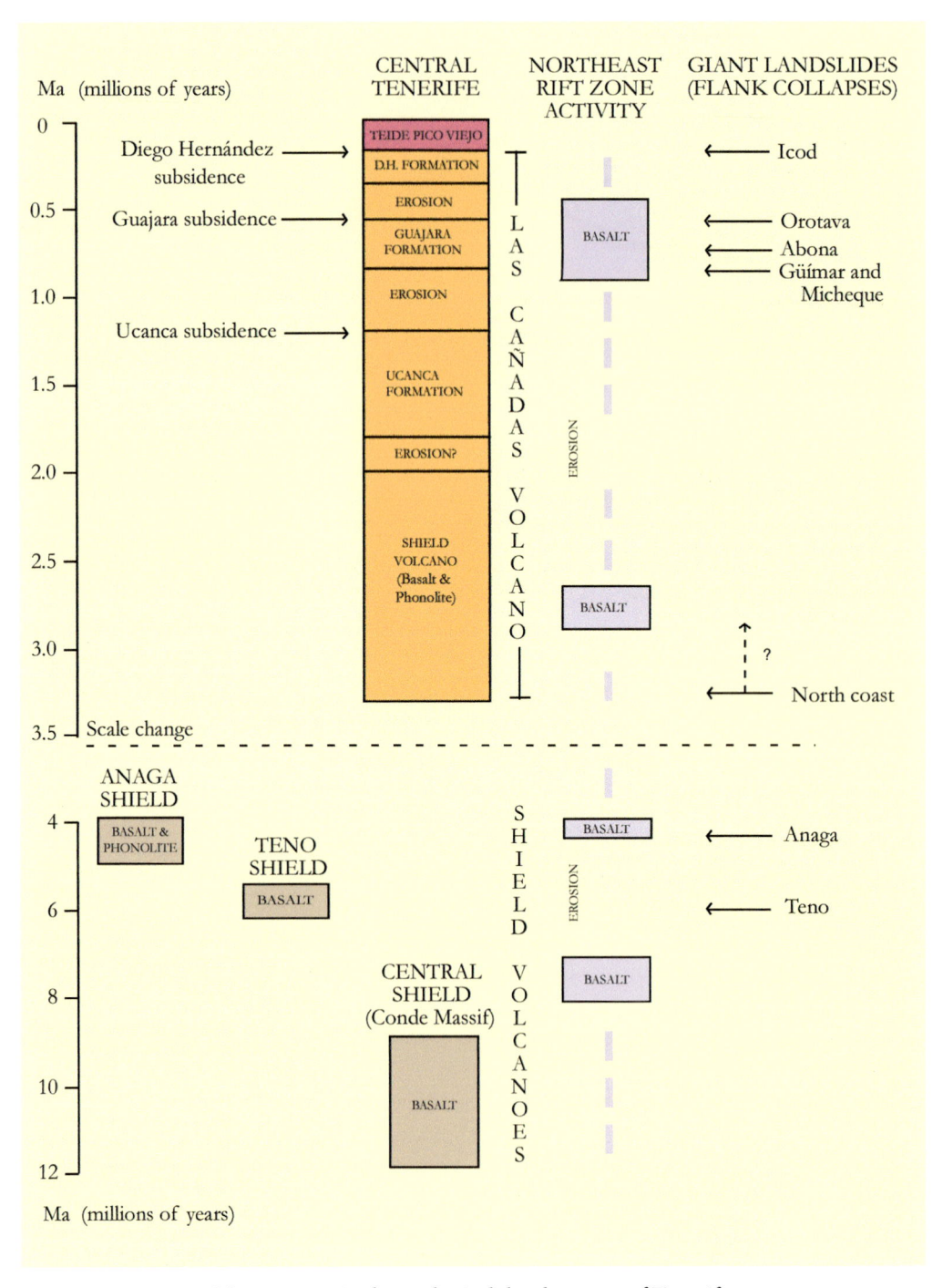

Major events in the geological development of Tenerife

years ago, with massive basaltic pyroclastic deposits and some lava flows. The oldest known rocks in Teno are dated at about 6.1 Ma, but they are at one of the highest points in the massif (1030 m) so much older strata are presumably buried beneath them. Around 6 Ma there was at least one major flank collapse (giant landslide) in which the whole of the northeast sector of the original Teno massif apparently slid into the sea (see diagram in next section). This left behind a high ridge running from north of Teno Alto southeastwards to Santiago del Teide and then northeast towards the Puerto de Erjos. The collapse embayment was immediately partially filled by new volcanic deposits, dipping towards the north. A period of erosion of the surviving original ridge was followed by massive emissions of basaltic lava flows dipping towards the southwest. Activity of the Teno Volcano continued until about 5.1 Ma.

Today, the Teno massif forms most of the northwest corner of Tenerife, mainly west of the TF82 road running northwards from Los Gigantes through Santiago del Teide to El Tanque. It has spectacular eroded topography, with deep barrancos and sea cliffs exceeding 500 m in height in some places. The pyroclastic deposits and lava flows are cut by dykes and a few pale-coloured phonolitic intrusive plugs; some of the later flows are of trachytic lava.

The impressive sea cliffs visible from Los Gigantes or Punta del Teno are composed of thin and brittle layers of lava, and are almost perfectly vertical; the continual pounding of the waves cuts into the lower part of the cliffs and eventually causes rock falls from above. When wave erosion creates major cliffs like this, it removes part of the supporting flank of the volcanic pile, and this may make oceanic islands more vulnerable to flank collapse than volcanoes on continents.

Santiago del Teide, on the southeastern border of the Teno massif, lies in a large basin formed when the lower part of an ancient eroded valley running up into the massif was obstructed by the growth of Las Cañadas Volcano. The blockage led to ponding of later lava flows from the Northwest Rift Zone and accumulation of sediments, and eventually to formation of a fertile plain around Santiago, in a similar way to the origin of La Vega de La Laguna on the edge of Anaga (see

Erosion by water has created the deeply incised Barranco de Masca

Los Gigantes cliffs show long series of thin lava flows cut by dykes; near the top is a reddish scoria cone

Northern coastline of Anaga, viewed from near Taganana,
showing the eroded roots of the ancient volcanoes

below). The eastern wall of the Tamaimo valley to the south of Santiago is formed of ancient rock of the Teno Shield, but is topped by historic lava flows.

On the north side of Teno relatively recent volcanic rocks form the land surface of the low-lying 'Isla Baja' near Buenavista, including both the area close to Punta de Teno and the long valley running southwards into the Teno massif at El Palmar and Las Portelas. However, older lava flows, which filled the embayment of the Teno lateral collapse almost six million years ago, are visible along the north coast in the impressive cliffs of Interián and Acantilado de La Culata. These stretch from Los Silos in the west past Garachico in the east, until they are truncated about a kilometre west of Icod at the edge of the Icod valley, which is the embayment of the relatively recent Icod lateral collapse (see below), now filled with debris and younger materials.

The end of volcanism in the west of Tenerife was quickly followed by the building of the Anaga Volcano between 4.9 and 4.0 Ma. Marine sediments found in boreholes near La Laguna suggest that the Anaga peninsula was once a separate island, though the volcano that formed it may have originated at the northeast corner of the older Central Shield. It became joined to the rest of Tenerife as a result of eruptions along the Northeast Rift Zone at a much later date. The axis of the ancient volcanism in Anaga ran from west to east well to the north of the present ridge, and at one time the land surface may have been as much as 2000 metres above the current eroded remains.

The formation of Anaga involved two phases, the first laying down mainly basaltic pyroclastics dipping towards the south, and the second involving basaltic lava flows dipping radially. The development of the peninsula was violently interrupted at some stage by the

The rubbly coastal cliffs of northern Anaga erode rapidly, leaving the hard intrusive phonolitic plugs exposed

Taganana giant landslide on the north side, which removed a large part of the eastern end of the ridge and determined the concave form of this part of the peninsula. The collapse embayment has thick deposits of breccia on its floor, suggesting that on this occasion the flank of the mountain fragmented rather than sliding into the sea in the form of coherent blocks.

Today, it is very evident that since its heyday as a major volcanic edifice, Anaga has been ravaged by millions of years of erosion. The coastline has moved far inland during the long period since the landslide, and the flanks of the peninsula are dissected by barrancos with sides towering hundreds of metres above the valley floors, carved out by runoff of the plentiful rainfall produced by the moisture-laden northeast trade winds. Skylines in the north of the peninsula are dominated by peaks representing the hard skeletal elements of the old volcano, including many dykes – constituting up to half of the rock outcrops in some places – and phonolitic plugs. The dykes and plugs often rise from slopes of rubbly material, since the rocks in many parts of Anaga are hydrothermally altered, probably by moderately heated groundwater derived from the high rainfall, which forms clay minerals and leaves surviving rocks weakened and much more prone to erosion. This is brought home by a visit to the north coast of Anaga in rough weather, when the sight of the waves clouded with silt shows that this is a place where erosion takes place on a timescale of days as well as of millennia.

The southern part of Anaga has generally harder rocks, mainly long sequences of basalt lava flows and scoria dipping to the south, well exposed in the long southern valleys and the cliffs along the coast east of Santa Cruz. Along the main ridge, roadcuts expose phonolitic lava and pyroclastic rocks that are among the few surviving elements of the latest (~4 Ma) eruptive sequences in Anaga, which once formed a thick cap above the basaltic base.

In the west of Anaga the plain around La Laguna forms a flat saddle between the peaks of the peninsula and the ridge of the Northeast (Dorsal) Rift Zone running up to Las Cañadas. Activity along the rift zone around 2.7 Ma produced lavas and pyroclasts that filled the sea passage between the old Anaga massif and the developing Las Cañadas Edifice; they also formed the apron close to sea level on which Santa Cruz de Tenerife is built. These volcanic materials were supplemented by debris from the once great eroded valley that ran down from Cruz del Carmen in Anaga to the present site of La Laguna. This blocked valley now levels out at Las Mercedes, to the west of which a marshy lagoon, enclosed roughly by the 600 m contour, survived until the time of the Castilian conquest but was then soon transformed into the agricultural area known as La Vega (fertile plain) de La Laguna (see photo in Chapter 1).

RIFT ZONES AND GIANT LANDSLIDES

As already implied, the development of Tenerife has been punctuated by a series of catastrophic flank collapses (giant landslides). Events of this type are a fundamental feature of the development of oceanic islands, and their massive nature is reflected in the accumulation of debris around the islands; to the north of Tenerife the deposits have a volume of more than 500 km^3. There is evidence of about nine collapses on the flanks of Tenerife, but the real number may be higher. The Teno shield volcano suffered at least one collapse about six million years ago, and the Anaga volcano suffered at least one during its construction rather later. Since then, after the basic structure of Tenerife was established some three million years ago, at least six more major landslides have removed large sections from most of the flanks of the island.

The occurrence of collapses on oceanic islands seems often to be related to the volcanic activity on rift zones radiating from the centre. Tenerife has two active rift zones, and also a Southern Volcanic Zone with widely scattered vents in the far south of the island, which has sometimes been considered as a third rift zone. The Northwest (Santiago del Teide) Rift Zone is oriented SE–NW from the centre of the island towards Teno, and has given rise to a series of eruptive vents including Volcán Chinyero just over a century ago. More obvious is the Northeast Rift Zone or 'Dorsal de la Esperanza', oriented roughly SW–NE, which built the dorsal ridge, running for some 30 km between the high centre of the island and La Laguna. Geologists now think that the NW and NE rift zones are long-lasting

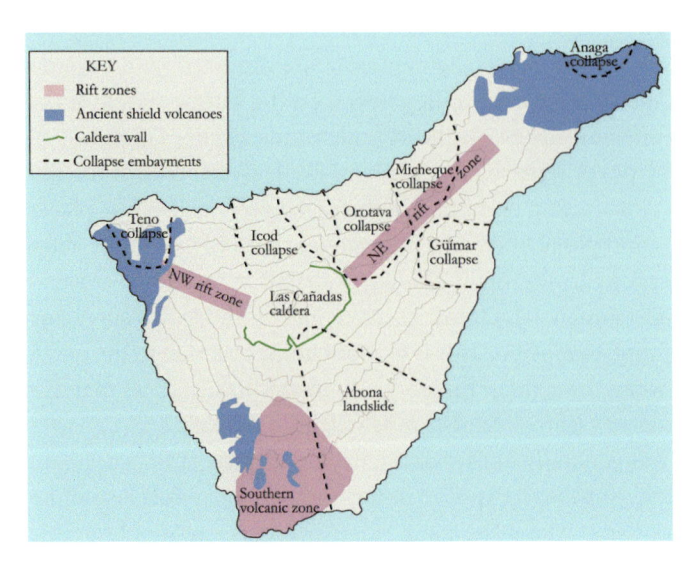

Ancient shield volcanoes, rift zones, Las Cañadas caldera and outlines of collapse embayments

features acting throughout the development of the island and determining its fundamental shape. They are effectively joined beneath Las Cañadas Edifice, forming a continuous axis of activity along which eruptions are concentrated and flank collapses originate.

The extraordinary amount of debris on the sea floor that forms an apron along the centre of the north coast, suggests a massive landslide there in the distant past; it may have occurred around 3-3.5 Ma, after a phase of rapid volcano growth. This collapse is thought to have removed a large part of the northern flank of Tenerife, from near Icod to Orotava. This may help to explain the basic concavity of the north coast as viewed from above, although the landslide embayment has been filled by later eruptive materials and erosion debris.

Several better known flank collapses have occurred in the last million years. Probably the first was the Micheque collapse on the northern flank of the island around 830 thousand years ago (830 ka), creating an enormous concavity between Santa Úrsula and Tacoronte. However, this collapse was undetected until recently, because the huge hollow left by the landslide was filled by massive eruptions of basalt, followed by eruptions of trachytic lava that built the Micheque stratovolcano, now heavily eroded and inconspicuous, southeast of Santa Úrsula. Evidence for these events came to light only through study of rocks penetrated by one of the 'galerías' constructed to collect water from distant parts of the island, which provide geologists with the kind of access to deeply-buried strata that they would normally only dream about.

The Micheque collapse on the north flank of the island seems to have been the result of intense volcanism in the NE Rift Zone, but when the collapse occurred it apparently destabilised the other side of the ridge, causing another collapse (perhaps almost immediately) which created the Güímar valley on the southeast flank of the island. In this event a mass of rock up to a kilometre thick slid into the ocean, leaving a 10 km x 10 km rectangular scar. This embayment was soon partially filled by lavas cascading in from the active ridge along its upper edge, and also from the southwest near the present Ladera de Güímar.

The next documented collapse was the Abona landslide about 733 ka, initiated by a violent eruption followed by an ignimbrite and ash fall deposit. The landslide probably started near where Guajara now stands; it did not gouge as deeply into the side of the volcano as the major flank collapses, but its debris covered some 90 km² of the southeast slopes of the island, and extended up to 50 km southwards over the ocean floor. Later pyroclastic flows followed the same route down the mountain, mainly between Granadilla de Abona and Arico, contributing to the Bandas del Sur ignimbrite deposits that blanket this part of the island.

The subsequent pause ended with the most recent collapse on the Northeast Rift Zone, roughly dated to between 780 and 566 ka. This

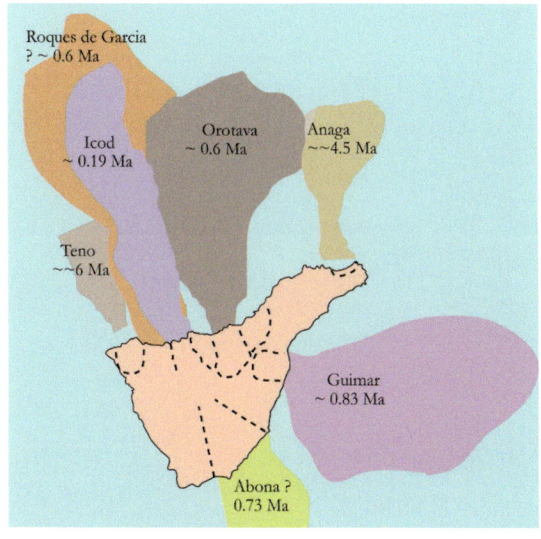

One tentative interpretation of the collapse debris aprons that fan out from the shores of Tenerife

created the Orotava valley, one of the most spectacular features of the island, some 9-12 km wide and 400-500 m deep. It seems to have involved two separate landslides in fairly quick succession, first in the east and then in the west, and is reckoned to have removed 130 km³ of rock. After the collapse there were some eruptions on the north side of the ridge, including some of trachytic and intermediate materials.

The last major landslide event on Tenerife was the Icod collapse, probably at 198 ka; this removed the northern flank of Las Cañadas Edifice, providing the foundation for the development of Las Cañadas and El Teide in their modern form.

Las Cañadas Edifice and the origin of Las Cañadas Caldera

The construction of the Teno and Anaga massifs between about six and four million years ago was followed by a quiet period during which erosive processes were dominant. Shortly after 3.5 Ma, however, a new eruptive centre became active between the three older massifs, and this led to the construction of Las Cañadas Edifice (or Las Cañadas Volcano) a massive shield volcano centred slightly to the east of the original Central Shield, underneath the highest part of the modern island. The base of Las Cañadas Edifice was formed mainly by layers of mafic lava. Exposed rocks from this period include the base of the caldera wall and also the Tigaiga massif in the north, the area west of Adeje, and Montaña de Guaza in the far south.

Activity of the Las Cañadas Edifice decreased about 2 Ma, and when it became active again about 1.8 Ma it took on a different character. A lower frequency of eruptions allowed time for differentiation (partial crystallisation) of the magma in the reservoir feeding the volcano, so that eruptions were of felsic magma with high viscosity, sometimes mixed with mafic material from deeper down. Rather than being emitted as slow moving lava that might create a steep-sided stratovolcano like Teide, the eruptive material was often ejected explosively.

Between about 1.8 Ma and 0.2 Ma Las Cañadas Edifice went through three cycles of this kind of volcanic activity, represented by the Ucanca, Gua-jara and Diego Hernández formations, each terminated by particularly violent explosive 'Plinian' eruptions in which vast quantities of volcanic materials were hurled into the air as pyroclastic fragments and deposited by pyroclastic density currents.

These volcanic cycles built up a massive and complex series of deposits of 'ignimbrite', a type of rock comprising thick beds of low density and usually pale coloured volcanic fragments that are barely welded together. The ignimbrites are sometimes interspersed with beds derived from the less dramatic ash falls that occurred early in each

Artificial caves cut into ignimbrite deposits of the Bandas del Sur at Los Derriscaderos, near Chimiche

Western section of Las Cañadas caldera, viewed from the crater of El Teide. The Llano de Ucanca in centre and right was formed by an early subsidence and is at a lower level than the caldera floor at the left, formed by a later subsidence; Los Roques de García mark the separation

eruptive event. All the major flows travelled on into the ocean, forming extensive submarine deposits off the south coast of the island. On land, the main concentrations of ignimbrite deposits form the Bandas del Sur, an enormous apron covering the southern part of the island between the western edge of the Güímar valley and the southwest coast. The total volume of pyroclastic materials deposited in this period is estimated at more than 130 km^3. The ignimbrite layers are easily viewed from the motorway between the Aeropuerto del Sur and Arico when passing through road cuttings.

The explosive eruption of an enormous volume of magma may partially or completely empty the magma chamber below the volcano, leading to its collapse. This may then cause vertical subsidence around the eruptive vent, often involving a ring fault and the formation of a roughly circular depression known as a 'caldera'. Subsidence of this kind seems to have affected Las Cañadas Edifice at least three times during the period concerned, at the end of each main volcanic cycle, but possibly as many as seven times. Subsidence may sometimes have been followed by edifice building, but it is likely that for much of the past million years (and perhaps earlier) the volcano took the form of a squat, truncated mountain with a large summit caldera, rather than a steep-sided cone like that of the modern Teide–Pico Viejo stratovolcano.

This scenario provides one plausible explanation for the origin of the modern Las Cañadas Caldera, a topic that has been debated for many years. The caldera is one of the most striking features of the topography of Tenerife. It is a depression some 16 x 9 km across and currently up to 500 m deep. The caldera has two obvious sections separated by the Roques de García, with the western depression now some 150 metres below the eastern one, which is itself made up of two parts, but with a less obvious boundary between them. Gravimetric studies of central Tenerife support the idea that the three sections resulted from the three major explosive eruptive events during the last million years. The newer sections of the caldera have formed progressively further east, suggesting a gradual eastward migration of the active centre of the volcano.

It has been suggested that the western depression around the Llano de Ucanca (or at least part of it) was formed by an explosion and vertical subsidence about 1.0 Ma. The central section (immediately east of Los Roques) may have formed in a similar crisis terminating the Guajara formation around 0.56 Ma, which contributed to the massive layers of ignimbrite around Granadilla de Abona. Finally, the easternmost section, south of the Centro de Visitantes and El Portillo, relates to the Abrigo event, a violent explosion around 0.19 Ma, marking the end of the Diego Hernández formation. On this hypothesis, all the cliffs and cañadas originated as the southern half of a large caldera, with three or more overlapping sections created by successive vertical subsidence events. Later eruptions of Teide largely filled the caldera sections, to different levels, but the original depressions may have been over a thousand metres deep.

Los Roques de García, one of the most striking features of Las Cañadas, are the summits of an ancient spur projecting northwards from the south wall of the caldera and extending thousands of metres below the modern caldera floor. Detailed study shows that the rocks forming the spur are geologically similar to the caldera wall and are representative of the lower part of Las Cañadas Edifice; they were laid down more than two million years ago, prior to the onset of explosive volcanism. The spur includes several phonolitic intrusions – notably La Catedral about half way along the ridge – that seem to represent ancient volcanic conduits for phonolitic magma. The structure of Los Roques supports the idea that independent vertical subsidence events formed the several sections of the caldera – and the surrounding cliffs – following separate major explosive eruptions.

Significantly, the results of the study of Los Roques are considered to be incompatible with the alternative hypothesis for the origin of Las Cañadas, which is that a single giant flank collapse created the entire Las Cañadas Caldera, leaving a simple gutter-shaped embayment with a headwall surviving as the cliffs forming the rim of the modern caldera, though in an eroded state. Further evidence against this idea is provided by a recent gravimetric study, which found evidence of a collapse headwall further north, now deeply buried beneath the Teide-Pico Viejo volcano. This semicircular north-facing headwall may have been in roughly the same position as the northern rim of the original caldera, formed by the subsidence events described above, but this rim would have been concave towards the south.

Although it now seems that the caldera owes its origin primarily to subsidence events, its modern form has been determined by one or more lateral flank collapses.. The evidence is best for the Icod collapse around 0.198 Ma, when a large section of the northern flank of Las Cañadas Edifice slid into the sea, gouging out a huge embayment (around 10 km wide, 19 km long and up to 2 km deep) from the steep northern flank of the mountain. The Icod-La

Guancha Valley is the lower part of the embayment, now partially filled with lavas, while the cone of Teide is built on the upper part of the embayment and the northern half of the original caldera.

The available information leaves open the possibility that the Icod lateral collapse comprised two or more separate events, separated in an east-west dimension. A recent study of marine deposits derived from the Icod collapse indicates that it occurred in seven stages, at intervals of at least several days. The first three stages were at least partially submarine and the other four originated above sea level, with the last ones presumably coming from high up the volcano. The fact that the succession of landslides started below sea level suggests the possibility that the explosive eruption terminating the Diego Hernández formation – which caused vertical subsidence (caldera formation) at the surface and may have involved disintegration of the magma chamber at or below sea level – also created shock waves through the base of the mountain that caused the Icod flank collapse. This would imply that the two events were simultaneous; however, neither of them has yet been precisely dated.

This lava flow spilled over from the Guajara section of Las Cañadas caldera down into Llano de Ucanca to the west

It is known that the Icod lateral collapse created an enormous raw and unstable scar, covered with a thick layer of debris, which doubtless suffered rapid erosion as a result of debris flows, rockfalls and rockslides from steep slopes. Furthermore, depressurisation resulting from removal of the flank of the mountain led to eruptions of large volumes of lava within the debris-strewn embayment. The eruptions continued over a long period and some of the pyroclastics and lava poured into the caldera depressions, eventually raising the floors to their current levels. One of the latest of the lava flows spilled over the lowest point in the wall formed by the Roques de García and flowed down into the Llano de Ucanca, 200 metres below.

The Teide–Pico Viejo stratovolcano

The majestic cone of El Teide, about 8 km across, rises some 1700 m above the relatively level depression of Las Cañadas Caldera, which itself lies at more than 2000 m above sea level. Below the sea surface the slopes of the volcano plunge another 3700 m to the sea floor, so just half of the Teide volcanic pile is visible. The caldera is largely surrounded by inward-facing cliff walls up to 650 m high, and the mountain therefore gives a striking impression of a conical candle – or firework – sitting in a saucer on a much broader pedestal. The metaphor is apt because the volcano is still active: at the summit there are fumaroles emitting steam and sulphur dioxide, and major eruptions of lava have occurred in historical times.

The Teide–Pico Viejo stratovolcano from the west, showing the pale and steep-sided flows of felsic lava from the Roques Blancos vent (top centre)

Teide–Pico Viejo is a twin-peaked stratovolcano that owes its origin to the major crisis described above, involving both explosive eruption leading to subsidence, and flank collapse. The Diego Hernández explosion and subsidence (Abrigo event) and the Icod flank collapse destroyed the northern flank of the older Las Cañadas Edifice and led immediately to the start of construction of the modern stratovolcano on the base of the earlier mountain. A long series of eruptions starting around 198 ka created the twin volcanoes of Teide (3718 m) and the younger Pico Viejo (3134 m); the summits of the two peaks are about 2.5 km apart, with a shallow saddle between them at ~3050 m. Both of the volcanoes are composed of many layers of lava, volcanic ash, scoria, blocks and bombs.

In parallel with the eruptions of Teide and Pico Viejo during the last 200 ka, there was extensive volcanic activity along the Northwest and Northeast Rift Zones, and it is suggested that the rifts and the central volcanoes act as an essentially continuous, interconnected system. However, the eruptive style and materials ejected tend to differ in different parts of the system. Eruptions along the rift zones are typically from fissures and are mainly 'effusive', involving small explosions and violent emission of fragments of magma, as well as flows of lava; the magma is usually mafic and relatively fluid.

Early in the development of the Teide Volcanic Complex the lava produced by Teide was similar to that along the rift zones, of mafic to intermediate composition. However, there was then a transition to felsic materials, especially in the infrequent later eruptions, which were from the central vent of Teide but also from Pico Viejo and numerous satellite vent systems on the flanks. Vents located between the two volcanoes and the rift zones produced materials of intermediate composition. The many satellite vents may have developed because the peak of Teide had become so high that only the highest pressure from below could lift magma to the summit vent; lateral emission through vents on the flanks was easier.

Viscous felsic (phonolitic) lava of the type emitted in the last phase of development of the Teide Volcanic Complex forms domes and thick, steep-sided flows, and generally hardens before spreading far, so the slopes are steep. During the building of the earlier Las Cañadas Edifice the high viscosity of the magma was one of the factors leading to the high frequency of violent 'Plinian' eruptions producing deposits of ignimbrite. However, the style of eruptions during the construction of the Teide Volcanic Complex was different, with many eruptions of pyroclasts and flows of lava, but apparently without major explosive eruptions until about 2000 years ago.

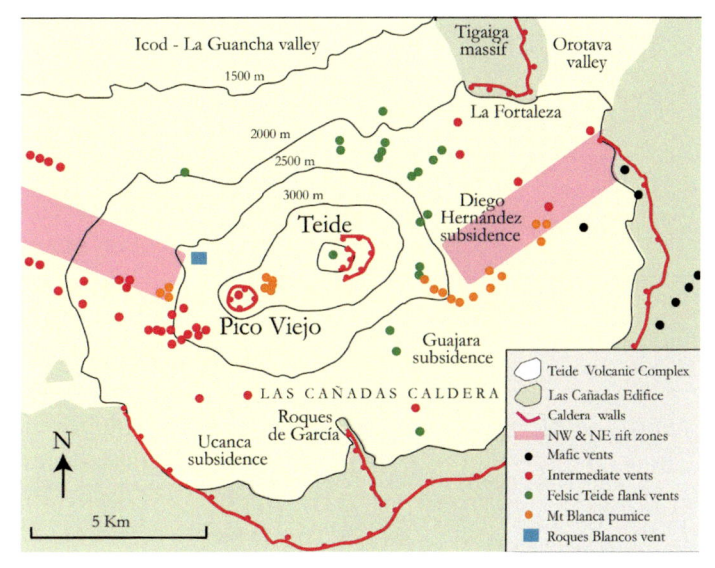

The Teide-Pico Viejo stratovolcano.
Simplified and redrawn from Ablay and Martí (2000)

At that time, however, there was a massive explosive event at Montaña Blanca, on the eastern flank of Teide. It started with emission of phonolitic lava from fissures, but this was followed by an explosive 'subplinian' phase lasting several hours, in which ~0.8 km³ of pale green phonolitic pumice (oxidising to pale reddish brown) and ash was discharged from a magma chamber some 4 km below; the event ended with dome-building activity. The pumice was drifted by strong southwesterly winds and fell mainly in the area between the vent and El Portillo, the latter receiving a blanket of pumice about 30 cm thick.

A period of relative calm terminated with a quite different eruption of Teide itself, dated to around 800 AD. This formed the steep-sided summit cone called El Pitón (or 'El Pilón de Azucar', meaning sugar loaf) which rises 220 m from its base near the modern teleférico, and has a shallow crater about 100 m across, tilted to the southwest. There are cracks and holes in the floor and sides of the crater from which sulphurous steam is emitted at a temperature of 60–85°C; this causes hydrothermal alteration of the volcanic materials and leaves the surface whitish, while there are also yellow deposits of sulphur. The eruption that created the cone also produced about 0.5 km³ of glassy phonolitic lava, the conspicuous Lavas Negras, blackish flows that travelled slowly down the north, east and south flanks of the volcano; they are a spectacular feature of the road between the Montaña Blanca access path and the teleférico car park. In the east the lava flowed over the Montaña Blanca pumice and also formed the 'huevos del Teide' lava balls.

El Teide, with the summit cone (El Pitón) and the Lavas Negras, products of the eruption of Teide about 1200 years ago. The black lava covered much of the pumice of the Montaña Blanca eruption

Pico Viejo became active on the southwestern flank of Teide about 27.5 ka, when eruptions on Teide itself had become rare. It lies at the eastern end of the Northwest Rift Zone and has generally produced lava of intermediate composition, as do the vents on its western and southern flanks (and also vents in the northeast of the caldera); this lava tends to be more fluid than that from Teide and therefore spreads more widely and produces some 'lavas cordadas' (ropy lava). However, the Roques Blancos dome and vent, high on the northwest flank of Pico Viejo, which were formed around 1800 years ago, produced the most voluminous flow of viscous felsic lava (evolved phonolite) of the Teide Volcanic Complex. This flow is also the longest known example of this type, with a total length of 14.7 km; it reached the north coast at Playa de San Marcos after creeping down the mountainside close to the older basaltic flow containing the Cueva del Viento.

The summit of Pico Viejo has an impressive crater 800 m wide, but also evidence of an old lava lake in the form of an enormous, slightly tilted, flat-surfaced block that probably once formed part of a caldera floor. A later subsidence, perhaps during the Roques Blancos episode, left this block isolated high on the southern rim of the modern crater. This main crater is truncated in the southwest by a smaller but deep crater probably caused by a steam-powered 'phreatomagmatic' explosive eruption. This type of event can occur when hot magma (either felsic or mafic) suddenly encounters water on a sufficiently large scale to generate large quantities of steam. The crater of Pico Viejo accumulates snow in winter, and the explosion doubtless occurred when magma (possibly associated with the Las Narices eruption of 1798) suddenly upwelled below the crater floor. An example of this kind of eruption on Teide itself is found in the pale brown deposits named Las Calvas del Teide (Teide's bald patches) on the very steep (~35°) slopes northwest of the peak, which were only recently recognised as phreatomagmatic. Elsewhere on Tenerife there are other examples of this type of activity, including Montaña Amarilla near the southernmost point of Tenerife, where interaction of basaltic magma with seawater has produced large volumes of yellowish rocks.

12 Geology

Pico Viejo from Teide crater; the flat-topped block on left of Pico Viejo is part of the floor of an earlier summit caldera; orange pumice deposits in middle distance are from the time of the Montaña Blanca eruption about 2000 years ago; black lava in foreground is the western part of the flow shown on page 372.

Garachico was rebuilt on an apron of 1706 lava, but is still overshadowed
by the eastern flank of the flows, which destroyed much of the town

Historical eruptions on Tenerife

Tenerife has seen several eruptions within historical times. By an extraordinary chance, Christopher Columbus was the first person to record observations of an eruption on Tenerife. On 24th August 1492, as he sailed from La Gomera on his first voyage to the Americas, he reported in the ship's log that '…they saw a great fire on the sierra of Tenerife' [transl.]. Columbus explained to his nervous crew that it was an eruption similar to those of Mount Etna and other volcanoes, with which they were familiar. The volcano on Tenerife was Boca Cangrejo, just outside the boundary of Teide National Park, below the Chío recreation area on the road running northwest from Boca Tauce. The eruption produced large flows of lava down the slopes above Arguayo and Las Manchas (photo in 'Volcán de Chinyero' in Chapter 6).

Another volcanic episode occurred in 1704-05. Small earthquakes were felt around Christmas in 1704 and were immediately followed by three small eruptions on the Northeast Rift Zone in the region of Izaña. First, Siete Fuentes volcano gave rise to a tiny set of cones and a short lava flow. This was followed by an eruption at Volcán de Fasnia, producing a group of basaltic cinder cones and also a significant lava flow. Finally, an eruption at 1500 m in the Güímar Valley formed a cinder cone 100 m high to the south of the Mirador La Crucita (at Km47 on the ridge road) and also a lava flow that threatened two towns far below: it split into two streams, one of which stopped just to the east of Güímar and another which passed close to Arafo and came to a halt only a kilometre from the coast. Activity continued through 1705, and on 5th May 1706 a much larger, damaging eruption started on the Northwest Rift Zone at Montaña Negra; this is generally known as the Garachico eruption. The lava flowed northwards down the mountainside for 5 km before reaching the steep Acantilado de La

Lava from the Narices del Teide eruption of 1798, both
in the foreground and on the slopes of Pico Viejo behind

Culata. It then completed its 1300 m descent by pouring over this cliff and largely filling the port of Garachico, which was one of the best on the island. A second flow of lava started on 12th May, overwhelming the town centre early on the following day. The Garachico eruption caused no casualties but led to immediate economic collapse of the town, which had been the main port in the Canaries for trade with South America. It also initiated fundamental changes in the economic geography of the island, including permanent reduction in the importance of Garachico and development of Santa Cruz de Tenerife as the island's main port and capital.

After a pause of nearly a century there was a spectacular eruption in 1798 from Las Narices del Teide (Teide's nostrils) high on the steep southwest slopes of Pico Viejo; this is sometimes referred to as the Chahorra eruption. Activity started on 9th June and continued until mid September; three vents were involved, the main one at the top involving explosions and venting of gases, the middle one producing scoria (lumps of molten rock) and lava, and the lowest emitting only lava. Streams of black lava of intermediate composition poured down the mountainside and spread over 4.9 km^2 of the floor of Las Cañadas; being confined by the rim of the caldera, it ponded in places to a thickness of 15–20 m. These impressive flows are crossed by the road immediately northwest of Boca Tauce on the edge of the caldera.

Finally, in 1909 a massive eruption on the Northwest Rift Zone created Volcán de Chinyero, close to Boca Cangrejo and 10 km northwest of Pico Viejo. Its lava flowed to the west and northwest, encircling Montaña Bilma to the east of Santiago del Teide, where it lies close to the flows from the earlier eruption of Boca Cangrejo. A visit to any of the areas where these recent eruptions occurred enables one to see the almost unaltered products of volcanoes and to get some idea of the violent events involved in creation of the island of Tenerife.

Teide: a volcano under surveillance

Active volcanic complexes present hazards to human life and property, but it seems that the risks associated with Teide are relatively small. Major earthquakes are improbable, mainly because the Canary Islands lie within a single tectonic plate rather than at an active plate junction. Few earthquakes of magnitude greater than 4 are recorded, although one of magnitude >5 occurred in May 1989. Seismic activity is concentrated in the area between the southeast coast of Tenerife and Gran Canaria, where there is an apparently live submarine volcano known as El Hijo de Tenerife.

The eruptions on Tenerife in recent centuries suggest that there are likely to be more in the future. Teide is one of 16 volcanoes around the world selected for special international study because they are close to populated areas and have had large, destructive eruptions relatively recently. One result of modern surveillance was the detection in spring 2004 of increased seismic activity in Las Cañadas Caldera, with a series of small earthquakes registered. However, minor episodes of this type would have gone unnoticed in the past, so it is not known whether they occur frequently.

The main danger to life from volcanoes comes from explosive eruptions, and these have been rare in the Canaries except in the distant past. Although the Montaña Blanca event (see above) was explosive and occurred just 2000 years ago, it may have been the only major eruption of its kind in the past 200,000 years. In general, eruptions in the Canaries produce lava, and most of those on Tenerife in the past 20,000 years have been either on the Northwest Rift Zone, with both mafic and felsic lava, or from flank vents on Pico Viejo, with intermediate (phonolitic) lava. The flows have been towards the west and north, and have 'resurfaced' most of the densely populated areas of north and west Tenerife during this period. However, although lava flows can be extremely destructive, they rarely cause danger to life. If there is major activity on Tenerife in future, modern monitoring techniques should provide adequate warning and ensure that appropriate safety measures are adopted.

Giant landslides (flank collapses) though potentially dangerous, are only a remote possibility, since there have been probably less than a dozen of these in the 12 million year history of Tenerife – an average of less than one every million years. However, the modern Teide–Pico Viejo stratovolcano is built on very insecure foundations. When the earlier volcano disintegrated at the time of the Icod collapse, it created a gigantic chute running down the side of the mountain; the floor of this embayment was left covered with a thick layer of landslide debris, and it was on top of this loose layer that the new Teide Volcanic Complex was built.

When a giant landslide occurs, the effect is potentially far-reaching, since a large segment of an island sliding suddenly into the sea can create a major tsunami, causing serious damage on distant coasts. The Güímar collapse of 830 ka produced a tsunami that deposited marine debris at over 100 m on slopes of the Agaete Valley, 70 km away on the northwest coast of Gran Canaria. Modelling suggests that the landslide triggered waves several hundred metres high, reaching Gran Canaria about 10 minutes later. However, the frequency of major tsunamis caused by landslides may be less than often assumed, as research on the Icod collapse shows that they sometimes occur in several stages, presumably with much reduced effects.

Raised beaches on Tenerife

Most beaches on Tenerife are composed of volcanic rocks or black volcanic sand, but there are also a few natural beaches of whitish shell sand, at El Médano and elsewhere. (Las Teresitas near Santa Cruz has siliceous sand imported from the Sahara.) In a few places there are also ancient shell sand deposits, some of them well above current sea level, where beach sand or dunes have been covered by volcanic rock or where the sand grains are cemented together, forming calcareous rock. The most striking example of such deposits is at Punta Gotera (locally known as La Mancha de la Laja) on the north coast near Bajamar, where a fossil sandy beach some 20 m above sea level is overlain by red 'almagre' (burnt soil) and a lava flow. It is dated about 130 ka, a time when worldwide sea level was somewhat higher than now. The sand may well have been blown some distance inland and uphill, but the bottom of the deposit contains shells of marine molluscs and was probably at shore level, although there are remains of landsnails nearer to the top.

The author of a recent study of Tenerife's raised beaches made the intriguing suggestion that the island has been uplifted asymmetrically in the last 300,000 years, with the north coast being raised by up to 10 m but the south coast around El Médano by as much as 45 m in the last 10,000 years. The evidence was mainly from fossil beach deposits, for instance at Punta Roja, just west of El Médano, where deposits with limpet shells can be seen about 11 m above present sea level. Shell sand deposits nearby have previously been dated at ~10,000 years ago, when worldwide sea level was tens of metres lower than now, so if the deposits were formed then, massive uplift of the land is implied. However, we are wary of this conclusion, since global sea level was rising rapidly around that time and the calculation is thus sensitive to errors in dating.

It is likely that the sand in deposits at various sites along the southeast coast was blown onshore mainly during the last Ice Age, when the shallow submarine platform just offshore – with its populations of molluscs with calcareous shells – was exposed to the wind by globally low sea levels.

Another relevant site is Montaña Amarilla on the Costa del Silencio. This is given official protection mainly for its underwater geological features, observable by divers. Some conspicuous whitish sloping banks of consolidated shell sand are now around modern sea level, but were probably formed above water when sand was blown up against the eroded edges of the tuff cone of Montaña Amarilla; the surviving deposits are doubtless remnants of a larger sandbank.

Fossil sandy beach at Punta Gotera, with shells of landsnails *Hemicycla* sp. from the upper section

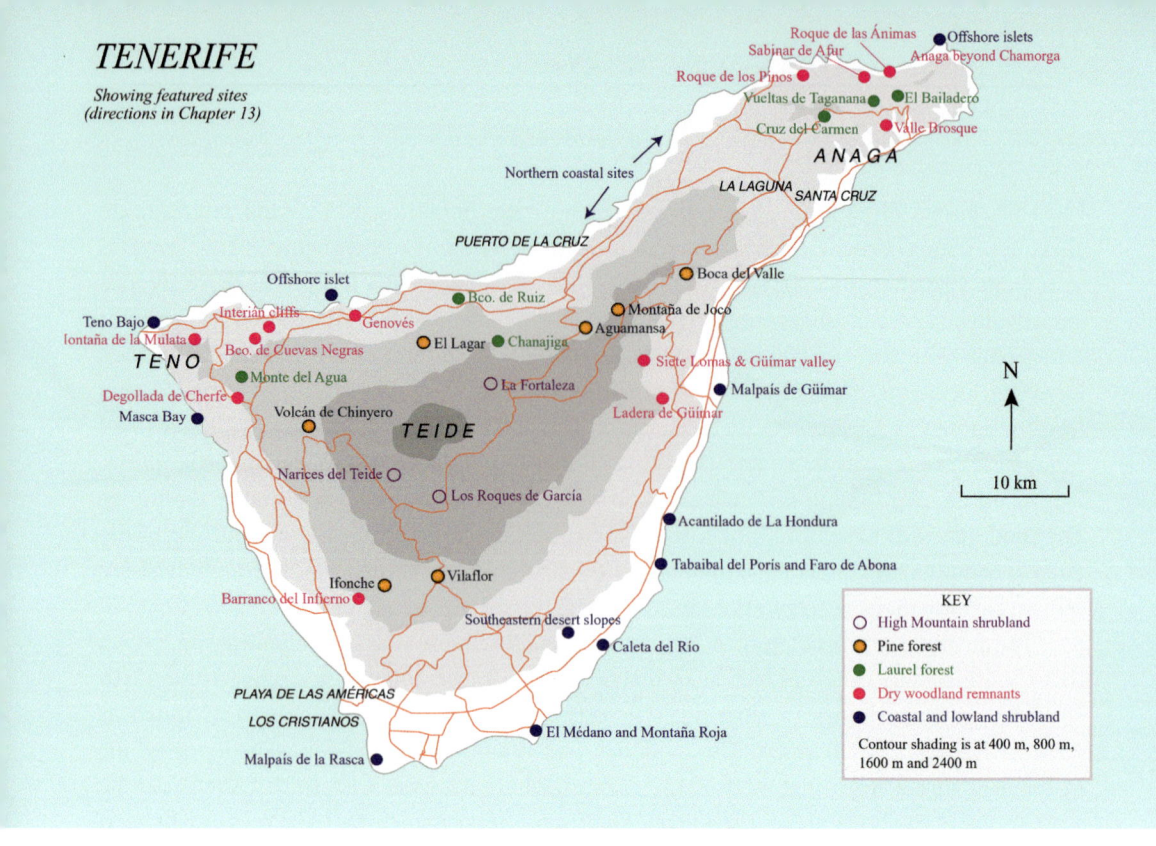

TENERIFE
Showing featured sites (directions in Chapter 13)

KEY
○ High Mountain shrubland
● Pine forest
● Laurel forest
● Dry woodland remnants
● Coastal and lowland shrubland

Contour shading is at 400 m, 800 m, 1600 m and 2400 m

13

Directions to Featured Sites

NOTE. Sites are listed alphabetically (ignoring definite article). The site name is followed by the page number, the name of the relevant ecosystem, and the region within the island.

Acantilado de la Hondura (p54)   Coastal and lowland shrubland: Southeast
Site of Scientific Interest. A rocky coastal strip of land exposed to strong onshore winds. It has an attractive community of salt-tolerant shrubs including the rare endemic shrub Piñamar *Atractylis preauxiana*.
Leave TF-1 at Exit 16 (Km36) signed for La Eras. Easy parking on the seaward side of the slope leading down to the village, and well marked walking tracks.

Aguamansa (p214)   Pine forest: North Central
Moist Canary Pine forest at about 1200 m with some areas including other tree species, and with varied undergrowth and rich invertebrate life.
Drive inland on TF-21 from Orotava near the north coast. Beyond Aguamansa take a small left turning signed to La Caldera. Many well marked trails.

Anaga beyond Chamorga (p118)   Dry woodland: Northeast
Part of Anaga Rural Park. A beautiful walk with steep sections, through a variety of habitats.
Chamorga is the most easterly village on the island and can be reached in two ways: either drive east from Santa Cruz on the TF11 to San Andrés and then turn inland on TF-12, joining TF-123 on the

main ridge; or find the TF-12 north of La Laguna and follow it along the long, narrow and windy ridge of Anaga to the junction with TF-123; take care to avoid going through the tunnel to Taganana. In either case, then follow TF-123 eastwards to the end at Chamorga, where there is limited parking and a bar. The path to the Faro de Anaga is well signed, and one can return up the Barranco de Roque Bermejo.

Barranco de Cuevas Negras, Los Silos (p146)   Dry woodland: Northwest

A fine barranco in the Teno Rural Park, cutting into the ancient cliffs of northwest Tenerife, which is famous for its special endemic plants.

Coming from the east from Puerto de la Cruz on TF-42 there is a roundabout as you get to Los Silos, with a sign left to Cuevas Negras. Follow this and drive as far as you are comfortable. There is then a well-marked but steep mule track PR TF-53 up to the hamlet of Cuevas Negras, where the track branches: to the left signed to Tierra del Trigo and straight on signed to Erjos.

Barranco del Infierno (p134)   Dry woodland: Southwest

Special Nature Reserve. A magnificent ravine cutting deep into the ancient rocks of the Conde massif at an altitude of about 400 m, terminating in vertical cliffs and a waterfall, and with a wealth of endemic plants.

Drive to Adeje, north of Playa de las Américas. The way into the gorge is from just beyond a restaurant at the top of the town. It has been officially closed, apparently because the path could do with some upkeep in places.

Barranco de Ruiz (p194)   Laurel forest: Central

Site of Scientific Interest and part of the Los Campeches, Tigaiga and Ruiz Protected Landscape. An extraordinarily deep ravine that contains a small area of laurisilva – one of the few surviving in the centre of the north coast. Access is easy on foot, and most of the trees typical of this habitat can be seen here, as well as both of the laurel pigeons and other endemic birds.

From the northern motorway TF-5 take the exit for Los Realejos; go steeply uphill in the town and eventually take a right turn on the TF-342, passing the Mirador El Lance and continuing to Icod el Alto. Near the western end of the town is a DISA petrol station and 0.3 km beyond this is a junction with TF-344 where you take the small right turn labelled Carretera Fajana and Casablanca Hotel Rural. After 0.2 km a sign points downhill to Barranco Ruiz and La Fajana; 0.5 km down this road a sign indicates a steep rough track down to the left for Barranco de Ruiz, and near it there is limited parking space. It is about 0.5 km walk to reach the laurel forest, and by continuing it is possible to cross the barranco. Access is also possible from far below, by a steep trail (signed to Icod el Alto) from the Barranco de Ruiz recreation area on the south side of the TF-5 at Km44+600 m).

Boca del Valle, Bosque de la Esperanza (p220)   Pine forest: Central

Near the eastern limit of the Corona Forestal Natural Park. Easy walking access to dry pine forest, with cliffs and rocky outcrops harbouring a wealth of unusual endemic plants, and magnificent views down to Candelaria and Arafo.

From the motorway TF-5 at La Laguna take the TF-24 signed to El Teide and drive through La Esperanza to Las Lagunetas (Km16). Pass the track to the right with a café and activity park, and after 200 m take the second rough but drivable track to the left. After 0.9 km you will reach a space with ample parking by a locked gate. Continue on foot, bearing right into Boca del Valle along a level track cut into the cliffs.

La Caleta del Río (p56)   Coastal & lowland shrubland: Southeast

A small area of wetland where the Barranco del Río reaches the sea. A wide variety of salt-tolerant plants and a chance to see shorebirds.

Leave the TF-1 motorway at Km49, Chimiche El Río, and turn towards the sea on a small road passing through the Urbanización Callao de Río. Where it reaches the coast there is a small area of marsh and saline lagoons.

Chanajiga (p192) — Laurel forest: Central
Within the Corona Forestal Natural Park. This site is high above Los Realejos on the western wall of the Orotava valley, created by a giant landslide half a million years ago. The track passes through modified but still interesting laurel forest. The area is often in cloud and it is worth waiting for a clear day if possible, since this is a good place to see a variety of butterflies when the sun is shining.
Leave TF-324 between Los Realejos and La Orotava on TF-326. After passing Palo Blanco at Km5 take a right turn on a road signed to 'Las Llanadas, Camino Agricola'. In Las Llanadas take a right turn signed to Chanajiga, reaching the recreation area after about 5 km; pass this and gain access to the laurel forest by using the level track, which in principle can be followed for many miles.

Degollada de Cherfe and the Masca road (p138) — Dry woodland: Northwest
Within Teno Rural Park, at around 1000 m. This area has spectacular eroded ridges and barrancos, with degraded but still rich and interesting remains of dry woodland, dominated by Retama blanca *Retama rhodorhizoides* and with many other plants characteristic of the Teno massif.
Drive to Santiago del Teide on TF-82 in the west of the island. Turn west in the town on TF-436, a very narrow, steep and twisty road, signed to Masca. It's wise to avoid peak times with tourist buses. Drive to the degollada (pass) where there is good parking; then continue to Masca village or beyond, through fascinating landscapes.

La Fortaleza trail (p266) — High mountain shrubland: Central
A track at about 2000 m in the Teide National Park, leading over ancient vegetated lava towards an isolated cliff which is a refuge for some rare plants; access to the cliff itself is not allowed.
The track starts from the visitor centre in the east of the park, which is on TF-21 from La Orotava 1 km south of its junction (at El Portillo) with TF-24 from La Laguna. Sun protection is advisable, since it is 4 km to La Fortaleza over unshaded terrain.

Genovés and El Guincho (p154) — Dry woodland remnants: Central
Genovés is a small but extraordinary site retaining one of the richest fragments of dry woodland to be found anywhere on Tenerife, with great variety of characteristic trees and shrubs. About 300 m immediately below it is El Guincho (the Osprey) and beyond it is Punta de la Sabina, which does indeed have junipers (Sabinas) right down by the shore.
The village of Genovés lies west of Icod de los Vinos, near Km 5 on TF-82, the high road to Santiago del Teide and Los Gigantes. The way to the site, which is high up on the cliffs that dominate this part of the coast, is just west of the Bar Arepera Las Venitas. Park by the bar and walk down the track under the power lines. El Guincho is on the main coast road TF-42 between Icod and Garachico. In the village take the small road past the church that leads down towards the coast.

Interián cliffs, Los Silos (p150) — Dry woodland remnants: Northwest
Site of Scientific Interest. A spectacular cliff with rich dry woodland and easy access along a water channel.
Los Silos is on the northern coast road TF-42 between Garachico and Buenavista. At the eastern end of the village (about 0.7 km east of the town hall) take a small, steep road signed to 'Hotel Rural, Finca la Hacienda' which winds up the cliff; on a hairpin bend after about 0.8 km there is a small parking place where the road crosses a water channel; one can walk in either direction along this, with access to a wonderful array of endemic plants. The road continues up to Tierra del Trigo, joining the TF-423, which traverses more dry woodland before reaching the TF-82.

Ladera de Güímar (p124) — Dry woodland remnants: Southeast
North-facing inland cliffs at an altitude of about 500 m, with fine views and a rich community of endemic plants, including several that are hard to find elsewhere.

Leave TF-1 at Exit 11 (Km23) signed to Güimar on TF-61. At Güimar turn south on TF-28. Continue for 3.1 km beyond Km29 to a large white building – Mirador de Don Martín. There are then three ways to approach these cliffs:
1. *Park at the mirador and walk back down the road for about 0.4 km until you reach a steep track (the Camino Real) up the cliff to the left.*
2. *Access the Camino Real from higher up; drive past the mirador for 0.5 km and turn steeply up to the right on a small paved road; after 0.2 km park opposite the second house. Take the rough track to the right immediately opposite this house (with a sign, recently broken, saying Camino Real) and go under the power lines. Follow the track for about 0.5 km eastwards through scrub to the cliff edge, where there is a Protected Area sign and an information board; one can then continue gradually down across the face of the cliff, with opportunities to view a wide range of endemic plants, and eventually reach TF-28 at the point described in 1.*
3. *Access the disused water channel along the face of the cliff higher up, although this is now deteriorated and rather dangerous. Drive or walk for another 0.3 km along the small paved road described in 2; when you see a large white concrete building on the right turn off and park beside it. Find your way round the building and walk through scrub across old terracing and several minor water channels. Turn left at the fourth wooden pole (above a small concrete building) and join the main water channel.*

El Lagar (p217)   Pine forest: North Central
Recreation area in the Corona Forestal Natural Park. Planted Canary Pine forest at 1000 m, with good understorey. A convenient place in the north of the island to see Blue Chaffinch, Great Spotted Woodpecker and other birds and insects of the pine forest.
Drive west from Puerto de la Cruz on the northern motorway TF-5, and after passing San Juan de la Rambla take the TF-351 up to La Guancha. In the centre of the town turn left at a small roundabout beside the Bank of Santander and then almost at once turn uphill to the right on Calle Los Pinos, following signs to Casa Forestal, Bco de la Arena and Aula de la naturaleza. After about 2 km turn right (at Km9) signed 'Área Recreativa El Lagar 5 km', and continue to the site.

Laurel forest of Anaga (p187)   Laurel forest: Northeast
Within the Anaga Rural Park. The peninsula of Anaga has the largest surviving area of laurel forest on Tenerife, at 600-1000 m, with examples of both laurisilva and fayal-brezal. The laurisilva here is the most diverse on the island and several rare or endangered species occur in the more remote places.
Drive along TF-12 eastward from La Laguna in the east of Tenerife, along the ridge through the laurel forest. There are numerous parking places, some marked trails and several unmarked ones. Stop for information at Cruz del Carmen, where there is an easily-missed visitor centre below the car park.

Malpaís de Güimar (p48)   Coastal and lowland shrubland: southeast
Special Nature Reserve. Coastal lava field of relatively recent geological date, with salt tolerant plants dominated by euphorbias. A beautiful place, especially in evening light.
Leave TF-1 at Exit 11 (Km23) signed for Puertito de Güimar and drive down to the port town. The Malpaís is at the northeast edge of the town. When you reach a 'no entry' sign and can see the sea in front of you, turn left and then right and then right again immediately. Follow the road to its end where you can park and walk along well marked trails. To access the malpaís from the north, leave TF-1 at Exit 10 (Km20) and drive down to the coast at El Socorro; park and walk south to reach the malpaís trails.

Malpaís de la Rasca (p64)   Coastal and lowland shrubland: Southwest
Special Nature Reserve featuring an arid coastal plain (including the southernmost point of Tenerife) from which several impressive cinder cones emerge. Easy walking on well-used tracks, with sparse but characteristic vegetation and opportunities to view shorebirds and seabirds as well as those of the malpaís.
Leave TF-1 at Km68 (taking care, as signage is confusing) and drive south past Guaza on TF-66. At the junction with TF-653 turn right towards Palm-Mar. Park on the roadside at the entrance to the

development where a sign indicates a specially protected area, and take the track to the south. Access is also feasible by driving and walking from Las Galletas in the south, but this is an unattractive approach.

Masca Bay (p79) Coastal and lowland shrubland: Southwest

Part of Teno Rural Park. An opportunity to explore the lower part of the Barranco de Masca ravine, with impressive scenery and endemic coastal plants (as well as some introduced species near the house).

The Barranco de Masca is a fine but challenging objective for walkers starting from Masca village 700 m above. However the mouth of the ravine and the lower section are easily accessed by boats that run from Los Gigantes several times each day.

El Médano & Montaña Roja (p58) Coastal and lowland shrubland: Southeast

Montaña Roja Special Nature Reserve. This is a fascinating area with an impressive cinder cone, a small saline lagoon and areas of wind-blown sand with many plants more typical of the eastern Canary Islands. The site is intensively used, with good marked paths but much disturbance to wildlife.

Leave TF-1 at Exit 22 (Km59) and drive towards El Médano on TF-64. After nearly 3 km turn right on TF-643, signed to Los Abrigos and marked with a blue boat. Montaña Roja is on the left, and two large car parks give access to the protected site as well as to Playa de la Tejita just beyond it. A small but interesting part of the site, with sand dunes, is conveniently reached by parking on the roadside where TF-643 comes close to the sea behind the Playa del Médano.

Montaña de Joco (p218) Pine forest: Central

Part of the Corona Forestal Natural Park. Fairly dense pine forest, with shrub understorey and some magnificent old trees, high on the north side of the dorsal ridge of the island.

Drive up the TF-24 ridge road from La Laguna towards El Teide. Just beyond Km24 on a bend there is an unsigned track to the right. This is driveable but not recommended. It leads to several tracks through the forest.

Montaña de la Mulata (Camino del Risco de Teno Alto) (p143)
Dry woodland remnants: Northwest

Part of Teno Rural Park. This is one of the richest areas for endemic plants in Tenerife. The route we suggest follows an old mule track leading uphill (and eventually reaching Teno Alto) but there is also access to the fascinating Barranco de Bujame immediately to the east. The track is now quite rough in places.

From Buenavista del Norte in the 'Isla Baja' region of northwest Tenerife take TF-445 westwards for about 2 km, to a point where the road is often restricted. Park by the roadside near three large water tanks and take the track up to the left, signed 'Camino de Risco, Teno Alto, Los Bailaderos 3.9 km'.

Monte del Agua (p197) Laurel forest: Northwest

This site forms the core of the eastern part of Teno Rural Park. Monte del Agua is the best area of laurel forest in northwest Tenerife, offering a chance to see both species of laurel pigeon and a wide variety of insects.

Take TF-82 north from Santiago del Teide or west from Icod de los Vinos. Just south of Erjos del Tanque (near the bar) turn off west on an unpaved track winding round the head of the enormous Barranco de los Cochinos that runs down to Los Silos. Driving is allowed only for the first 0.75 km (with parking near some masts) but the track eventually reaches the Buenavista to Masca road (TF-436) at Las Portelas.

Las Narices del Teide & Chavao (p269) High mountain shrubland: Central

This site in the Teide National Park features one of the most recent lava flows on the island (eruption in 1798). Fascinating volcanic landscapes and views of special plants on the caldera wall.

At the western end of the national park TF-38 runs northwest for 3 km straight across the Narices del Teide lava flow, from Boca Tauce to the Mirador de Chío. There is limited roadside parking near Boca

Tauce and more at the Mirador. At Boca Tauce, just to the west of the junction with TF-21 from Vilaflor, is the house of Juan Évora and beside it is the start of a track that follows the caldera wall past the Roques de Chavao to the Mirador de Chío; it also gives access to a trail that goes through a gap in the rim of the caldera and down into fine old pine forests to the southwest of the park.

Northern coastal sites (p72) Coastal and lowland shrubland: Central
We do not give directions for these sites, but several of them are well worth a visit if one has plenty of time and preferably a small car for exploring minor roads and tracks.

Offshore islets (p80) Coastal and lowland shrubland: Northwest & North
These islets can only be reached by boat (or in one case along a rocky ridge at low tide) and access is permitted only for special purposes.

Pine forest and mountains near Ifonche (p234) Pine forest: Southwest
The mountains of the Conde massif are in the geologically oldest part of Tenerife, from the head of Barranco del Infierno south to the massive Roque del Conde. Some surviving pine forest and fine communities of mountain plants.

Turn off TF-1 on to TF-28 near Playa de las Américas and follow it eastwards for about 4 km. Turn left on TF-51, passing through Arona and on to its junction with TF-565; then turn left on TF-567 signed to Ifonche. From there, a track to the northwest skirts around the dark and often mist-shrouded head of the Barranco del Infierno, and another heads south, giving access to the mountains of the Conde massif.

Pine forest near Vilaflor (p230) Pine forest: Southwest
An austere but beautiful area with dry pine forest, both planted and natural, and interesting geological features.

Drive uphill from Vilaflor on TF-21 to the start of a forest road just beyond Km66; or walk up from Vilaflor on the Camino de Chasna (PR TF-72) to intercept the forest road. This is a long and fairly level track that eventually stops at the Barranco del Río (cars are not allowed beyond Madre del Agua); it gives access to different types of pine forest and to the well known Paisaje Lunar.

Roque de las Ánimas, Taganana (p115) Dry woodland remnants: Northeast
This very special site within the Anaga Rural Park is an enormous mass of phonolitic rock towering nearly 400 m above the rapidly eroding coast east of Taganana; a wide variety of endemic plants grow on the rock faces and ledges out of reach of goats.

Access is from two directions. You can drive east from Santa Cruz on the TF11 to San Andrés and then drive inland on TF-12 almost to the ridge, where there is a sharp right turn which leads to a tunnel and towards Taganana on TF-134. Alternatively, drive from La Laguna on TF-12 along the ridge of Anaga, ignore the left turn on TF-123 to El Bailadero and Chamorga and continue to the next left on TF-134. Almost 1 km down from the tunnel is a viewpoint, and below it on a hairpin bend an inconspicuous, steep and rocky track goes downhill, skirting the eastern slopes of Roque de Enmedio and leading to Roque de las Ánimas.

Roque de los Pinos, Chinamada (p112) Dry woodland remnants: Northeast
A fascinating remnant of dry woodland on an isolated rock, with the only wild Canary Pines in Anaga, as well as many endemic plants.

Drive east from La Laguna on TF12 along the Anaga ridge; about 2 km beyond Cruz del Carmen take a left turn on TF-145. Ignore a right turn off this (signed to Taborno) and continue to Las Carboneras, where a small road turns off to the left to Chinamada. Park near the restaurant and take the track towards Punta del Hidalgo, which starts behind the church. After 300 m the Roque de los Pinos can be seen on the left, and a faint and very steep path leaves the main track, dropping down into the valley and giving access to the rock.

Los Roques de García (p272) — High mountain shrubland: Central
Eroded remnants of an ancient crater wall at about 2200 m, with dramatic rock formations separating upper and lower segments of the caldera, and both pāhoehoe and ʻaā lava. There is rich high mountain vegetation and an impressive view down into a large cañada – the Llano de Ucanca.
In the west of Teide National Park, opposite the entrance to the Parador, turn off TF-21 into the car park. A popular area, but a trail along the rocks to the right enables one to get away from the crowds, especially early or late in the day when the light is also best.

Sabinar de Afur (p108) — Dry woodland remnants: Northeast
Within the Anaga Rural Park. An area of fine juniper scrub, with many endemic plants, in a broad dry amphitheatre open to the west. A visit requires several hours.
Drive east from La Laguna on TF-12 along the Anaga ridge; about 4 km beyond Cruz del Carmen take a left turn on TF-136, passing Roque Negro and continuing down the narrow road to Las Casas de Afur. The path to the sabinar is not well marked, but it starts up steps beside some red houses a couple of hundred metres back up the hill from Afur. The path leads on upwards, past smallholdings with dogs and goats on left, but where it turns right to a last house a small path to the left leads across a barranco and then up to the right along a ridge. A short way up this a 15 cm metal water pipe runs off to the left accompanied by a small path leading across a couple of barrancos to the area dominated by junipers.

Siete Lomas & barrancos of the Güímar valley (p129) — Dry woodland remnants: Southeast
The Siete Lomas Protected Landscape is a relatively moist area with woodland near the otherwise dry southern slopes, and a biodiversity hotspot with many plants that do not occur elsewhere on the island.
The ridges and ravines of this area are interesting but confusing and it is easy to get lost. Each map seems to have a different set of names, and when driving much time is spent on steep hills in first gear, with safe turning places rare. The westernmost major ridge in the Güímar valley can be reached by taking the TF-28 just west of Güímar, and turning right just after a petrol station and bridge over the barranco at the edge of the town, on a small road signed to 'Güímar de Arriba, San Juan'. Continue upwards until you reach the Plaza in San Juan, and then go on up on Camino de los Zarzales and then the Camino del Hidro, to the end of the road. Limited parking, and then a very steep walk, giving access in principle to barrancos on both sides. Tracks often go up ridges, but there are also several barrancos with walking access. We have enjoyed especially the Barranco de Añavingo and the nearby Barranco de las Saletas; they can be reached by following a road straight on up from the top of the town of Arafo through agricultural plots, and eventually reaching a point where the track deteriorates just beyond a high wall on the left, beside which you can park (avoid parking in the turning area). The track leads downhill and then one path heads straight on into the Barranco de la Saletas, passing under an aqueduct, while a right fork (marginally driveable) goes on up and eventually reaches Barranco de Añavingo.

Southeastern desert slopes (p76) — Coastal and lowland shrubland: Southeast
Los Derriscaderos Natural Monument. Arid slopes on the porous rocks of the Bandas del Sur formation, formed by massive falls of pale volcanic particles. Diverse scrub vegetation, many goats, and birdlife including Stone Curlews.
Leave the TF-1 motorway at Km49, Chimiche El Río, and turn uphill on a small road that runs parallel with the Barranco del Río and eventually reaches the Carretera general, the winding old main road that runs along the whole of the southeast side of Tenerife. Access to the barranco and nearby bluffs is from either of the roads, on foot over rough ground.

Tabaibal del Porís and Punta de Abona (p51) — Coastal and lowland shrubland: Southeast
Site of Scientific Interest. A small coastal strip of volcanic terrain and sea cliffs with varied salt-tolerant plants, including the rare Piñamar *Atractylis preauxiana*.
Leave TF-1 at Exit 17 (Km 39), signed 'El Porís-Villa de Arico'. Drive on TF-625 towards Porís and turn right at a blue and red boat. Go left in about 200 m, then right around the roundabout to the road

alongside the large restaurant – Casa Blanca. Drive along road and turn right before the dead end; drive seaward along a windy road through a housing estate. Park at the end, close to the shore. The entrance to Tabaibal del Porís has been partially blocked by a rubbish dump, but there are well marked paths, probably used mainly by fishermen. The Punta Abona is reached by going to the southern edge of the town and following the road that runs to the south behind the beach; the scrub to the west of this road is especially interesting.

Teno Bajo (p68) — Coastal and lowland shrubland: Northwest

An important part of Teno Rural Park. Steep north facing slopes, cliffs and gullies, as well as more gentle slopes, near sea level in an ancient massif, with extremely rich endemic flora. Fine views of the Los Gigantes cliffs from the end of the road.

From Buenavista del Norte in the far northwest of Tenerife take TF-445 towards Punta de Teno, passing the precipitous Punta del Fraile and going through a tunnel. The road runs close to some of the most spectacular cliffs on the island. There are few places where it is safe to pull off the road, but parking is possible at Luz Teno sheds, opposite six wind turbines. From here the uphill track PR TF-51 provides easy access to an area where nearly all the plants are either Tenerife endemics or at least Canary endemics. Parking is also possible, though crowded, at the Punta de Teno itself where the road ends.

Valle Brosque (p122) — Dry woodland remnants: Southeast

An easily accessible remnant of dry woodland in the south of the Anaga peninsula.

Drive northeast along the coast from Santa Cruz on TF-11; about 1 km beyond Valle Seco turn left on the narrow road signed to Dos Barrancos. This eventually forks, giving access to Valle Crispin to the west and Valle Brosque to the east. From the road end at Valle Brosque (with very limited parking) there is again a choice of left or right tracks, both of which eventually reach the ridge road TF-12. The right hand track gives access to some interesting fragments of dry woodland.

Volcán de Chinyero and nearby habitats (p224) — Pine forest: Northwest

Special Nature Reserve. This volcano last erupted in 1909, covering large areas of land with cinders and sending massive lava flows to the west. The surrounding areas show pine forest in various stages of development, on varied volcanic terrain.

Ideally, approach from the north, using the high road TF-375 south from Santiago del Teide to Chio via Arguayo, so as to see lava from Boca de Cangrejo and Chinyero at Las Manchas. You can also reach Chinyero by going to Chío from Guia de Isora on TF-82. In either case turn uphill on TF-38 to Las Cañadas, passing an interesting lava tube by the road between Km19 and Km18. About 0.5 km after passing Km15 a rough track goes left towards Chinyero, but the gate may be locked on some days, and walking is restricted to marked tracks. A quite different approach to Chinyero is from the north from TF-373, which leaves TF-82 just west of Icod and rejoins it just south of Erjos del Tanque. Leave TF-373 south of La Montañeta on a small road signed to the Arenas Negras recreation area, where you can park; from there it is a walk of nearly 3 km southwards to Chinyero.

Further Reading

The literature on Tenerife is enormous and we list only a few basic references – some of them in Spanish. Most topics and species can be readily pursued online.

Books

Bramwell, David, & Bramwell, Zoë (2001) *Flores silvestres de las Islas Canarias*. 4th edition. Editorial Rueda, Madrid. ISBN 84-7207-128-6, 4th edition. (The out-of-print English version – *Wild flowers of the Canary Islands* – is of the 1974 edition.)
   Basic source for information on the endemic plants of the archipelago, but does not include non-endemic species.

Carracedo, J C & Troll, V R (eds.) (2013) *Teide Volcano. Geology and eruptions of a highly differentiated oceanic stratovolcano*. ISBN 978-3-642-25892-3 and ISBN 978-3-642-25893-0 (eBook).
   Recent technical accounts in English of aspects of the geology of the Tenerife volcano, by specialists from Spain and around the world.

Clarke, T & Collins, D (1996) *A birdwatchers' guide to the Canary Islands*. Prion Ltd, Cley next the sea, Norfolk. ISBN 1-871104-06-8.
   A straightforward account aimed at visitors to the islands, including 25 pages about places on Tenerife that are especially good for birds.

Hilbers, D & Woutersen, K (2015) *Canary Islands - II Tenerife and La Gomera - Spain.* Crossbill Guides. ISBN 978-94-9164-8069. Crossbill Guides Foundation, Arnhem, The Netherlands.
Modern nature guide, well illustrated and with routes and site descriptions.

Fernández-Palacios, J M & Martín, J L (eds.) (2001) *Naturaleza de las Islas Canarias.* Turquesa, Santa Cruz de Tenerife. ISBN 84-95412-187.
Multi-author book in Spanish on the natural history of the Canaries, edited by two eminent local ecologists.

Lamdin-Whymark, S (2013) *Tenerife Nature Walks.* Flintwork Publications, Oxford. ISBN 978-0-9575486-0-2.
Well-organised modern walk guide, with maps and photos of conspicuous plants.

Martín, A & Lorenzo, J A (2001) *Aves del archipiélago canario.* Francisco Lemus, Editor, La Laguna, Tenerife.
Comprehensive treatment in Spanish of the birds of the Canary Islands, by naturalists with unparalleled local knowledge.

Rodríguez, B, Siverio, F, Siverio, M, Rodríguez, A & Barone, R (2014) *Los vertebrados terrestres de Teno. Catálogo ilustrado y comentado.* Gohnic, Buenavista del Norte, Tenerife. ISBN 978-84-616-8670-4.
Modern and beautifully illustrated account in Spanish of the vertebrates of the westernmost part of Tenerife, including most of the species breeding on the island.

Schönfelder, P & Schönfelder, I (2012) Die Kosmos-Kanarenflora. Franckh-Kosmos Verlags-GmBH & Co., Stuttgart. ISBN 978-3-440-12607-3.
Comprehensive guide in German to the flora of all the Canary Islands.

ONLINE RESOURCES

Oromí, P, Zurita, N, Morales, E & López, H (2015) *Diversidad de artrópodos terrestres en las Islas Canarias.* ISSN 2386-7183. http://www.sea-entomologia.org/IDE@/revista_4.pdf
Concise analysis in Spanish of the terrestrial arthropod fauna of the Canary Islands, with useful bibliography.

http://www.floradecanarias.com
Comprehensive treatment of the flora of the archipelago, with photographs, local names and brief accounts (in Spanish) of most of the plants of Tenerife, including ferns.

http://www.gobiernodecanarias.org/medioambiente/piac/descargas/Biodiversidad/Listas-Especies/Lista_Especies_Silvestres.pdf
Authoritative lists (by scientific name) of the terrestrial fungi, plants and animals of the Canary Islands.

INDEX

The index is in several sections: first comes a list of English names of birds and other vertebrate animals, and this is followed by one for butterflies, hawkmoths and dragonflies. English names for all of these groups are used in the run of the book.

The main index, however, is of all the scientific names in the book, which are mainly of plants and invertebrates. These are the names used internationally, and anyhow English names are established for very few Tenerife plants. With regret, we have not found space to index Spanish names, but many of them (especially for trees and other well known plants) are given in captions to photographs or are mentioned in the text.

Page numbers in Bold type indicate photos.

For general topics and sites to visit, please see the Contents list and Chapter 13.

English Name Index of Birds and Other Vertebrates

Alpine Chough 21, 301
Alpine Swift 293
Arctic Skua 295
Arctic Tern 296

Barbary Falcon 126, 133, 298
Barbary Partridge 50, 78, 137, 210, 264, 268, 291, 298
Barn Owl 311, 317, 318
Bat **186**, 210, 264, 311, 316, 317, 318
Bee-eater 297
Berthelot's Pipit **59**, 65, 78, 210, 232, 264, 268, 271, 273, 290, 291, **300**, 304
Black Kite 292
Blackbird 185, 209, 264, 290, 307
Blackcap 137, 186, 290, 291, 306, 307
Black-headed Gull 295
Black-tailed Godwit 295
Black-winged Stilt 295
Blue Chaffinch 20. 209, 217, 218, 219, 223, 232, 264, 268, 289, 290, 291, **300**, 303, 330, 381
Blue Tit 78, 185, 209, 217, 264, 271, 290, 291, **300**, 305
Bolle's Pigeon 107, 133, 185, 189, 193, 197, 198, 209, 290, 296, 297, **301**
Budgerigar 310
Bulwer's Petrel 65, 80, 82, 309
Buzzard 19, 137, 185, 209, 223, 265, 290, 292

Canary 186, 209, 210, 232, 290, **300**, 304
Canary Chiffchaff 78, 185, 209, 223, 264, 271, 290, **300**, 306
Cat 17, 18, 21, 101, 210, 264, 299, 309, 310, 313, 319, 320, 321
Cattle Egret 296
Chaffinch 20, 185, **209**, 217, 223, 290, **300**, 303
Chough 21, 301
Cockatiel 310

Collared Dove 292, 297
Collared Pratincole 294
Common Myna 306
Common Sandpiper 295
Common Tern 296
Common Waxbill 302
Coot 288, 299
Corn Bunting 302
Cory's Shearwater 7, **8**, 65, 80, 137, 309
Cow 312
Cream-coloured Courser 65, 291, 294

Dark-tailed Pigeon 107, 133, 185, 189, 193, 197, 198, 209, 290, 296, 297, **301**
Dog 16, 17, 21, 58, 62, 101, 122, 142, 164, 165, 310, 313, 318, 319, 320
Dolphin 314, 315, 316
Donkey 318, 319
Dromedary 311
Dunlin 295

Egyptian Vulture 265, 290, 291, 292
European Wigeon 293

Ferret 242, 313, 316
Fish 6, 7, 8, 10, 38, 49, 286, 288, 315, 323, 324
Frog 79, 286, 288, 323

Gannet 308
Gecko 26, **67**, 81, 210, 264, 320, 321
Giant Rat **18**, 46, 186, 278, 311, 319
Goat 17, **46**, **78**, 188, 204, 206, 207, 240, 242, 245, 247, 268, 274, 292, 312
Goldcrest 185, 209, 223, 232, 290, 306
Goldfinch 303
Great Skua 295
Great Spotted Woodpecker **205**, 209, **210**, 217, 223, 232, 265, 290, **301**, 308, 381

Green Sandpiper 295
Greenfinch 20, 209, 210, 303
Greenshank 295
Grey Heron 82, 288, 296
Grey Plover 294
Grey Wagtail 209, 265, 290, 304l

Hedgehog 210, 264, 318
Helmeted Guineafowl 298
Hen Harrier 292
Herring Gull 294
Hoopoe 50, 65, 133, 210, 232, 265, 298
Horse 18, 19, 122, **199**, 318, 319
Houbara Bustard 21, 290, 299
House Martin 394
House Mouse 17, 210, 264, 311, 319

Kentish Plover 63, 194, 291
Kestrel 19, 78, 82, 137, 185, 209, 261, 264, 271, 290, 298

Lapwing 294
Laughing Dove 297
Laurel Pigeon **107**, 133, 185, 189, 193, 197, 198, 216, 290, 297, **301**
Lesser Short-toed Lark 65, 290, 291, 299
Lesser Black-backed Gull 294
Linnet 186, 290, 302
Little Bittern 291, 296
Little Egret 65, 82, 291, 296
Little Ringed Plover 63, 294
Lizard 16, **18**, **19**, 50, **80**, 81, **105**. 210, **264**, 278, 313, 320, 321
Long-eared Owl 50, 185, 209, 265, 290, 310
Long-legged Bunting 20, 278, 290, 291, 302

Macaronesian Shearwater 309
Madeiran Storm-petrel 65, 80, 308
Manx Shearwater 65, 291, 309
Marsh Harrier 292
Meadow Pipit 304
Monk Parakeet 310
Montagu's Harrier 292
Moorhen 288, 299
Mouflon 37, 242, 245, 247, 251, 254, 264, 269, 271, 312

Nanday Conure 310
Night Heron 291, 296

Orange-cheeked Waxbill 302
Osprey 156, 292, 380

Pallid Swift 293
Pekin Robin 308
Pheasant 299
Pied Flycatcher 217, 305
Pig 17, 23, 311
Plain Swift 82, 209, 264, 290, 293, **201**

Pochard 293
Pomarine Skua 295
Purple Glossy Starling 306
Purple Heron 296

Quail 298

Rabbit 37, 50, 54, 61, 67, 77, 82, 99, 122, 137, 155, 207, 210, 242, 245, 254, 264, 268, 269, 313, 318, 319
Rat **18**, 21, 46, 107, 164, 178, 186, 192, 297, 299, 309, 311, 313, 319, 320, 321
Raven 26, 93, 210, 264, 265, 290, 291, 299, 301
Red Avadavat 302
Red Bishop 305
Red Kite 265, 291, 292
Redshank 295
Red-whiskered Bulbul 306
Ring Ouzel **265**, 308
Ringed Plover **65**, 294
Ring-necked Parakeet 310
Robin 185, 209, 265, 290, **300**, 307
Rock Dove 82, 210, 264, 292, 296, 297, 298
Rock Sparrow 142, 291, 305
Ruddy Shelduck 291, 293

Sand Martin 304
Sanderling 295
Sandwich Tern 296
Sardinian Warbler 137, 186, 265, 290, 307
Senegal Parrot 310
Serin 291, 304
Sheep 17, 21, 28, **46**. 83. 125, 138, 188, 191, 207, 242, 245, 247, 264, 266, 268, 269, 274, 312, 313
Shrew 318
Skink **16**, 81, 210, 264, 321
Skylark 299
Slender-billed Greenfinch **20**, 278, 290, 291, 302
Snake 322
Snipe 295
Southern Grey Shrike 19, 65, 78, 210, 264, 290, **300**, 304
Spanish Sparrow 78, 305
Sparrowhawk 185, 209, 217, 265, 290, 292, 308
Speckled Pigeon 296
Spectacled Warbler 50, 78, 210, 264, 268, 290, 307
Spotted Redshank 295
Starling 306
Stone Curlew 21, 65, 78, 290, 293, 384
Storm Petrel 65, 80, 308
Swallow 304
Swift 291, 293

Teal 293
Tenerife Blue Tit 78, 185, 209, 217, 264, 271, 290, 291, **300**, 305
Tenerife Giant Rat **18**, 46, 186, 278, 311, 319
Tortoise 9, 19, 323

Trumpeter Finch 50, 65, 290, 291, 302
Turnstone 295
Turtle (and terrapin) 322
Turtle Dove 210, 232, 265, 292, 297

Whale 314, 315, 316
Wheatear 307
Whimbrel 65, 295
White Wagtail 305

White-tailed Laurel Pigeon **107**, 133, 185, 189, 193, 197, 198, 216, 290, 297, **301**
Willow Warbler 306
Wood Sandpiper 295
Woodcock 185, 209, 292, 295

Yellow Wagtail 305
Yellow-crowned Amazon 310
Yellow-legged Gull 65, 80, 82, 290, 291, 294

English Name Index of Butterflies, Hawkmoths & Dragonflies

African Grass Blue **326**, 330
African Migrant **327**, 332
American Painted Lady 198, **326**, 332

Bath White **46**, 47, **67**, 193, 198, 211, 262, **327**, 333
Blue Emperor 336
Broad Scarlet 337

Canary Blue 133, 193, 198, 211, **262**, 271, 273, 320
Canary Brimstone 145, 184, 193, 198, 211, **327**, 333
Canary Islands Large White 211, **327**, 333
Canary Red Admiral 184, **185**, 193, 198, 211, **326**, 332
Canary Speckled Wood **184**, 193, 198, 211, **327**, 332
Cardinal 184, **193**, 198, 211, **326**, 330
Clouded Yellow **133**, 193, 198, 211, **327**, 333
Common Zebra Blue 329
Convolvulus Hawkmoth 191, 334

Death's Head Hawkmoth 334
Diadem 331

Epaulet Skimmer 337

Geranium Bronze **328**, 329
Grayling 193, 211, 262, **327**, 331
Green-striped White 211, **262**, **327**, 333

Hummingbird Hawkmoth 262, 335

Island Darter 337

Large White 211, **327**, 333
Lesser Emperor 223
Long-tailed Blue 126, 211, **326**, 329
Lulworth Skipper 185, **193**, 198, 211, **326**, 329

Meadow Brown **193**, 198, 211, **327**, 331
Monarch (Milkweed) 311, **326**, 331

Painted Lady **184**, 193, 198, 262, **326**, 332
Plain Tiger **47**, **326**, 330, 331

Queen of Spain Fritillary 185, **326**, 331

Red Admiral 184, **185**, 193, 198, 211, **326**, 332
Red-veined Darter 337
Red-veined Dropwing 338
Ringed Cascader 338

Sahara Bluetail 337
Senegal Bluetail 337
Silver-striped Hawkmoth 334
Small Copper **47**, 193, 198, 211, **326**, 330
Small White **193**, 198, 207, 211, 262, 333
Southern Brown Argus 193, 211, **215**, **326**, 329
Spurge Hawkmoth **47**, 50, 335
Striped Hawkmoth 335

Tenerife Grayling 193, 211, 262, **327**, 331

Vagrant Emperor 337

Scientific Name Index

Accipiter nisus 185, 290, 292, 308
Acherontia atropos 334
Acridotheres tristis 306
Acrostira tenerifae **106**, 343
Actitis hypoleucos 295
Adenocarpus foliolosus **36**, 133, 173, 193, 197, **203**, 206, **207**, 215, 220, 221
Adenocarpus viscosus 206, 218, 219, 227, 232, **250**, 251, 261, 262, 266, 274
Adiantum capillus-veneris 79, 135, 181, **287**, 288
Adiantum reniforme 104, 126, 149, 180, **181**
Aelurillus lucasi 264

Aeonium arboreum 74, **91**, 126, **132**, 137, 142, 155
Aeonium canariense 71, 102, **103**, **109**, 110, **111**, 121, 123, 142, 182, 218
Aeonium ciliatum 149, **178**, 182
Aeonium cuneatum **179**
Aeonium haworthii **141**
Aeonium lindleyi **109**, **116**
Aeonium pseudourbicum 137, **139**, 236
Aeonium smithii 102, **222**, 256, 258
Aeonium spathulatum 142, 208, **218**, 223, 228, **229**, 258

Aeonium tabulaeforme 71, 102, 111, **116**, 145, 147, 153, 182
Aeonium urbicum 78, 123, **128**, 137, 149, **175**
Aeonium volkerii 81
Agave americana 26, 91, 93, 99, 123, 140, 145, 152, 156
Ageratina adenophora 26, 153, 178, 196, 288
Agrius convolvuli 191, 334
Aichryson laxum 133, **179**, 182, **228**
Aichryson pachycaulon ssp. *immaculatum* **179**
Aichryson parlatorei 258
Aichryson punctatum 182
Aizoon canariense 51, 55, 56, **57**
Alauda arvensis 299
Alectoris barbara 137, 298
Alevonota outereloi 283
Allagopappus canariensis 49, 63, **102**, 126, 133, 134, **151**, 153
Allagopappus dichotomus 49
Allium sp. 123, 156
Alucita canariensis 262
Amandava amandava 392
Amazona ochrocephala 310
Amicta cabrerai 82, 334
Ammotragus lervia 207
Anagyris latifolia **89**, 131
Anas crecca 293
Anas penelope 293
Anataelia canariensis 264, **276**, 344
Anax ephippiger 336, 337
Anax imperator 287, 336,
Anax parthenope 336
Andryala pinnatifida 111, 117, 145, 149, **179**, 189, **251**, 257
Anguilla anguilla 285, 323
Anisantha 260
Anogramma leptophylla 181
Anolis carolinensis 321
Anthus berthelotii 290, **300**, 304
Anthus pratensis 304
Apis mellifera 243, 261, 346
Apollonias barbujana 93, 113, **114**, 146, 153, 156, 161, 162, **169**, 170, 171, 190, **191**, 196, 198, 320
Apus apus 293
Apus melba 293
Apus pallidus 293
Apus unicolor 82, 290, 293, **301**
Aradus canariensis 210
Aratinga nenday 310
Arbutus canariensis 15, 93, **101**, **130**, 131, 132, 162, 165, **167**, 171, 195, 197, 215
Ardea cinerea 82, 288, 296
Ardea purpurea 296
Arenaria interpres 294, 295
Argynnis pandora 185, **193**, 211, **326**, 330
Argyranthemum adauctum 319

Argyranthemum foeniculaceum 133
Argyranthemum frutescens **42**, 43, 67, 71, 72, 77, 79, **127**, 133, 143, 156
Argyranthemum gracile 136, **137**, 235
Argyranthemum teneriffae 219, **250**, 251, 262, 267
Argyranthemum vincentii 222, **223**
Argyrodes argyrodes 340
Ariagona margaritae 343
Aricia agestis cramera 329
Aricia cramera 193, 211, **215**, **326**, 329
Arisarum simorrhinum 118, 196
Arminda 343
Arrhenatherum calderae 254, 268
Artemisia 85
Artemisia ramosa **67**
Artemisia reptans 62, **63**
Artemisia thuscula 71, 77, 92, 108, 116, **120**, 121, 123, 126, 133, 142, 145, 149, 152, 155
Artogeia rapae 333
Arundo donax 79, 108, 123, 142, 145, 149, 153, 288
Ascotis fortunata 328
Asio otus 185, 290, 310, 311
Asparagus arborescens **49**, 81
Asparagus fallax **178**, 190
Asparagus plocamoides 208
Asparagus scoparius 89, **90**, **147**, 152, 155
Asparagus sp. 49, 71, 85, 111, 126, 145
Asparagus umbellatus 82, 91, 123, **131**, **155**
Asphodelus ramosus 77, 93, **109**, 110, 121, 145, 156
Asphodelus sp. 46
Asplenium aureum 181
Asplenium filare 236
Asplenium hemionitis **180**, **236**
Asplenium onopteris 180
Asplenium trichomanes 181
Asteriscus aquaticus 67, **77**, 126, 235
Astydamia latifolia 43, **45**, 49, 52, **54**, 55, 63, 71, 74
Atalanthus capillaris 66, 96, 102, **103**, 134, **137**, 139
Atalanthus microcarpus 77, 126
Atalanthus pinnatus 71, **75**, 113, 144, **145**, 149, 152, 156
Atalanthus sp. 235
Atelerix algirus 264, 318
Athyrium filix-femina 181
Atractylis preauxiana 52, **53**, 55, 61, 378, 384
Atriplex glauca **60**
Attalus 339
Aythya farina 293

Balaenoptera borealis 314
Balaenoptera edeni 314
Balaenoptera physalus 314
Barbastella barbastellus **186**, 210, 316, 317
Bembecia vulcanica 262, 334
Bembix flavescens 63
Bencomia caudata 93, **129**, 132, 137, 166, 172, 190, 193, 215, 235

Bencomia exstipulata 101, 254, **255**, **270**, **271**
Bidens spp. 178
Bifiditermes rogeriae 50
Bituminaria bituminosa 111, 121, 123, 133, 137, 142, 145, **147**, 156
Blechnum spicant **189**
Blepharopsis mendica **135**, 342
Bombus canariensis **133**, 261
Bos taurus 312
Bosea yervamora 88, **90**, 104, 123, 133, 137, 146, 152, 156
Brachycarenus tigrinus 263
Bromus rigidus 225
Bryonia verrucosa 104, 113, **120**, 121
Bubulcus ibis 296
Bucanetes githagineus 290
Bufonia paniculata 259
Bulweria bulweri 80, 82
Bunochelis spinifera 264, 340
Bupleurum salicifolium 89, **90**, 96, 119, 131, **140**, 191
Buprestis bertheloti 211, **132**
Burhinus oedicnemus 21, 290, 293
Buteo buteo 19, 137, 185, 290, 292
Bystropogon canariensis **132**, 198
Bystropogon odoratissimus 113, **114**
Bystropogon origanifolius 225, 249
Bystropogon plumosus 208

Cacyreus marshalli **328**, 330
Cakile maritima **61**
Calandrella rufescens 290, 299
Calathus 183, 345
Calendula arvensis 235
Calidris alba 295
Calidris alpina 295
Calliphona konigi 184, **325**, 343
Calliptamus plebeius 261, 343
Calliteara fortunata 211, 262, **263**, 318, 328, 330
Callophrys rubi 325
Calonectris diomedea 7, **8**, 80, 82, 137, 309
Camelus dromedarius 311
Camponotus hesperius 82
Campylanthus salsoloides 43, **49**, 66, 71, 92, 126, **127**
Campylopus pilifer **260**
Canarichelifer teneriffae 339
Canaricoris 345
Canariella 163, 338
Canarina canariensis 93, 104, 108, 118, 133, 149, 153, 172, **173**, 190, 191
Canariola nubigena **343**
Canariola willemsei **343**
Canariomys bravoi **18**, 186, 278, 319
Canarionesticus 340
Canarivitrina 183
Canis familiaris 313

Capra hircus 312
Carabus faustus 184, **343**
Carassius auratus 288. 324
Carduelis aurelioi **20**, 278, 290, 302
Carduelis cannabina 186, 290, 302
Carduelis carduelis 302, 303, 304
Carduelis chloris 20, 303
Carduus clavulatus 91, **120**, 121, 131
Caretta caretta 322
Carex paniculata ssp. *calderae* 246
Carlina salicifolia 89, 78, **90**, 96, 108, 113, 121, 123, 126, 133, 140, 149, 189, 193, 198, **219**, 222, 235
Carlina xeranthemoides 219, 232, 256, **257**
Carpinus 22, 23, 85
Castanea sativa 23, 26, 214
Casuarina 26
Catopsilia florella **327**, 332
Ceballosia fruticosa 70, 70, 92, 133, **136**, 145, 152
Cedronella canariensis **176**, 177, 193, 198, 208
Cedrus atlantica 245
Celerio lineata 355
Centaurium erythraea **73**, 75
Centaurium sp. 111
Centranthus calcitrapae 133
Centruroides gracilis 26, 339
Cerastium sventenii 259
Ceratonia siliqua 117, 122, **123**
Ceropegia dichotoma 47, 69, 70, **71**, 92, 111, 113, 116, 123, 145, **147**
Ceropegia fusca 43, 49, 62, 67, **77**, 137, **236**
Chalcides coeruleopunctatus 16
Chalcides sexlineatus 16
Chalcides viridanus **16**, 81, 82, 105, 210, 264, 321
Chamaecytisus proliferus 137, 140, 193, **202**, 206, 207, 211, 215, 218, **220**, 225, 232, **234**, 254, 268, 303
Charadrius alexandrinus 294
Charadrius dubius 294
Charadrius hiaticula 294
Cheilanthes pulchella 225, **236**
Cheiracanthium pelasgicum 185
Cheirolophus burchardii **70**
Cheirolophus metlesicsii **130**, 133
Cheirolophus tagananensis 74, 117
Cheirolophus teydis **212**, **251**, 262, 263
Cheirolophus webbianus 195
Chelonia mydas 322
Chenoleoides tomentosa **52**, 53, 63, 71
Chenopodium coronopus 82
Chlamydotis undulata 21, 290, 299
Christella dentata 181
Cicer canariense 207
Cionus griseus 262
Circus aeruginosus 292
Circus cyaneus 292
Circus pygargus 292

Cistus chinamadensis 112
Cistus monspeliensis 78, 93, 94, 99, 117, 126, 131, 137, 142, 193, 206, **207**, 215, 220, **234**, **236**
Cistus symphytifolius 113, 133, 137, 198, 206, **207**, 216, **217**, 218, **219**, 221, 227, **231**, 232, 234, 255
Cladycnis 340
Cleora fortunata 328
Colias crocea **133**, 193, 198, 211, **327**, 333
Columba bollii 107, 133, 185, **186**, 189, 193, 290, 296, **301**
Columba guinea 296
Columba junoniae **107**, 133, 185, 189, 193, 290, 297, **301**
Columba livia 82, 297
Columba trocaz bollii 296
Convolvulus canariensis 131, **160**, 165, **173**, 190
Convolvulus cf *fruticulosus* 81
Convolvulus floridus 89, 111, 123. 116. 121, 133, **134**, 137, 145, 149, 153
Corvus corax 26, 93, 265, 290, 299
Cosentinia vellea 236
Coturnix coturnix 298
Crambe arborea 125, **128**
Crambe scaberrima 137, **141**
Crambe strigosa 133, **176**, 177, 190, 198
Crassula tillaea 155
Crithmum maritimum **42**, 43, 52, 69, 74
Crocidura russula 318
Crocothemis erythraea 287, 336, 337
Cryptotaenia elegans 108, **177**
Ctenolepisma 264
Cupressus macrocarpa 26
Cursorius cursor 294
Cuscuta sp. **125**, **144**
Cyanistes teneriffae 133, 183, 290, **300**, 305
Cyanistes ultramarinus 305
Cyclirius webbianus 198, **326**,
Cynthia virginiensis 332
Cyperus 111, 288
Cyphocleonus armitagei **261**, 262
Cyprinus carpio 288, 324
Cyrtomium falcatum 181
Cyrtophora citricola 50, 340
Cystopteris fragilis 180

Dactylopius coccus **25**, 143
Danaus chrysippus 47, **326**, 330, 331
Danaus plexippus **326**, 331
Daphne gnidium 133, 165, 173, 196, 208, **215**
Datura sp. 56, 77, 79
Davallia canariensis 104, 111, 114, **121**, 126, 133, 142, 148, 156, 180, 228, **229**
Delairea odorata 104
Delichon urbica 304
Delphinus delphis 314
Dendrocopos major 290, **301**, 308

Dermochelys coriacea 322
Descurainia bourgeauana 119, 211, 219, 247, **248**, 255, 260, 262, 266, 274, 333
Descurainia gonzalezii 247
Descurainia lemsii **132**, 133, 219
Descurainia millefolia 91, **114**, 119, 137, 147, 235
Dicallomera fortunata 330
Dicheranthus plocamoides 71, 142, **151**, 153
Dicladispa occator **216**
Diplazium caudatum **180**, 190
Dipsosphecia vulcania 334
Dittrichia viscosa 121
Dolichoiulus 264, 339, 341
Domene vulcanica **283**
Dorycnium 131
Dorycnium spectabile 148, 149
Dracaena draco 61, **86**, 94, **97**, 101, 102, 119, 121, **126**, 136, 148, 152, 153
Dracunculus canariensis 93, 118, **120**, 128, 133, 144
Drimia hesperia 46
Drimia maritima 46
Dryopteris affinis 181
Dryopteris guanchica **180**
Dryopteris oligodonta 178, 180
Dysdera 210, 263, 284, 339, 340
Dysdera gollumi 263, **284**
Dysdera levipes **284**

Echium aculeatum 71, **98**, 99, **155**, 137, 140, **144**, 198, 225
Echium auberianum **244**, 248, **249**, 260
Echium giganteum 83, 91, **149**, 153, 155, 193, **196**
Echium leucophaeum 123
Echium plantagineum 111, 145
Echium simplex 81, 91, 111, **118**, 119
Echium strictum 88, **90**, 121, 128, **133**, 137, 178
Echium virescens 96, 114, 133, **136**, 137, 140, **208**, 222, 235, 236
Echium wildpretii 119, **231**, 232, 243, **244**, **245**, 248, **249**, 255, 257, 253, 267, **273**
Egretta garzetta 82, 296
Emberiza alcoveri 20, 278, 290, 302
Emberiza calandra 302
Ephedra 85, 258,
Ephedra fragilis 79, **88**, 99, 125, 131, 155
Ephedra major **258**
Epilobium sp. 79
Equus africanus asinus 318
Equus caballus 319
Eretmochelys imbricata 322
Erica arborea 23, 93, **98**, 108, 113, 118, 123, 128, 131, 142, 149, 156, 159, **161**, **164**, 189, **191**, 193, 196, 197, 200, 203, 212, 215, 217, 218, 220, 227, 234, 236, 296, 306
Erica platycodon 15, **161**, **164**, 345
Erigeron calderae **250**, 251

Erithacus rubecula 185, 290, **300**, 307
Erucastrum sp. 128
Erysimum bicolor 137, **148**, 149, 156
Erysimum scoparium 248, **249**, 260, 263, 266, **273**
Eschscholzia californica 235
Estrilda astrild 302
Estrilda melpoda 302
Eubalaena glacialis 314
Eucalyptus 26, 214
Euchloe belemia 211, **262**, **327**, 333, 334
Euphorbia aphylla 68, 69, 75
Euphorbia atropurpurea 69, 70, 79, 91, 95, 99, **124**, 137, **139**, 145, **234**, **235**, 236, **358**
Euphorbia balsamifera **41**, **43**, **48**, 49, **51**, **55**, 61, 62, 66, 69, 74, 76, 81, 82, 134, 145, 153, **156**
Euphorbia canariensis **43**, 49, 50, 66, **69**, 77, 82, 92, 106, 111, **116**, 123, **124**, 125, 134, 142, 149, 153
Euphorbia cf *peplus* 137
Euphorbia lamarckii **43**, 47, 49, 50, 66, 69, 77, 92, 93, **106**, **108**, 113, **117**, 119, 123, 133, 134, **137**, 142, 143, 145, 149, 153, **155**, 156
Euphorbia mellifera **163**, 198
Euphorbia paralias **52**, 61
Euplectes orix 305
Eurydema lundbladi 263
Eusimonia wunderlichi **263**, 340

Fagonia cretica **63**,
Falco pelegrinoides 126, 133, 298
Falco tinnunculus 19, 82, 137, 185, **256**, 290, 298
Felis catus 313
Ferula linkii 255, **268**
Ficedula hypoleuca 305
Foeniculum vulgare 121, 123, 133, 142, 198, 235
Forsskaolea angustifolia 49, 55, 137, 142
Frankenia capitata 51, **73**, 74
Frankenia ericifolia **73**, 74
Frankenia laevis 67
Frankenia pulverulenta 56, **64**
Fringilla coelebs 20, 185, **209**, 290, **300**, 303
Fringilla teydea 20, **209**, **289**, 290, **300**, 303
Fulica atra 288, 299
Furcraea foetida 26

Galium 178
Gallinago gallinago 295
Gallinula chloropus 288, 299
Gallotia galloti **18**, 19, 26, 50, **80**, 81, 82, **105**, 210, **264**, 265, 320
Gallotia goliath **18**, 19, 278, 313, 320
Gallotia intermedia **19**, 313, 321, 321
Gambusia affinis 288, 324
Geochelone burchardi 9, 19, 323
Geranium reuteri 174, **176**, 193, 197
Gesnouinia arborea 163, **169**, **189**, 190, 191, 195
Gibbulinella 338
Gladiolus italicus 145, **177**, 178

Glareola pratincola 294
Globicephala macrorhynchus 315
Globularia salicina **88**, 96, 99, 108, 114, 117, 123, 128, 131, 137, 140, 145, 149, 152, 156, 165, 173, 198
Gnaphalium sp. 79
Gobius paganellus 286
Gonepteryx cleobule **145**, 184, 193, 195, 198, 211, **327**, 333
Gonepteryx eversi 328
Gonospermum fruticosum 92, **102**, 104, **109**, 110, 128, 142, 145, 149, 165
Gonospermum revolutum 81, **110**, 111, **116**, 117
Grampus griseus 315
Greenovia aizoon 102, **222**
Greenovia aurea 102, 208, **215**, 218, **221**, 223, 258
Greenovia dodrentalis 141
Grimmia trichophylla 224
Gryllomorpha canariensis 216, 343
Gryllotalpa africana 63, 343
Gryllus bimaculatus 311, 343
Guanchia 216, 344
Guanchia cabrerae **183**, 184
Guerrina 183
Gymnocarpos decandrus **54**, 55

Habenaria tridactylites **123**, 131, 145, 156, 213
Haematopus meadewaldoi 291
Heberdenia excelsa 86, 87, 93, 121, 146, **151**, 152, **153**, 156, 162, 164, **168**, 171, **186**, 198
Hedera canariensis 172, **173**, 190
Hegeter 345
Hegeter amaroides 81, 82
Hegeter lateralis 262
Helianthemum broussonetii **109**, 110
Helianthemum canariense 63, **77**, 137
Helianthemum juliae 252, **253**, 255
Helianthemum teneriffae 128
Heliotropium ramosissimum **63**, 77,
Hemicycla 338, 377
Hemicycla bidentalis 81, 82, 183
Hemidactylus turcicus 26, 320
Herniaria canariensis 63
Herse convolvuli 334
Himantopus himantopus 295
Hipparchia wyssii 193, 211, **327**, 331
Hippotion celerio 334
Hirschfeldia incana 56
Hirundo rustica 304
Hydrobates pelagicus 80, 308
Hydroporus pilosus 286
Hyla meridionalis **286**, 288, 323
Hyles livornica 335
Hyles tithymali **47**, 50, 335
Hymenophyllum tunbrigense 180, **181**
Hyoscyamus albus **73**, 74

Hypericum canariense 87, **90**, 93, **99**, 114, 123, 128, 131, 142, 145, 149, 152, 165, 173, 191, 196
Hypericum glandulosum 104
Hypericum grandifolium 133, **176**, 190, 193, 198
Hypericum reflexum **102**, 104, 123, **125**, 128, 133, 137, 142, 144, 149, 193
Hypochaeris oligocephala **71**, 104
Hypolimnas misippus 331
Hypsugo savii 264, 317

Ilex 22
Ilex canariensis 17, 35, 108, 111, 113, 117, 131, 145, 149, 156, 155, 162, 164, 165, **168**, 170, 171, 111, 193, 197, 203, 215,
Ilex perado 162, **166**, 170, 171, 190, 215
Insulivitrina **183**, 338
Ischnura saharensis 288, 337
Ischnura senegalensis 288, 337
Isoplexis canariensis 118, 133, **174**, 197
Issoria lathonia 185, **326**, 331
Ixanthus viscosus 173, **174**, 190, 198
Ixobrychus minutus 296

Jasminum 85
Jasminum odoratissimum 88, **89**, 99, 108, 114, 116, 119, 123, 133, 137, 144, **145**, 149, 152, 156, 165, 173
Juncus 79, 288
Juniperus 22, 85
Juniperus cedrus 191, 132, 355, **265**, **268**, 308
Juniperus turbinata 81, **84**, **86**, **92**, **94**, **96**, **99**, 102, **110**, 113, 116, **117**, 121, 123, 125, 128, 133, 136, 148, 153, **154**, **156**, 234, 236
Justicia hyssopifolia 70, 92, 134, **135**, 145, **151**, 152

Kickxia sagittata **52**, 53, 62, 67, 137
Kickxia scoparia 67, 137
Kleinia neriifolia 43, **45**, 66, 69, 77, 79, 92, 113, 123, 128, 137, 142, 145, 152, 156, 236
Kunkeliella retamoides **131**,
Kunkeliella subsucculenta **73**, 75

Lampides boeticus 126, 211, **326**, 329, 330
Lamprotornis purpureus 306
Lanius excubitor 304
Lanius meridionalis koenigi 19, 290, **300**
Laparocerus 13, 16, 184, 263, 339, 345
Laparocerus crassus **13**
Laparocerus fernandezi **13**
Laparocerus inaequalis **13**
Laparocerus undatus 161
Laphangium sp. 56
Laphangium teydeum 159, **260**
Larus argentatus 294
Larus cachinnans atlantis 294
Larus fuscus 294
Larus marinus 294
Larus michahellis atlantis 80, 82, 290, 294

Larus ridibundus 295
Launaea arborescens **42**, 43, 59, 67, 69, 76, 134
Laurobasidium lauri 146
Laurus novocanariensis 35, 113, 131, 146, 152, 156, 159, 161, 164, 165, **169**, 171, 182, 189, 190, 193, 195, 196, 197, 215, 319
Lavandula buchii **81**, 145
Lavandula canariensis 49, 74, 77, 111, 145, **235**
Lavandula minutollii **141**
Lavandula pinnata 133
Lavandula sp. 43, 71, 79, 116, 137
Lavatera acerifolia 70, 79, **90**, 91, 121, 125, **128**, 136, **143**, 144, 152, **153**
Leiothrix lutea 308
Lepidochelys kempii 322
Lepromoris gibba **50**
Leptotes pirithous 329
Limnebius punctatus 286
Limonium arborescens 91, 102, **103**, 146
Limonium imbricatum **73**, 74, 82
Limonium macrophyllum **103**, 111, 119, **120**
Limonium pectinatum **42**, 43, **51**, 55, 63, 67, 71, 74, 76
Limosa lapponica 295
Limosa limosa 295
Lindbergopsallus 345
Lithobius pilicornis 341
Lithobius sp. 264
Lithobius speleovulcanus 341
Loboptera 384, 342
Loboptera subterranea 283
Loboptera troglobia **277**
Lobularia canariensis 123, 133, **149**, 222
Locusta migratoria 343
Lotus campylocladus **207**, 217, 220, 227, 235, 255, 334
Lotus maculatus **75**, 81
Lotus sessilifolius 51, 53, **55**, 62, 128
Lotus sp. 116
Lotus tenellus **73**, 75, 156
Lugoa revolutum 104
Luzula 178, 332
Luzula canariensis 178
Lycaena phlaeas **47**, 193, 198, 211, **326**, 330
Lycium intricatum 46, 52, 63, **66**, 67, 304

Macaronesia fortunata 330,
Macroglossum stellatarum 262, 335
Maiorerus randoi 340
Maniola jurtina **193**, 198, 211, **327**, 331
Mantis religiosa 216, 342
Marcetella moquiniana **85**, 66, **90**, **95**, 114, 136, 146, **150**, **151**, **152**
Marrubium vulgare 235
Maytenus canariensis 87, 102, 121, 136, 146, **148**, 152, 163, **168**, 171
Megaptera novaeangliae 314

Meloe flavicomus **105**, 106
Melopsittacus undulatus 310
Mercurialis annua 178
Mercurialis canariensis 177, **179**
Mercurialis sp. 133
Merops apiaster 297
Mesembryanthemum crystallinum 43, 56, **57**, 71
Mesembryanthemum nodiflorum 56, **57**
Mesoplodon densirostris 316
Micromeria 45, 104, 121, 123, 177,
Micromeria lasiophylla 250
Micromeria hyssopifolia 55, 77, 207, **208**, 219, 235, 236
Micromeria lachnophylla 219, 250
Micromeria teneriffae 104, 128
Micromeria varia 71, 74, **91**, 111, 145
Miliaria calandra 302
Milvus migrans 292
Milvus milvus 292
Monanthes 71, 102, 147
Monanthes brachycaulos 128, **223**, 258
Monanthes laxiflora 102, **103**, **109**, 116, 118, **120**, 145, 149, 153
Monanthes pallens 71, **141**, 145, 236
Monanthes polyphylla 116, **147**, 149
Monilearia 338,
Morella faya 22, 35, 93, 108, 113, 131, 142, 146, 159, 162, 164, 165, **167**, 169, 171, 193, 195, 196, 197, 203, 212, 215, 217, 220, 328
Morus bassanus 308
Motacilla alba 305
Motacilla cinerea 290, 304
Motacilla flava 305
Mus musculus 17, 264, 319
Mustela putorius furo 313
Myopsittacus monachus 310
Myosotis 177
Myosotis latifolia 193
Myrica faya 162
Myrmeleontidae 345
Myrmeleon alternans 271

Nandayus nenday 310
Napaeus 183, 338
Napaeus variatus 183
Navaea 85
Navaea phoenicea **90**, 91, 148, **149**,
Neochamaelea pulverulenta **42**, 43, 67, 71, 134
Neophron percnopterus 290, 292
Nepeta teydea 213, **244**, **248**, 255, **271**
Nesotes 345
Nesotes conformis 161
Nicotiana glauca 26, 56, **59**, 62, 77, 81, 135
Normania nava 172
Notholaena marantae ssp. *subcordata* 181
Numenius arquatus 295
Numenius phaeopus 295

Numida meleagris 298
Nyctalus leisleri 210, 264, 317
Nycticorax nycticorax 296
Nymphicus hollandicus 310

Oceanodroma castro 80, 82, 308
Ocotea foetens **162**, **169**, 171, 190, 195, 198, 297
Oedipoda canariensis 343
Oenanthe oenanthe 307
Olea 85
Olea cerasiformis 78, 81, 84, **86**, 94, **96**, 99, 102, 113, 116, **117**, 121, 123, 125, **127**, 131, 144, 153, 156
Olea europaea 96
Olios canariensis 210, **211**
Ononis serrata 63
Ononis tournefortii **62**
Opuntia 93
Opuntia dillenii 26, 40, 50, 62, 70, 82
Opuntia maxima **25**, 26, 40, 77, 91, 99, 108, 114, 123, 128, 135, 140, 143, 146, 149, 152, 156, 235
Orchestia guancha 183
Orchis patens 222
Orcinus orca 315
Origanum vulgare 177
Orthetrum chrysostigma **105**, 336, 337
Oryctolagus cuniculus 155, 207, 245, 252, 264, 318
Osyris lanceolata **83**, 88, **89**, 148, 155
Ovis aries 312, 252
Ovis musimon 312
Ovis orientalis 245, 252, 264, 312
Oxalis pes-caprae 26, 91, 196

Pachydema obscura 230, **321**
Pancratium canariense 70, **71**, 91, 116, 137, 145
Pandion haliaetus 292
Pandoriana pandora 330
Pararge aegeria 332
Pararge xiphioides **184**, 193, 195, 198, 211, **327**, 332
Parascleropilio 340
Parentucellia viscosa 111
Parietaria filamentosa 128, 332
Parolinia intermedia **70**
Paronychia canariensis 71, 74, 92, 111, 114, 119, **120**, 123, 133, 142, 149, 153, 155
Parus caeruleus teneriffae 305
Parus major 305
Parus teneriffae teneriffae 305
Passer domesticus 305
Passer hispaniolensis 305
Patellifolia patellaris 56, **57**, 74
Pelophylax perezi 288, 323
Pericallis appendiculata 121, **175**, 195
Pericallis cruenta 121, 123, **175**, 193, 222
Pericallis echinata 149
Pericallis lanata **102**, 104, 126, **127**, 133, **222**
Pericallis tussilaginis 174, **175**
Periplaneta americana 342

Periplaneta australasiae 342
Periploca laevigata 44, **45**, 49, 67, 71, 77, 79, 92, 111, 113, 123, 128, 137, 142, 145, 152, 156
Persea indica 17, 161, **169**, 171, 190, 193, 195, 197, 319
Petronia petronia 142, 305
Phagnalon cf *saxatile* 236
Phagnalon purpurascens 222
Phagnalon saxatile **109**, 236
Phaneroptera sparsa **123**
Phasianus colchicus 299
Phoenix canariensis 23, 79, 64, 85, 86, **87**, 94, 108, 137, 142, 145, 149, **152**
Phyllis nobla **109**, 110, 131, 133, **176**, 177, 186, 189, 193, 196
Phyllis viscosa 79, **103**, 104, 140, 145, 153, 198
Phyllodromica brullei 184, 267, 342, **343**
Phylloscopus canariensis 185, 209, 290, **300**, 306
Phylloscopus collybita 306
Phylloscopus trochilus 306
Physeter macrocephalus 316
Picconia excelsa 17, **130**, 131, 146, 162, **166**, 185, 191, **195**, **196**, 198
Pieris cheiranthi 195, 211, **327**, 333
Pieris daplidice 333
Pieris rapae **193**, 195, 198, 211, **327**, 333
Pimelia 345
Pimelia ascendens **261**, 262
Pimpinella anagodendron 121
Pimpinella cumbrae **251**, 255, **271**
Pimpinella dendrotragium 131, 208, 221
Pinus canariensis 22, 35, **139**, 132, 136, 199
Pinus halepensis 205
Pinus insignis 205
Pinus pinaster 295
Pinus radiata 192, 205, 245, 273, 303
Pipistrellus kuhlii 317
Pipistrellus maderensis 186, 210, 264, 316, 317
Pipistrellus savii 317
Pisidium casertanum 338
Pistacia 85, 96
Pistacia atlantica **85**, **86**, 96, **97**, 99, 113, 119, 125, 145, **150**, 152
Pistacia lentiscus **96**, 97
Plantago arborescens 91, **109**, 110, 114, 119, 123
Plantago coronopus **73**, 75
Plantago webbii **257**, 268
Plecotus teneriffae 210, 264, 316, 317
Pleiomeris canariensis 93, 113, **114**, 152, 156, 162, **168**, 171, 189, 190, 195
Plocama pendula 44, 49, 55, 62, 71, 74, **75**, 77, 79, 123, 134, 142, 320
Plutonia sp. 159, **183**, 338
Pluvialis squatarola 294
Poa 260
Poecilia reticulata 288, 324

Poicephalus senegalus 310
Polycarpaea carnosa 71, **79**
Polycarpaea divaricata **73**, 74
Polycarpaea latifolia 123
Polycarpaea nivea **61**, 77
Polycarpon tetraphyllum 56
Polygonum balansae **61**
Polygonum salicifolium 79
Polygonum spp. 288
Polyommatus celina 325
Polyommatus icarus 325
Polypodium macaronesicum 104, **123**, 156, 180, 236
Polystichum setiferum **180**
Pontia daplidice **46**, 47, **67**, 193, 198, 211, 262, **327**, 333
Portulaca sp. 63
Procambarus clarkii 288, 341
Promachus vexator **216**, 330
Prunus avium 132
Prunus lusitanica 162, 163, 164, **167**, 171, 190
Pseudorca crassidens 315
Pseudoyersinia subaptera **50**
Pseudoyersinia teydeana 263, 342
Psittacula krameri 310
Pteridium aquilinum 108, 135, 178, 193, 195, 196
Pteris incompleta 180, **181**
Pteris vittata 181
Pterocephalus dumetorus **124**, 125, **128**, 122, 133
Pterocephalus lasiospermus 133, 208, 219, 225, **246**, 247, 260, 263, 267, 269, 273, 274
Pterocephalus virens **74**, 117
Puffinus assimilis baroli 309
Puffinus baroli 80, 82, 309
Puffinus holeae 21
Puffinus olsoni 21
Puffinus puffinus 309
Punica granatum 123
Pycnonotus jocosus 306
Pyrrhocorax graculus 21, 301
Pyrrhocorax pyrrhocorax 21, 301

Quercus 22, 23, 26, 85,
Quercus cf *ilex* 152
Quercus suber 132

Ramphotyphlops braminus 322
Ranunculus cortusifolius 174, **176**, 189, 190, 198
Rattus norvegicus 18, 319
Rattus rattus **18**, 186, 264, 297, 319
Reduvius personatus 63
Regulus regulus 185, 290, 306
Reichardia crystallina 56, **57**
Reichardia ligulata **73**, 74, 104, 136, **137**, 153, 156
Reseda scoparia 44, 63, **66**, 67, 77, 111, 137
Retama rhodorhizoides 86, **87**, 91, 93, 94, **95**, **106**, 131, 133, 138, **139**, 155, **277**, 236, 380

Rhamnus crenulata **88**, 111, 119, 123, 125, 133, 137, 144, 146, 152, 156
Rhamnus glandulosa 108, 113, 118, 131, 163, 165, **166**, 169, 170, 171, 185, 189, 190, 191, 215
Rhamnus integrifolia **88**, 89, 101, 104, 136, 254, 255, **256**, 265, 274
Rhodopechys githagineus 302
Rhyparia rufescens 328
Ricinus communis 26, 56
Riparia riparia 304
Roccella 101, 115, **116**, 128, 145, 153, 156
Romulea columnae **109**
Rosa canina 265, **274**
Rousettus aegyptiacus 316
Rubia fruticosa 44, **45**, 71, 77, 79, 92, 108, 113, 117, 121, 128, 133, 137, 142, 145, 149 152, 320
Rubia peregrina 108, 190, 196
Rubus bollei 193, 196
Rubus fruticosa 153
Rubus palmensis 172
Rubus sp. 79, 108, 121, 121, 123, 133, 135, 145, 149, 170, 172, 190, 197
Rumex lunaria **45**, 46, 70, 79, 111, 117, 121, **124**, 133, 137, 142, 145, 149, 153, 156, 174, 198, 235
Rumex maderensis 200, 227, **228**, **229**
Ruta pinnata 88, 90, 137, 152, 155

Salix canariensis **95**, 123, 135, 142, 153, 163, **167**, 170, 171, 195, 196, **288**, 306
Salsola divaricata 55, **60**, 74, 156
Salvia broussonetii **102**
Salvia canariensis 91, 93, 111, 123, 128, **130**, 137, 142
Sambucus nigra ssp. *palmensis* 163, **166**, 190
Sardinus pilchardus 7
Satchelliella binunciolata 286
Scabiosa atropurpurea 117
Scarites buparius 63
Schistocerca gregaria 5, **9**, 343
Schizogyne sericea 43, 44, **45**, **48**, 49, 51, 55, 62, **63**, 67, 71, 74, 76, 79, 156
Scilla haemorrhoidalis 49, 145, **155**, 156
Scilla latifolia 71, 123, 145
Scilla spp. 46
Scolopax rusticola 185, 295
Scolopendra morsitans 341
Scolopendra valida 210, 341
Scrophularia glabrata 200, 208, 213, 225, 227, **248**, 250, 269, 274
Scrophularia smithii 174, 193
Scutigera coleoptrata 342
Selaginella denticulata 180
Semele androgyna 133, 149, **172**, **190**
Senecio glaucus 63
Senecio palmensis 101, **254**, 255
Serinus canaria 186, 290, **300**, 304
Serinus serinus 304

Sideritis brevicaulis 79, 96, 140, **141**
Sideritis canariensis 133, 177, 193
Sideritis cretica 70, 145, **148**, 149, 153
Sideritis dendro-chahorra 104, 111, 114, 116, **117**, 121
Sideritis eriocephala 249, **256**
Sideritis infernalis 104, **136**
Sideritis kuegleriana 156, **104**
Sideritis oroteneriffae **218**, 221, **222**
Sideritis soluta 133, 268
Sideroxylon canariense 87, 102, **113**, 136, 152, 156, 163, **168**, 170, 171
Silene berthelotiana 133, 208, 255, **256**, 259
Silene lagunensis 104, **121**
Silene nocteolens **259**, 268
Silene sp. 128, 235
Simulium ruficorne 187
Smilax aspera 91, 104, 114, **154**, 155, 196
Smilax canariensis **172**, 189, 190, 193
Solanum nava 172
Solanum vespertilio 91, 114, 119, **120**, 123, 149
Sonchus acaulis 111, 116, 118, **119**, 128, 131, 133, 139, 145, 149, 153, 156, 213, **221**
Sonchus asper **219**
Sonchus canariensis 139, **140**, **117**, 236
Sonchus congestus 133, **147**, 174, **175**, 182
Sonchus fauces-orci **137**
Sonchus gummifer **125**, 126, 131, 133, **222**
Sonchus radicatus 121, 145, 147
Sonchus tectifolius 102, **103**, 123
Sorbus aria 265
Spartocytisus filipes **89**, 152, **154**, 155, 207
Spartocytisus supranubius 211, **212**, 232, 243, **246**, 257, 260, 261, 266, 273, 329
Spergula fallax **73**, 74
Sphingonotus willemsei 261, **343**
Spilopelia senegalensis 297
Stemmacantha cynaroides 251, 252, **253**, 262
Stenella coeruleoalba 315
Stenella frontalis 315
Steno bredanensis 315
Stenochrus portoricensis 340
Stercorarius parasiticus 295
Stercorarius pomarinus 295
Stercorarius skua 295
Stereocaulon vesuvianum 224
Sterna hirundo 296
Sterna paradisaea 296
Sterna sandvicensis 296
Streptopelia decaocto 297
Streptopelia risoria **107**,
Streptopelia roseogrisea 297
Streptopelia senegalensis 297
Streptopelia turtur 297
Sturnus vulgaris 306
Sula bassana 308

Suncus etruscus 318
Sus domesticus 311
Sylvia atricapilla 137, 186, 290, 306
Sylvia conspicillata orbitalis 290, 307
Sylvia melanocephala 137, 186, 290, 307
Sympetrum fonscolombii **67**, 336, 337
Sympetrum nigrifemur 337
Sympetrum striolatum nigrifemur 337

Tadarida teniotis 210, 164, 316
Tadorna ferruginea 293
Tamarix canariensis **56**, 61
Tamus edulis **90**, 91, 114, 121, 128, 133, 149, 172, 173, 189
Tarentola 19, 321
Tarentola chazaliae 321
Tarentola delalandii **67**, 81, 82, 105, 210, 264, 321
Tarentola mauritanica 321
Teline canariensis 118, 119, **139**, 140, 173, **175**, 189, 190, 329
Teline osyrioides 207, **221**
Teline pallida 101, 115, **116**
Teline stenopetala 207, 215
Temnoscheila coerulea 232
Tetraena fontanesii **42**, 43, 49, 51, 55, 56, 60
Teucrium heterophyllum **88**, 89, 92, 116, 121, 144, 152
Thymelicus christi 185, **193**, 198, 211, **326**, 329
Tilapia tholloni 323
Tinguarra cervariaefolia **125**, 126, 131, 133
Tipula macquarti 184
Todaroa aurea 71, 119, 123, 142, 149, 155
Todaroa montana **220**, 221
Tolpis cf *crassiuscula* 136
Tolpis lagopoda 136
Tolpis webbii 219, **257**, **262**, 265
Trachemys scripta elegans 322
Traganum moquinii **60**
Tragopogon porrifolius 222
Trichoferus roridus 261
Trichomanes speciosum 180
Trifolium sp. 221
Trifolium squarrosum **109**, 110
Tringa glareola 295
Tringa nebularia 295
Tringa ochropus 295
Tringa totanus 295
Trithemis arteriosa **286**, 287, 336, 338
Troglohyphantes oromii **283**
Turdus merula 185, 290, 307
Turdus torquatus **265**, 308
Tursiops truncatus 315
Typha domingensis 288
Typha sp. **287**
Tyto alba 311

Umbilicus gaditanus **219**
Upupa epops 133, 298
Urtica 177, 328, 332
Urtica morifolia **179**, 189, 196, 332,

Vandenboschia speciosa 180, **181**
Vanellus vanellus 294
Vanessa atalanta 184, **185**,193, 198, 211, **326**, 332
Vanessa cardui **184**, 193, 198, **326**, 332
Vanessa indica 332
Vanessa virginiensis 198, **326**, 332
Vanessa vulcania 184, **185**, 193, 195, 198, 211, **326**, 332
Venezillo tenerifensis **283**
Viburnum rigidum 23, 131, 146, 163, **166**, 171, 189, 193, 195, 196, 197, 215, 319
Vicia cirrhosa 119, **120**, 131
Vieria laevigata 70, **71**, 79, 102, 142, 145, **151**, 153
Viola anagae **178**
Viola cheiranthifolia **258**, 259, **274**
Visnea mocanera 87, **93**, **99**,113, 125, **127**, **130**, 131, 146, 152, 162, 165, **167**, 171, 190, 198
Vitis vinifera 24
Volutaria canariensis **62**
Vulpia 260

Wahlenbergia lobelioides **109**, 110
Withania aristata 71, 74, **87**, 119, 152, 156
Woodwardia radicans 158, 178, **180**, 190

Xerotricha nubivaga 262

Yponomeuta gigas **135**, 335

Ziphius cavirostris 316
Zizeeria knysna **326**, **328**, 329, 330
Zygonyx torridus 336, 338